3
mf

PESTICIDES *in the* DIETS OF INFANTS AND CHILDREN

Committee on Pesticides in the Diets of
Infants and Children

Board on Agriculture
and
Board on Environmental Studies and Toxicology

Commission on Life Sciences

National Research Council

D0171584

NATIONAL ACADEMY PRESS
Washington, D.C. 1993

NATIONAL ACADEMY PRESS • 2101 Constitution Avenue • Washington, DC 20418

NOTICE: The project that is the subject of this report was approved by the Governing Board of the National Research Council, whose members are drawn from the councils of the National Academy of Sciences, the National Academy of Engineering, and the Institute of Medicine. The members of the committee responsible for the report were chosen for their special competences and with regard for appropriate balance.

This report has been reviewed by a group other than the authors according to procedures approved by a Report Review Committee consisting of members of the National Academy of Sciences, the National Academy of Engineering, and the Institute of Medicine.

Support for this project was provided by the U.S. Environmental Protection Agency, Contract No. 68D-80101, with contributions from International Life Sciences Institute and Health and Welfare Canada. In addition, support for this project was provided by the Kellogg Endowment Fund of the National Academy of Sciences and the Institute of Medicine.

Library of Congress Cataloging-in-Publication Data

National Research Council (U.S.). Committee on Pesticides in the Diets of Infants and Children.
  Pesticides in the diets of infants and children/Committee on Pesticides in the Diets of Infants and Children, Board on Agriculture and Board on Environmental Studies and Toxicology, Commission on Life Sciences, National Research Council.
      p.   cm.
  Includes bibliographical references and index.
  ISBN 0-309-04875-3
    1. Pediatric toxicology. 2. Pesticide residues in food—United States. 3. Pesticides—Toxicology. 4. Infant formulas—Contamination. 5. Food contamination. I. Title.
    [DNLM: 1. Pesticides. 2. Diet—in infancy & childhood. 3. Food Contamination. WS 115 N277p 1993]
  RA1225.N38 1993
  615.9'54—dc20
  DNLM/DLC
  for Library of Congress

Any opinions, findings, conclusions, or recommendations expressed in this publication are those of the author(s) and do not necessarily reflect the view of the organizations or agencies that provided support for this project.

Additional copies of this book are available from the National Academy Press, 2101 Constitution Avenue, NW, Box 285, Washington, DC 20055. Call 800-624-6242 or 202-334-3313 (in the Washington Metropolitan Area).

Printed in the United States of America

Technical Advisers

EMMANUEL AKPANYIE, Environmental Systems International, Vienna, Va.

SHERYL BARTLETT, Health and Welfare Canada, Ottawa, Ontario

JUDY HAUSWIRTH, Jellinek, Schwartz, Connolly and Freshman, Washington, D.C.

JOHN P. WARGO, Yale University, New Haven, Conn.

RICHARD WILES, Center for Resource Economics, Washington, D.C.

COMMISSION ON LIFE SCIENCES

The National Academy of Sciences is a private, nonprofit, self-perpetuating society of distinguished scholars engaged in scientific and engineering research, dedicated to the furtherance of science and technology and to their use for the general welfare. Upon the authority of the charter granted to it by the Congress in 1863, the Academy has a mandate that requires it to advise the federal government on scientific and technical matters. Dr. Frank Press is president of the National Academy of Sciences.

The National Academy of Engineering was established in 1964, under the charter of the National Academy of Sciences, as a parallel organization of outstanding engineers. It is autonomous in its administration and in the selection of its members, sharing with the National Academy of Sciences the responsibility for advising the federal government. The National Academy of Engineering also sponsors engineering programs aimed at meeting national needs, encourages education and research, and recognizes the superior achievements of engineers. Dr. Robert M. White is president of the National Academy of Engineering.

The Institute of Medicine was established in 1970 by the National Academy of Sciences to secure the services of eminent members of appropriate professions in the examination of policy matters pertaining to the health of the public. The Institute acts under the responsibility given to the National Academy of Sciences by its congressional charter to be an adviser to the federal government and, upon its own initiative, to identify issues of medical care, research, and education. Dr. Kenneth I. Shine is president of the Institute of Medicine.

The National Research Council was organized by the National Academy of Sciences in 1916 to associate the broad community of science and technology with the Academy's purposes of furthering knowledge and advising the federal government. Functioning in accordance with general policies determined by the Academy, the Council has become the principal operating agency of both the National Academy of Sciences and the National Academy of Engineering in providing services to the government, the public, and the scientific and engineering communities. The Council is administered jointly by both Academies and the Institute of Medicine. Dr. Frank Press and Dr. Robert M. White are chairman and vice-chairman, respectively, of the National Research Council.

Preface

Iₙ 1988, the U.S. Congress requested that the National Academy of
Sciences establish a committee within the National Research Council
to study scientific and policy issues concerning pesticides in the diets
of infants and children. The Committee on Pesticide Residues in the Diets
of Infants and Children appointed to undertake this study was charged
with responsibility for examining what is known about exposures to pesticide residues in the diets of infants and children, the adequacy of current
risk assessment methods and policies, and toxicological issues of greatest
concern. The committee operated under the joint aegis of the Board on
Agriculture (BA) and the Board on Environmental Studies and Toxicology
(BEST).

The committee first met in October 1988 and held its last meeting in
January 1993. Several full committee meetings were held each year, and
subgroups of the committee were convened on a number of occasions to
address such topics as the physiology of infants and children, the age-
specific patterns of children's diets, the measurement of residue levels,
and the mathematical modeling of risks. The expertise represented on
the committee included pediatrics, toxicology, epidemiology, biostatistics,
food science and nutrition, analytical chemistry, and child growth and
development. When required, advice was obtained from experts outside
the committee on a variety of topics.

Critical assessment of potential risks to health resulting from exposures
to toxicants in the environment has been the focus of several recent studies
conducted by BEST and BA. Many of the approaches to risk assessment
used in this report trace their origins to the reports on *Drinking Water and*

Health developed since 1977. Of particular value was Volume 6 in that series. The committee also found useful *Risk Assessment in the Federal Government: Managing the Process* (1983), *Biologic Markers in Reproductive Toxicology* (1989), *Biologic Markers in Immunotoxicology* (1992), and *Environmental Neurotoxicology* (1992). The analysis in this volume draws conceptually from the 1987 report from the Board on Agriculture called *Regulating Pesticides in Food: The Delaney Paradox*—an examination of the process by which levels of pesticide residues in foods are regulated by the U.S. Government.

The Committee on Pesticides in the Diets of Infants and Children was greatly assisted by many individuals and groups who provided information on food consumption patterns and on pesticide residue concentrations in the U.S. diet. The groups include the U.S. Department of Agriculture, the U.S. Food and Drug Administration, the National Food Processors Association, the Gerber Products Company, and the Infant Formula Council. Many other food manufacturers as well as pesticide manufacturers also provided useful data to the committee either individually or through various organizations.

The committee is grateful for the assistance of the National Research Council (NRC) staff in the preparation of this report. In particular the committee wishes to acknowledge Frances Peter, project manager; Richard Thomas, principal staff scientist (BEST); Sandi Fitzpatrick, senior program assistant (BEST); James Reisa, director of BEST; and Susan Offutt, executive director of BA. Other staff members who contributed to this effort include Shelley A. Nurse, senior project assistant (BEST); Ruth P. Danoff, project assistant (BEST); Craig Cox, senior staff officer (BA); Mary Lou Sutton, administrative assistant (BA); Carla Carlson, director of communications (BA); Barbara J. Rice, editor (BA); Janet Overton, associate editor (BA); Lee R. Paulson, program director for information systems and statistics (BEST); Bernidean Williams, information specialist (BEST); and Dawn M. Eichenlaub, production manager, and Richard E. Morris, editor, National Academy Press. Thanks are also due to Richard Wiles and Charles Benbrook, formerly of the BA staff. The interest in this report shown by the Executive Office of the National Research Council, especially by the Deputy Executive Officer Mitchel Wallerstein, is greatly appreciated. These individuals provided invaluable support to the committee throughout its deliberations.

As consultant to the committee, John Wargo of the Yale University School of Forestry and Environmental Studies developed numerous innovative approaches to the analysis of highly complex data. His pellucid presentations permitted clear understanding of issues that previously had been opaque. Valuable assistance was also provided to the committee by Emmanuel Akpanyie, Sheryl Bartlett, and Judy Hauswirth, who served as technical advisers.

Last, but by no means least, the work of all the members of the committee is greatly appreciated. We are also grateful to the U. S. Environmental Protection Agency, Health and Welfare Canada, the International Life Sciences Institute, and the Kellogg Endowment Fund of the National Academy of Sciences and the Institute of Medicine, whose financial support made the study possible.

PHILIP J. LANDRIGAN, M.D., M.Sc.
Chairman

Contents

Executive Summary

PESTICIDES ARE USED WIDELY in agriculture in the United States. Their application has improved crop yields and has increased the quantity of fresh fruits and vegetables in the diet, thereby contributing to improvements in public health.

But pesticides may also cause harm. Some can damage the environment and accumulate in ecosystems. And depending on dose, some pesticides can cause a range of adverse effects on human health, including cancer, acute and chronic injury to the nervous system, lung damage, reproductive dysfunction, and possibly dysfunction of the endocrine and immune systems.

Diet is an important source of exposure to pesticides. The trace quantities of pesticides that are present on or in foodstuffs are termed residues. To minimize exposure of the general population to pesticide residues in food, the U.S. Government has instituted regulatory controls on pesticide use. These are intended to limit exposures to residues while ensuring an abundant and nutritious food supply. The legislative framework for these controls was established by the Congress through the Federal Insecticide, Fungicide, and Rodenticide Act (FIFRA) and the Federal Food, Drug, and Cosmetic Act (FFDCA). Pesticides are defined broadly in this context to include insecticides, herbicides, and fungicides.

Tolerances constitute the single, most important mechanism by which EPA limits levels of pesticide residues in foods. A tolerance is defined as the legal limit of a pesticide residue allowed in or on a raw agricultural commodity and, in appropriate cases, on processed foods. A tolerance must be established for any pesticide used on any food crop.

Tolerance concentrations are based primarily on the results of field trials conducted by pesticide manufacturers and are designed to reflect the highest residue concentrations likely under normal conditions of agricultural use. Their principal purpose is to ensure compliance with good agricultural practice. Tolerances are not based primarily on health considerations.

This report addresses the question of whether current regulatory approaches for controlling pesticide residues in foods adequately protect infants and children. The exposure of infants and children and their susceptibility to harm from ingesting pesticide residues may differ from that of adults. The current regulatory system does not, however, specifically consider infants and children. It does not examine the wide range of pesticide exposure patterns that appear to exist within the U.S. population. It looks only at the average exposure of the entire population. As a consequence, variations in dietary exposure to pesticides and health risks related to age and to such other factors as geographic region and ethnicity are not addressed in current regulatory practice.

Concern about the potential vulnerability of infants and children to dietary pesticides led the U.S. Congress in 1988 to request that the National Academy of Sciences (NAS) appoint a committee to study this issue through its National Research Council (NRC). In response, the NRC appointed a Committee on Pesticide Residues in the Diets of Infants and Children under the joint aegis of the Board on Agriculture and the Board on Environmental Studies and Toxicology.

The committee was charged with responsibility for examining scientific and policy issues faced by government agencies, particularly EPA, in regulating pesticide residues in foods consumed by infants and children. Specifically, the committee was asked to examine the adequacy of current risk assessment policies and methods; to assess information on the dietary intakes of infants and children; to evaluate data on pesticide residues in the food supply; to identify toxicological issues of greatest concern; and to develop relevant research priorities. Expertise represented on the committee included toxicology, epidemiology, biostatistics, food science and nutrition, analytical chemistry, child growth and development, and pediatrics.

The committee was not asked to consider toxicities resulting from exposures to microorganisms (bacteria and viruses) or from other naturally occurring potential toxins. It was not asked to weigh the benefits and risks to be derived from a plentiful and varied food supply against the potential risks resulting from pesticide exposure. It was not asked to assess the overall safety of the food supply.

In this report, the committee considered the development of children from the beginning of the last trimester of pregnancy (26 weeks) through

18 years of age, the point when all biological systems have essentially matured.

CONCLUSIONS

Age-Related Variation in Susceptibility and Toxicity

A fundamental maxim of pediatric medicine is that children are not "little adults." Profound differences exist between children and adults. Infants and children are growing and developing. Their metabolic rates are more rapid than those of adults. There are differences in their ability to activate, detoxify, and excrete xenobiotic compounds. All these differences can affect the toxicity of pesticides in infants and children, and for these reasons the toxicity of pesticides is frequently different in children and adults. Children may be more sensitive or less sensitive than adults, depending on the pesticide to which they are exposed. Moreover, because these processes can change rapidly and can counteract one another, there is no simple way to predict the kinetics and sensitivity to chemical compounds in infants and children from data derived entirely from adult humans or from toxicity testing in adult or adolescent animals.

The committee found both quantitative and occasionally qualitative differences in toxicity of pesticides between children and adults. Qualitative differences in toxicity are the consequence of exposures during special windows of vulnerability—brief periods early in development when exposure to a toxicant can permanently alter the structure or function of an organ system. Classic examples include chloramphenicol exposure of newborns and vascular collapse (gray baby syndrome), tetracycline and dysplasia of the dental enamel, and lead and altered neurologic development.

Quantitative differences in pesticide toxicity between children and adults are due in part to age-related differences in absorption, metabolism, detoxification, and excretion of xenobiotic compounds, that is, to differences in both pharmacokinetic and pharmacodynamic processes. Differences in size, immaturity of biochemical and physiological functions in major body systems, and variation in body composition (water, fat, protein, and mineral content) all can influence the extent of toxicity. Because newborns are the group most different anatomically and physiologically from adults, they may exhibit the most pronounced quantitative differences in sensitivity to pesticides. **The committee found that quantitative differences in toxicity between children and adults are usually less than a factor of approximately 10-fold.**

The committee concluded that the mechanism of action of a toxicant— how it causes harm—is generally similar in most species and across age and developmental stages within species. For example, if a substance is cytotoxic in adults, it is usually also cytotoxic in immature individuals.

Lack of data on pesticide toxicity in developing organisms was a recurrent problem encountered by the committee. In particular, little work has been done to identify effects that develop after a long latent period or to investigate the effects of pesticide exposure on neurotoxic, immunotoxic, or endocrine responses in infants and children. The committee therefore had to rely mostly on incomplete information derived from studies in mature animals and on chemicals other than pesticides.

The committee reviewed current EPA requirements for toxicity testing by pesticide manufacturers, as well as testing modifications proposed by the agency. In general, the committee found that current and past studies conducted by pesticide manufacturers are designed primarily to assess pesticide toxicity in sexually mature animals. Only a minority of testing protocols have supported extrapolation to infant and adolescent animals. Current testing protocols do not, for the most part, adequately address the toxicity and metabolism of pesticides in neonates and adolescent animals or the effects of exposure during early developmental stages and their sequelae in later life.

Age-Related Differences in Exposure

Estimation of the exposures of infants and children to pesticide residues requires information on (1) dietary composition and (2) residue concentrations in and on the food and water consumed. **The committee found that infants and children differ both qualitatively and quantitatively from adults in their exposure to pesticide residues in foods.** Children consume more calories of food per unit of body weight than do adults. But at the same time, infants and children consume far fewer types of foods than do adults. Thus, infants and young children may consume much more of certain foods, especially processed foods, than do adults. And water consumption, both as drinking water and as a food component, is very different between children and adults.

The committee concluded that differences in diet and thus in dietary exposure to pesticide residues account for most of the differences in pesticide-related health risks that were found to exist between children and adults. Differences in exposure were generally a more important source of differences in risk than were age-related differences in toxicologic vulnerability.

Data from various food consumption surveys were made available to the committee. In analyzing these data, the committee found it necessary to create its own computer programs to convert foods as consumed into their component raw agricultural commodities (RACs). This analytic approach facilitated the use of data from different sources and permitted evaluation of total exposure to pesticides in different food commodities. For processed foods, the committee noted that effects of processing on residue concentrations should be considered, but that information on these effects is quite limited. Processing may decrease or increase pesticide residue concentrations. The limited data available suggest that pesticide residues are generally reduced by processing; however, more research is needed to define the direction and magnitude of the changes for specific pesticide-food combinations. The effect of processing is an important consideration in assessing the dietary exposures of infants and young children, who consume large quantities of processed foods, such as fruit juices, baby food, milk, and infant formula.

Although there are several sources of data on pesticide residues in the United States, the data are of variable quality, and there are wide variations in sample selection, reflecting criteria developed for different sampling purposes, and in analytical procedures, reflecting different laboratory capabilities and different levels of quantification between and within laboratories. These differences reflect variations in precision and in the accuracy of methods used and the different approaches to analytical issues, such as variations in limit of quantification. There also are substantial differences in data reporting. These differences are due in part to different record-keeping requirements, such as whether to identify samples with multiple residues, and differences in statistical treatment of laboratory results below the limit of quantification.

Both government and industry data on residue concentrations in foods reflect the current regulatory emphasis on average adult consumption patterns. The committee found that foods eaten by infants and children are underrepresented in surveys of commodity residues. Many of the available residue data were generated for targeted compliance purposes by the Food and Drug Administration (FDA) to find residue concentrations exceeding the legal tolerances established by the EPA under FFDCA.

Survey data on consumption of particular foods are conventionally grouped by broad age categories. The average consumption of a hypothetical "normal" person is then used to represent the age group. However, in relying solely on the average as a measure of consumption, important information on the distribution of consumption patterns is lost. For example, the high levels of consumption within a particular age group are especially relevant when considering foods that might contain residues capable of causing acute toxic effects. Also, geographic, ethnic, and other differences may be overlooked.

To overcome the problems inherent in the current reliance on "average" exposures, the committee used the technique of statistical convolution (i.e., combining various data bases) to merge distributions of food consumption with distributions of residue concentrations. This approach permits examination of the full range of pesticide exposures in the U.S. pediatric population. As is described in the next section, this approach provides an improved basis over the approach now used for assessing risks for infants and children.

A New Approach to Risk Assessment for Infants and Children

To properly characterize risk to infants and children from pesticide residues in the diet, information is required on (1) food consumption patterns of infants and children, (2) concentrations of pesticide residues in foods consumed by infants and children, and (3) toxic effects of pesticides, especially effects that may be unique to infants and children. If suitable data on these three items are available, risk assessment methods based on the technique of statistical convolution can be used to estimate the likelihood that infants and children who experience specific exposure patterns may be at risk. To characterize potential risks to infants and children in this fashion, the committee utilized data on distributions of pesticide exposure that, in turn, were based on distributions of food consumption merged with data on the distribution of pesticide residue concentrations. The committee found that age-related differences in exposure patterns for 1- to 5-year-old children were most accurately illuminated by using 1-year age groupings of data on children's food consumption.

Exposure estimates should be constructed differently depending on whether acute or chronic effects are of concern. Average daily ingestion of pesticide residues is an appropriate measure of exposure for assessing the risk of chronic toxicity. However, actual individual daily ingestion is more appropriate for assessing acute toxicity. Because chronic toxicity is often related to long-term average exposure, the average daily dietary exposure to pesticide residues may be used as the basis for risk assessment when the potential for delayed, irreversible chronic toxic effects exists. Because acute toxicity is more often mediated by peak exposures occurring within a short period (e.g., over the course of a day or even during a single eating occasion), individual daily intakes are of interest. Examining the distribution of individual daily intakes within the population of interest reflects day-to-day variation in pesticide ingestion both for specific individuals and among individuals.

Children may be exposed to multiple pesticides with a common toxic effect, and estimates of exposure and of risk could therefore be improved by accounting for these simultaneous exposures. This can be accomplished by assigning toxicity equivalence factors to each of the compounds having a common mechanism of action. Total residue exposure is then estimated by multiplying the actual level of each pesticide residue by its toxicity equivalence factor and summing the results. This information may be combined with data on consumption to construct a distribution of total exposure to all pesticides having a common mechanism of action. To test this multiple-residue methodology, the committee estimated children's acute health risks resulting from combined exposure to five members of the organophosphate insecticide family. This was accomplished by combining actual food consumption data with data on actual pesticide residue levels.

Through this new analytical procedure, the committee estimated that for some children, total organophosphate exposures may exceed the reference dose. Furthermore, although the data were weak, the committee estimated that for some children exposures could be sufficiently high to produce symptoms of acute organophosphate pesticide poisoning.

Compared to late-in-life exposures, exposures to pesticides early in life can lead to a greater risk of chronic effects that are expressed only after long latency periods have elapsed. Such effects include cancer, neurodevelopmental impairment, and immune dysfunction. The committee developed new risk assessment methods to examine this issue.

Although some risk assessment methods take into account changes in exposure with age, these models are not universally applied in practice. The committee explored the use of newer risk assessment methods that allow for changes in exposure and susceptibility with age. However, the committee found that sufficient data are not currently available to permit wide application of these methods.

RECOMMENDATIONS

On the basis of its findings, the committee recommends that certain changes be made in current regulatory practice. Most importantly, estimates of expected total exposure to pesticide residues should reflect the unique characteristics of the diets of infants and children and should account also for all nondietary intake of pesticides. Estimates of exposure should take into account the fact that not all crops are treated with pesticides that can be legally applied to those crops, and they should consider the effects of food processing and storage. Exposure estimates should recognize that pesticide residues may be present on more than one food commodity consumed by infants and children and that more than one pesticide may be present on one food sample. Lastly, determinations of safe levels of exposure should take into consideration the physiological factors that can place infants and children at greater risk of harm than adults.

• *Tolerances.* Tolerances for pesticide residues on commodities are currently established by the EPA under FIFRA and FFDCA. A tolerance concentration is defined under FFDCA as the maximum quantity of a pesticide residue allowable on a raw agricultural commodity (RAC) (FFDCA, Section 408) and in processed food when the pesticide concentrates during processing (FFDCA Section 409). Tolerance concentrations on RACs are based on the results of field trials conducted by pesticide manufacturers and are designed to reflect the highest residue concentrations likely under normal agricultural practice. More than 8,500 food tolerances for pesticides are currently listed in the Code of Federal Regulations (CFR). Approximately 8,350 of these tolerances are for residues on raw commodities (promulgated under section 408) and about 150 are for residues known to concentrate in processed foods (promulgated under section 409).

The determination of what might be a safe level of residue exposure is made by considering the results of toxicological studies of the pesticide's effects on animals and, when data are available, on humans. Both acute and chronic effects, including cancer, are considered, although acute effects are treated separately. These data are used to establish human exposure guidelines (i.e., a reference dose, RfD) against which one can compare the expected exposure. Exposure is a function of the amount and kind of foods consumed and the amount and identity of the residues in the foods (i.e., Theoretical Maximum Residue Contributions, TMRCs). If the TMRCs exceed the RfD, then anticipated residues are calculated for comparison with the proposed tolerance. The percent of crop acreage treated is also considered. If the anticipated residues exceed the RfD, then the proposed tolerance is rejected, and the manufacturer may recommend a new tolerance level.

Although tolerances establish enforceable legal limits for pesticide residues in food, they are not based primarily on health considerations, and they do not provide a good basis for inference about actual exposures of infants and children to pesticide residues in or on foods.

Tolerances constitute the only tool that EPA has under the law for controlling pesticide residues in food. **To ensure that infants and children are not exposed to unsafe levels of pesticide residues, the committee recommends that EPA modify its decision-making process for setting tolerances so that it is based more on health considerations than on agricultural practices. These changes should incorporate the use of improved estimates of exposure and more relevant toxicology, along with continued consideration of the requirements of agricultural production. As a result, human health considerations would be more fully reflected in tolerance levels. Children should be able to eat a healthful diet**

containing legal residues without encroaching on safety margins. This goal should be kept clear.

• *Toxicity testing.* The committee believes it is essential to develop toxicity testing procedures that specifically evaluate the vulnerability of infants and children. Testing must be performed during the developmental period in appropriate animal models, and the adverse effects that may become evident must be monitored over a lifetime. Of particular importance are tests for neurotoxicity and toxicity to the developing immune and reproductive systems. Extrapolation of toxicity data from adult and adolescent laboratory animals to young humans may be inaccurate. Careful attention to interspecies differences in pharmacokinetics and metabolism of pesticides and the relative ages at which organ systems mature is essential. It is also important to enhance understanding of developmental toxicity, especially in humans, during critical periods of postnatal development, including infancy and puberty.

• *Uncertainty factors.* For toxic effects other than cancer or heritable mutation, uncertainty factors are widely used to establish guidelines for human exposure on the basis of animal testing results. This is often done by dividing the no-observed-effect level (NOEL) found in animal tests by an uncertainty factor of 100-fold. This factor comprises two separate factors of 10-fold each: one allows for uncertainty in extrapolating data from animals to humans; the other accommodates variation within the human population. Although the committee believes that the latter uncertainty factor generally provides adequate protection for infants and children, this population subgroup may be uniquely susceptible to chemical exposures at particularly sensitive stages of development.

At present, to provide added protection during early development, a third uncertainty factor of 10 is applied to the NOEL to develop the RfD. This third 10-fold factor has been applied by the EPA and FDA whenever toxicity studies and metabolic/disposition studies have shown fetal developmental effects.

Because there exist specific periods of vulnerability during postnatal development, the committee recommends that an uncertainty factor up to the 10-fold factor traditionally used by EPA and FDA for fetal developmental toxicity should also be considered when there is evidence of postnatal developmental toxicity and when data from toxicity testing relative to children are incomplete. The committee wishes to emphasize that this is not a new, additional uncertainty factor but, rather, an extended application of a uncertainty factor now routinely used by the agencies for a narrower purpose.

In the absence of data to the contrary, there should be a presumption of greater toxicity to infants and children. To validate this presumption, the sensitivity of mature and immature individuals should be studied systematically to expand the current limited data base on relative sensitivity.

• *Food consumption data.* The committee recommends that additional data on the food consumption patterns of infants and children be collected within narrow age groups. The available data indicate that infants and children consume much more of certain foods on a body weight basis than do adults. Because higher exposures can lead to higher risks, it is important to have accurate data on food consumption patterns for infants and children. At present, data are derived from relatively small samples and broad age groupings, making it difficult to draw conclusions about the food consumption patterns of infants and children. Because the composition of a child's diet changes dramatically from birth through childhood and adolescence to maturity, "market basket" food consumption surveys should include adequate samples of food consumption by children at 1-year intervals up to age 5, by children between the ages of 5 and 10 years, and by children between 11 and 18 years. Food consumption surveys should be conducted periodically to ascertain changes in consumption patterns over time.

• *Pesticide residue data.* To maximize the utility of pesticide residue data collected by various laboratories, the committee recommends the use of comparable analytical methods and standardized reporting procedures and the establishment of a computerized data base to collate data on pesticide residues generated by different laboratories. Reports on pesticide residue testing should describe the food commodity analyzed (whether processed or raw), the analytical methods used, the compounds for which tests were conducted, quality assurance and control procedures, and the limit of quantification of the tests. All findings should be reported, whether or not the residue sought is found.

–In its surveillance of pesticide residues, FDA should increase the frequency of sampling of the commodities most likely to be consumed by infants and children. The residue testing program should include all toxic forms of the pesticide, for example, its metabolites and degradation products.

–Food residue monitoring should target a special "market basket" survey focused toward the diets of infants and children.

–Pesticide field trials currently conducted by pesticide manufacturers in support of registration provide data on variation in residue concentrations associated with different rates and methods of application. Such data should be consulted to provide a basis for estimating potential maximum residue levels.

–More complete information is needed on the effects of food processing on levels of pesticides—both the parent compound and its metabolites—in specific food-chemical combinations potentially present in the diets of infants and children.

• *Risk assessment.* All exposures to pesticides—dietary and nondietary—need to be considered when evaluating the potential risks to infants and children. Nondietary environmental sources of exposure include air, dirt, indoor surfaces, lawns, and pets.

–Estimates of total dietary exposure should be refined to consider intake of multiple pesticides with a common toxic effect. Converting residues for each pesticide with a common mechanism of action to toxicity equivalence factors for one of the compounds would provide one approach to estimating total residue levels in toxicologically equivalent units.

–Consumption of pesticide residues in water is an important potential route of exposure. Risk assessment should include estimates of exposure to pesticides in drinking water and in water as a component of processed foods.

Given adequate data on food consumption and residues, the committee recommends the use of exposure distributions rather than single point data to characterize the likelihood of exposure to different concentrations of pesticide residues. The distribution of average daily exposure of individuals in the population of interest is most relevant for use in chronic toxicity risk assessment, and the distribution of individual daily intakes is recommended for evaluating acute toxicity. Ultimately, the collection of suitable data on the distribution of exposures to pesticides will permit an assessment of the proportion of the population that may be at risk.

Although the committee considers the use of exposure distributions to be more informative than point estimates of typical exposures, the data available to the committee did not always permit the distribution of exposures to be well characterized. Existing food consumption surveys generally involve relatively small numbers of infants and children, and food consumption data are collected for only a few days for each individual surveyed. Depending on the purpose for which they were originally collected, residue data may not reflect the actual distribution of pesticide residues in the food supply. Since residue data are not developed and reported in a consistent fashion, it is generally not possible to pool data sets derived from different surveys. Consequently, the committee recommends that guidelines be developed for consumption and residue data permitting characterization of distributions of dietary exposure to pesticides.

The committee identified important differences in susceptibility to the toxic effects of pesticides and exposure to pesticides in the diet with age.

For carcinogenic effects, the committee proposed new methods of cancer risk assessment designed to take such differences into account. Preliminary analyses conducted by the committee suggest that consideration of such differences can lead to lifetime estimates of cancer risk that can be higher or lower than estimates derived with methods based on constant exposure. However, underestimation of risk assuming constant exposure was limited to a factor of about 3- to 5-fold in all cases considered by the committee. Because these results are based on limited data and specific assumptions about the mechanisms by which carcinogenic effects are induced, the applicability of these conclusions under other conditions should be established.

Currently, most long-term laboratory studies of carcinogenesis and other chronic end points are based on protocols in which the level of exposure is held constant during the course of the study. To facilitate the application of risk assessment methods that allow for changes in exposure and susceptibility with age, it would be desirable to develop bioassay protocols that provide direct information on the relative contribution of exposures at different ages to lifetime risks. Although the committee does consider it necessary to develop special bioassay protocols for mandatory application in the regulation of pesticides, it would be useful to design special studies to provide information on the relative effects of exposures at different ages on lifetime cancer and other risks with selected chemical carcinogens.

In addition to pharmacodynamic models for cancer risk assessment, the committee recommends the development and application of physiologically based pharmacokinetic models that describe the unique features of infants and children. For example, differences in relative organ weights with age can be easily described in physiologic pharmacokinetic models; special compartments for the developing fetus may also be incorporated. Physiologically based pharmacokinetic models can be used to predict the dose of the proximate toxicant reaching target tissues, and may lead to more accurate estimates of risk.

In summary, better data on dietary exposure to pesticide residues should be combined with improved information on the potentially harmful effects of pesticides on infants and children. Risk assessment methods that enhance the ability to estimate the magnitude of these effects should be developed, along with appropriate toxicological tests for perinatal and childhood toxicity. The committee's recommendations support the need to improve methods for estimating exposure and for setting tolerances to safeguard the health of infants and children.

1

Background and Approach to the Study

PESTICIDES ARE USED widely in agriculture in the United States. When effectively applied, pesticides can kill or control pests, including weeds, insects, fungi, bacteria, and rodents. Chemical pest control has contributed to dramatic increases in yields for most major fruit and vegetable crops. Its use has led to substantial improvements over the past 40 years in the quantity and variety of the U.S. diet and thus in the health of the public (see, for example, Block et al., 1992).

On the negative side, many pesticides are harmful to the environment and are known or suspected to be toxic to humans. They can produce a wide range of adverse effects on human health that include acute neurologic toxicity, chronic neurodevelopmental impairment, cancer, reproductive dysfunction, and possibly dysfunction of the immune and endocrine systems.

The diet is an important source of exposure to pesticides. The trace quantities of pesticides and their breakdown products that are present on or in foodstuffs are termed residues. Residue levels reflect the amount of pesticide applied to a crop, the time that has elapsed since application, and the rate of pesticide dissipation and evaporation. Pesticide residues are widespread in the U.S. diet. They are consumed regularly by most Americans, including infants and children.

To protect the U.S. public against dietary pesticides and their potentially harmful effects, the U.S. Congress has enacted legislation to regulate residue exposures and to ensure that the food supply is safe as well abundant and nutritious. The two principal components of the legislative framework—the Federal Insecticide, Fungicide, and Rodenticide Act (FIFRA) and the Federal Food, Drug, and Cosmetic Act (FFDCA)—have provided the foundation for a comprehensive regulatory system.

Concern has arisen in recent years that the current pesticide regulatory system, which is intended to minimize health risk to the general population, may not adequately protect the health of infants and children. The traditional system assesses dietary pesticide risk on the basis of the average exposure of the entire U.S. population. However, it does not consider the range of exposures that exists within the population, nor does it specifically consider exposures of infants and children. The exposure of infants and children and their susceptibility to harm from ingesting pesticide residues may differ considerably from that of adults.

Concern about this uncertainty led the U.S. Congress in 1988 to request that the National Academy of Sciences (NAS) appoint a committee to study scientific and policy issues concerning pesticides in the diets of infants and children through its National Research Council (NRC). The committee was specifically charged with examining

- what is known about exposures to pesticide residues in the diets of infants and children;
- the adequacy of current risk assessment methods and policies; and
- toxicological issues of greatest concern and in greatest need of further research.

PESTICIDE USE

A pesticide is defined under FIFRA as "any substance or mixture of substances intended for preventing, destroying, repelling, or mitigating any insects, rodents, nematodes, fungi, or weeds, or any other forms of life declared to be pests, and any substance or mixture of substances intended for use as a plant regulator, defoliant, or desiccant."

Pesticides have been used by humankind for centuries. Their use was recorded as early as the eighth century BC when the application of fungicides was documented in Homeric poems (Mason, 1928; McCallan, 1967). From then until the present, numerous mixtures have been developed to control fungi, insects, weeds, and other pests.

In the 19th century, sulfur compounds were developed as fungicides, and arsenicals were used to control insects attacking fruits and vegetables. Those compounds were highly toxic and consequently were replaced by chlorinated organic pesticides such as DDT and benzenehexachloride (BHC), which were developed during the 1930s and became widely used in the 1950s and 1960s. Chlorinated hydrocarbon insecticides such as DDT, BHC, dieldrin, aldrin, and toxaphene were enthusiastically adopted by farmers who hoped to control previously uncontrolled insects with what were believed to be relatively safe compounds with long environmental persistence. These chemicals were also used widely in the control of malaria and other insectborne diseases. By 1955, more than 90% of all pest control chemicals used in U.S. agriculture were synthetic organic compounds, and in 1961 DDT was registered for use on 334 crops. Phenoxy herbicides such as 2,4-dichlorophenoxyacetic acid (2,4-D), 2,4,5-trichlorophenoxyacetic acid (2,4,5-T), and ethylenebisdithiocarbamates (EBDCs) and dicarboximide fungicides also gained widespread use during that time.

Beginning in the late 1960s, the potential of the chlorinated hydrocarbons for bioaccumulation and long-term toxicity became widely recognized. Also, pest resistance to chlorinated pesticides became increasingly evident and problematic throughout the 1960s, leading many farmers to substitute organophosphates and carbamates for DDT and other chlorinated compounds. Public pressure to end the use of chlorinated pesticides contributed to the creation of the Environmental Protection Agency (EPA) in 1970 and the ultimate administrative revocation in 1972 of the use of DDT on all food sources in the United States. By the end of the 1980s, most food uses of chlorinated compounds were discontinued in this country, although heavy application continues in other nations.

Since the late 1960s, a decline has occurred in insecticide use on major commodities such as corn, soybeans, cotton, and wheat. This decrease was primarily the result of pest management programs, which led to an approximately 50% reduction in pesticide application to cotton crops nationwide. Another important factor was the development and widespread adoption of synthetic pyrethroid compounds, which are applied in gram quantities rather than pounds per acre. During this period, fungicide use on peanuts and wheat declined, but because of the continued application of fungicides to fruits and vegetables and the increasing acreage of those crops under cultivation, the overall volume of fungicides used has remained steady.

In contrast, the use of herbicides has increased dramatically. In 1955 approximately 3% of all acreage planted with corn and soybean crops were treated with a herbicide; by 1985 that figure had increased to more than 95%, primarily because of the development of effective herbicides that were applied before the crop was planted. Herbicides now account for approximately 66% of all agricultural pesticides, but for a lower percentage of dietary exposure than is attributed to fungicides and insecticides, which are applied directly to the food closer to, or even after, its harvest. More than 90% of all herbicides are applied to just four crops: corn, soybeans, cotton, and wheat.

Today, most pesticides are synthetically produced organic and inorganic chemicals or microbial agents. Some of these pesticides have been found naturally and have been synthetically reproduced for commercial use. The variety and amounts of pesticides now used are far greater than at any previous time in human history. Approximately 600 pesticides are currently registered with the EPA (P. Fenner-Crisp, EPA, personal commun., 1993).

The most common food-use pesticides fall into three classes: insecticides, herbicides, and fungicides. In 1991, an estimated 817 million pounds of active pesticide ingredients were used for agricultural application in the United States. Of this total, herbicides accounted for 495 million pounds; insecticides, 175 million pounds; fungicides, 75 million pounds; and other pesticides, 72 million pounds (EPA, 1992). "Other" pesticides were defined as rodenticides, fumigants, and molluscicides but do not include wood preservatives, disinfectants, and sulfur.

- *Insecticides*. Insecticides control insects that damage crops through a variety of modes. Some work as nerve poisons, muscle poisons, desiccants, sterilants, or pheromones; others exert their effects by physical means such as by clogging air passages. The classes of insecticides most commonly used today are chlorinated hydrocarbons, organophosphates, and carbamates, and of these, the organophosphates are the most widely used. Typically they are very acutely toxic, but they do not persist in the environment. Well-known organophosphate pesticides include parathion, dichlorvos, malathion, chlorpyrifos, and azinphos-methyl. The toxicity to humans resulting from exposure to these compounds can differ markedly from chemical to chemical.

The carbamate insecticides are also very widely used in the United States today. They too are highly toxic, e.g., aldicarb. Other insecticides such as the synthetic pyrethroids, e.g., permethrin, are valued because of their fast action and relatively low toxicity to mammals.

- *Herbicides*. Herbicides are used to control weeds, which compete with crop plants for water, nutrients, space, and sunlight. By reducing the weed population, the need for farm labor is decreased and crop quality is enhanced. Herbicides work through a variety of modes of action. Some damage leaf cells and desiccate the plant; others alter nutrient uptake or photosynthesis. Some herbicides inhibit seed germination or seedling growth. Others are applied to foliage and kill on contact, thereby destroying leaf and stem tissues. Some of the most widely used herbicides are 2,4-D [(2,4-dichlorophenoxy)acetic acid], atrazine, simazine, dacthal, alachlor, metolachlor, and glyphosate.

- *Fungicides*. Fungicides control plant molds and other diseases. They include compounds of metals and sulfur as well as numerous synthetics. Some fungicides act by inhibiting the metabolic processes of fungal organisms and can be used on plants that have already been invaded and damaged by the organism. Other fungicides protect plants from fungal infections and retard fungal growth before damage to plants can occur. Fungicides frequently provide direct benefit to humans by retarding or eliminating fungal infections that can produce toxicants such as aflatoxins. Fungicides that have been used heavily over the years include benomyl, captan, and the EBDC family of fungicides such as mancozeb.

In addition to their agricultural applications, pesticides are also used for many nonagricultural purposes, e.g., in homes and public buildings to kill termites and other pests; on lawns and ornamental plantings to kill weeds, insects, and fungi; and on ponds, lakes, and rivers to control insects and weeds. Therefore, humans are exposed to pesticides from a variety of sources other than the diet, for example, through the skin or by inhalation. Some of these exposures are especially important when considering total exposures of infants and children.

PESTICIDE CONTROL LEGISLATION

The societal response to the dual nature of pesticides—to their combination of benefits and toxicity—has been to develop a comprehensive regulatory system that seeks to make possible the beneficial use of pesticides while minimizing their hazards to public health and the environment. This regulatory system originated with the enactment of FIFRA in 1947. The legislation regulating pesticides in the United States now consists of FIFRA, its comprehensive amendments of 1972, 1975, 1978, 1980, and 1988, and certain provisions of the FFDCA, which was enacted in 1954 and later amended.

FIFRA is intended by Congress to be a "balancing" or risk-benefit statute. It states that a pesticide when used for its intended purpose must not cause "unreasonable adverse effects on the environment." This balancing process must take into account "the economic, social, and environmental costs as well as the potential benefits of the use of any pesticide" [7 USC $\int$ 136(a) (1978)]. Wilkinson (1990, p. 11) has commented: "While use of the term 'unreasonable risk' implies that some risks will be tolerated under FIFRA, it is clearly expected that the anticipated benefits will outweigh the potential risks when the pesticide is used according to commonly recognized, good agricultural practices."

Under FIFRA, pesticide use is controlled through a registration process. This process is administered by EPA. A given pesticide may have several different uses, and each use is required to have its own registration. EPA registration of a pesticide use and approval of a label detailing the legally binding instructions for that use are required before a pesticide can be legally sold.

For a pesticide to be registered, manufacturers must submit to EPA the data needed to support the product's registration, including substantiation of its usefulness and disclosure of its chemical and toxic properties, its likely distribution in the environment, and its possible effects on wildlife and plants.

Pesticides that are to be registered for use on food crops must be granted a *tolerance* by EPA. These tolerances constitute the principal mechanism by which EPA limits levels of pesticide residues in foods. A tolerance concentration is defined under FFDCA as the maximum quantity of a pesticide residue allowable on a raw agricultural commodity (RAC) (FFDCA, Section 408) and in processed food when the pesticide has concentrated during processing (FFDCA, Section 409). A tolerance must be defined for any pesticide used on food crops. Tolerance concentrations on RACs are based on the result of field trials conducted by pesticide manufacturers and are designed to reflect the highest residue concentrations likely under normal agricultural practice. Thus, tolerances are based on good agricultural practice rather than on considerations of human health.

The determination of what might be a safe level of residue exposure is made by considering the results of toxicological studies of the pesticide's effects on animals and, when data are available, on humans. Both acute and chronic effects, including cancer, are considered, although currently, acute effects are treated separately. These data are used to establish human exposure guidelines (i.e., reference dose, RfD) against which one can compare the expected exposure. Exposure is a function of the amount and kind of foods consumed and the amount and identity of residues in the foods (i.e., Theoretical Maximum Residue Contributions, TMRCs). If the TMRCs exceed the RfD, then anticipated residues are calculated and compared with the proposed tolerance. The percent of crop acreage treated is also considered. If the anticipated residues exceed the RfD, then the proposed tolerance is rejected, and the manufacturer may recommend a new level.

Tolerances are the single most important tool by which the U.S. Government regulates pesticide residues in food. More than 8,500 food tolerances for all pesticides are currently listed in the *Code of Federal Regulations* (CFR). Approximately 8,350 of these tolerances are for residues on raw commodities (promulgated under section 408) and about 150 are for residues known to concentrate in processed foods (promulgated under Section 409). Table 1-1 shows the number of tolerances established for insecticides, herbicides, and fungicides in the mid-1980s for purposes of comparison.

APPROACH TO THE STUDY

Infants and children are unique. They are undergoing growth and development. Their metabolic rates are rapid. Their diets and their pat-

TABLE 1-1   Food Tolerances Established Under
Sections 408 and 409 of the Federal Food, Drug, and
Cosmetic Act

	Number of Tolerances Under:	
Type of Pesticide	Section 408	Section 409
Insecticides	3,654	63
Herbicides	2,462	39
Fungicides	1,256	20
Total	7,372	122

NOTE: This table does not include feed-additive tolerances
listed in the CFR.

SOURCE: NRC, 1987.

terns of dietary exposure to pesticide residues are quite different from
those of adults.

To determine whether the current regulatory system in the United
States adequately protects infants and children against dietary residues
of pesticides, the committee considered two main issues—susceptibility
and exposure:

• *Susceptibility*: Are infants and children more or less susceptible (sensitive) than adults to the toxic effects of pesticides? Is there a uniform and
predictable difference in susceptibility, or must each pesticide (and each
toxic response) be considered separately? Does susceptibility increase during periods of rapid growth and development? Does high metabolic activity lead to more rapid excretion of xenobiotic compounds and thus to
reduced susceptibility? Is the ability to repair damaged tissues and organs
greater in childhood, thus leading to apparently lower sensitivity? In what
fashion does the potentially long life span of infants and children affect
their susceptibility to diseases with long latent periods?

• *Exposure*: What foods do infants and children eat? How much of these
foods do they eat? How much variation in diet is there among children
in the United States? How much, and what, residues are found in or on
the food eaten by infants and children? What are the nonfood sources of
pesticide exposure? How important are they? What data are available on
exposure? Are there adequate, frequently collected food consumption data
categorized by age, sex, and race that can serve as a basis for computations
of intakes by potentially more sensitive subgroups in the population?
What are the proper measures of exposure?

The committee examined current procedures for *toxicity testing* of pesticides to learn whether these approaches provide sufficient information on toxicity in the young. Specific questions posed by the committee included: How are toxic effects identified? If they are determined by experiments in laboratory animals, what problems exist in transferring the results to humans? To infants and children? What information on toxicity is needed? For example, is information on mechanisms of action needed to establish risks to children? Are animal studies on weanlings and older animals adequate to estimate toxicity in infants and children at relatively earlier stages of development? Are there toxicities unique to some species of laboratory animals? To humans? How can exposures of animals to toxicants late in life predict responses in humans exposed early in life?

The committee reviewed approaches to pesticide *risk assessment* to assess whether these approaches adequately consider the effects of exposure in young age groups. Specific issues included: How is exposure to pesticide residues associated with response? If special consideration needs to be given to childhood exposures that result in risk, how can laboratory data from lifetime animal studies be used to develop meaningful estimates? Does risk accumulate faster during the early years of life? When exposure to a pesticide leads to more than one toxic responses, how can, or should, the total toxicity be described or evaluated?

Two final issues that the committee considered were:

• How can the lifetime risks associated with exposures to pesticides and other chemicals during infancy and childhood be assessed?

• How can methods for assessing and controlling these risks be improved?

In this report, the committee considers the development of children from the last trimester of gestation (26 weeks) through adolescence— approximately 18 years of age. Twenty-six weeks of gestation is considered the beginning of infancy because this age coincides closely with the earliest point at which an infant can survive outside the uterus. All major organ systems can function independently at that point, and the lungs have developed to the degree that reasonable exchanges of oxygen and carbon dioxide can take place.

Chapters 2, 3, and 4 of the report consider the susceptibility of infants and children to pesticides. Chapter 2 examines current evidence on the impact of children's exposures to pesticides and other toxicants in light of the special demands imposed by their rapid development, their special nutritional requirements, and their rapid metabolism. Chapter 3 explores current data on perinatal and pediatric toxicity. In Chapter 4, the committee reviews EPA's current and proposed toxicity testing requirements for pesticide registration and tolerance setting.

Chapters 5, 6, and 7 assess the dietary exposure of infants and children to pesticides. The committee began this examination by reviewing in Chapter 5 the food consumption patterns of this age group and exploring the ways that the patterns differ from those of adults—not only in the types and amounts of food and water consumed, but also in the proportion of the diet comprising certain foods. Then in Chapter 6 the committee reviews the data available on pesticide residues in food and gives particular attention to sampling of the foods consumed most by infants and children. In Chapter 7, the committee ties together the information on dietary patterns and residue levels from the two preceding chapters and provides examples for estimating the dietary pesticide exposures of infants and children. This linking of the data on dietary patterns of infants and children with data on pesticide residue levels was accomplished by applying a computer-based technology that enabled the committee to examine and quantify the full range of dietary pesticide exposures. This methodologic innovation obviates the need to study the average exposure of the hypothetical "normal" child and focuses instead on the full distribution of exposures.

In Chapter 8, the committee focuses on risk assessment. Using the data developed in Chapters 5, 6, and 7 on exposure levels, the committee presents a new method that can be used by government regulatory agencies to assess the health risks to infants and children resulting from exposures to pesticide residues in the diet. Like the exposure assessment method developed in Chapter 7, this risk assessment method permits examination of the full range of risks across the entire pediatric population.

This report embodies three unique features:

• It is the first assessment of dietary exposures to pesticides that has focused specifically on infants and children. It makes the case that children are different from the rest of the population, both in their vulnerability to toxicants as well as in their patterns of dietary exposure to pesticide residues. Children therefore deserve specific attention in the risk assessment and regulatory processes.

• It considers the total distribution of dietary exposures to pesticides among infants and children. It does not focus merely on average exposure, nor does it simply use summary statistics to examine the pesticide exposures of a hypothetical "average" child. Instead, through the use of newly applied statistical techniques, the committee was able to examine and quantify the entire range of exposures confronting the pediatric population of the United States. In this way, the committee was able to develop improved estimates of the numbers of children with high levels of dietary exposure to pesticides. This approach should be of considerable value to the government regulatory agencies, especially EPA, as they continue their efforts to use risk assessment methodologies to safeguard the health of the U.S. population.

• It proposes new cancer risk assessment methods that take into account temporal patterns of exposure to pesticide residues in the diet of infants and children, as well as tissue growth and changes in cell kinetics with age. Because of their greater consumption of certain foods relative to body weight, children may be at greater risk than adults from pesticides with carcinogenic potential. Infants and children are subject to rapid tissue growth and development, which will have an impact on cancer risk.

This report indicates how such variations in exposure with age can be accommodated in the Moolgavkar-Venzon-Knudson model of carcinogenesis (Moolgavkar et al., 1988), along with data on tissue growth and changes in cell kinetics. The methods proposed here can be adapted and extended, based on the availability of appropriate data on dietary exposure to pesticides and on tissue growth and cell kinetics, to arrive at improved estimates of lifetime cancer risks that may be posed by dietary exposure to pesticides.

REFERENCES

Block, G., B. Patterson, and A. Subar. 1992. Fruit, vegetables, and cancer prevention: A review of the epidemiological evidence. Nutr. Cancer 18:1-29.

EPA (U.S. Environmental Protection Agency). 1992. Pesticide Industry Sales and Usage. 1990 and 1991 Market Estimates. Office of Pesticide Programs. Washington, D.C.: U.S. Environmental Protection Agency.

Mason, A.F. 1928. Spraying, Dusting, and Fumigating of Plants. New York: Macmillan.

McCallan, S.E.A. 1967. History of fungicides. In Fungicides: An Advanced Treatise, Vol. 1, D.C. Torgeson, ed. New York: Academic.

Moolgavkar, S.H., A. Dewanji, and D.J. Venzon. 1988. A stochastic two-stage model for cancer risk assessment. I. The hazard function and the probability of tumor. Risk Anal. 8:383-392.

NRC (National Research Council). 1987. Regulating Pesticides in Food: The Delaney Paradox. Washington, D.C.: National Academy Press. 288 pp.

Wilkinson, C.F. 1990. Introduction and overview. Pp. 5-33 in Advances in Modern Environmental Toxicology: The Effects of Pesticides on Human Health, Vol. XVIII, S.R. Baker and C.F. Wilkinson, eds. Princeton, N.J.: Princeton Scientific. 438 pp.

2

Special Characteristics of Children

Because they are growing and developing, infants and children are different from adults in composition and metabolism as well as in physiological and biochemical processes. In a period of 26 weeks, or about 6 months, the human conceptus grows from microscopic size to recognizable human form weighing almost 500 g (1 pound). At that time, its organs and body systems (cardiovascular, pulmonary, genitourinary, gastrointestinal, neurological, hematological, immunologic, endocrine, and musculoskeletal) are sufficiently mature that extrauterine existence is possible—but survival is very risky. After 3 more months of intrauterine growth (38 weeks of gestation), the average fetal weight increases to 3.5 kg (7.5 pounds), and the organs and body systems become mature enough that adaptation to life outside the uterus is relatively assured. From birth through adolescence, physical growth and functional maturation of the body continue. The rates of physical growth and functional development vary from system to system, organ to organ, and tissue to tissue during this time. Thus, not only do infants and children differ from adults, but at any point during maturation, the individual differs in structure and function from herself or himself at any other age.

This chapter summarizes what is known about these differences. Several aspects of physical growth (structure) and development (functional maturation) are considered in terms of the implications they may have for evaluating the impacts of pesticides on children's health.

Physical development of the body (overall growth), nervous and digestive systems, liver and kidneys, and the proportions of body water and body fat are of special concern in the study of developmental toxicology. Prior to full maturation, damage to an organ or organ system, such as the central nervous system, could permanently prevent normal physical maturation. Also of concern are the physical properties of the toxic substances. For example, water-soluble compounds will be more diluted when the proportion of water in the body is higher, as in infancy, and lipid-soluble substances will be more concentrated in fatty or adipose tissue when the proportion of fat in the body is lower, as may also occur in infancy.

Functional development involves changes in the operational reactions that constitute the process of living, such as the ability to digest and absorb substances in the gastrointestinal tract, to alter their composition in the liver, to develop new metabolic pathways, and to excrete chemicals in the urine. These functions also develop at differing rates, so the organism may respond to chemicals in different ways at different ages. For example, because the filtering function of the kidney develops at a rate different from that of the reabsorptive and secretory functions, the kidney's overall ability to rid the body of toxic substances and the relative ability to excrete substances that are partially reabsorbed or are secreted will vary in a nonlinear fashion with increasing age.

The development of the functions of digestion, absorption, distribution, metabolic alteration, and elimination is of major importance in studies of developmental toxicology. This importance results in part from the possibility that toxicity to a functional system (e.g., glucose homeostasis) prior to its full development may permanently affect that system, resulting in altered function (e.g., glucose metabolism) in the mature animal, and in part from differences in the rates of absorption, metabolism, and excretion and therefore differences in susceptibility to toxicity at different ages prior to maturation.

Behavioral development includes the maturational changes in physical and mental activities associated with the relation of the individual to the environment. Behavioral development has four interrelated aspects: (a) gross motor and fine motor activities; (b) cognitive ability; (c) emotional development; and (d) social development. Alteration in one of these domains can affect the development of each of the other three.

Because of the dependence of behavioral development on physical and functional development, toxic effects occurring before maturation may permanently alter behavioral development. The most commonly encountered and well-known toxicants that can permanently change all four of the components of behavioral development are bilirubin toxicity in the newborn and lead toxicity in the infant or young child. All four aspects of behavioral development are important in studies of developmental toxicology, but much more attention has been given to the first two because they are easier to measure.

The committee's conclusions regarding how infants and children differ from adults with respect to susceptibility to toxicity are based on an extensive literature review. The few available epidemiological studies of chemical toxicity in humans, usually adults, are supplemented by experimental studies in animals. However, much less experimental work has been done on the relationship between body systems and pesticide toxicity in immature organisms than in mature ones. Therefore, data on the relationship between differences in structure and function of the young as they pertain to pesticide toxicity are supplemented with data on the effects of other toxic substances on immature organisms.

Although infant and adult nonhuman animals differ in much the same way that human infants and adults differ, there are substantial interspecies differences among the young. For example, the newborn mouse or rat more nearly resembles the human fetus in the third trimester of gestation than the human infant at birth. On the other hand, the rate of maturation and growth of the mouse or rat after birth is relatively more rapid than that of the human. Thus, cross-species comparisons of potential toxicity for pesticides in the very young animal, although helpful, cannot be used in the same manner that cross-species comparisons are used with adult animals because of differences in developmental patterns. At birth, the dog's brain reaches approximately 8% of its mature weight, the hamster 8%, the rat 15%, the mouse 22%, humans 24%, and the monkey 60% (Himwich, 1973). Subsequent rates of growth are shown in Figure 2-1. Examining the data in another way, as percent of mature weight, human brain weight at 15 to 20 months of age is similar to rat brain weight at 13 to 17 days of age. The rat brain from birth to 26 days is like the hamster brain from 3 to 17 days. For a different variable, the development of γ-aminobutyric acid (GABA) in the cat between the fetal ages of 40 and 44 days is like that of the dog between birth and 10 days of postnatal age (Himwich, 1973).

In some parts of this report, the term *development* is used to include both functional and behavioral development. Thus when reference is made to "growth and development," the term *growth* refers to the physical or structural changes associated with the process of maturing, and *development* refers to the functional and/or behavioral changes that occur during maturation.

GROWTH

Physical growth is a regulated process that represents the sum of the processes of growth of individual cells, tissues, organs, and body systems. These components do not grow at the same rate, but each component has its own rate characteristics. Thus, it is possible to predict the composition of the body from one time to the next; however, the overall composition

% of Mature Brain Weight

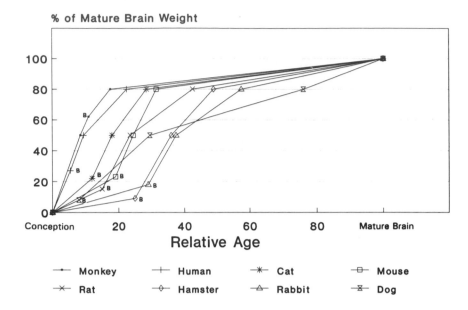

FIGURE 2-1 Differences in growth rate of the brain in several animal species. The relative-age scale depends on the age for mature brain size in each species. At birth, for example, the monkey brain will be 60% of its mature size, the human and mouse brains will be 24 and 22% of their mature size, and the hamster and dog brains will be less than 10% of mature size. Body weight does not reach 50% of its adult value until after 10 years of age, but by about 6 months of age brain weight is half of adult brain weight. Skeletal muscle grows more slowly in mass than total body weight before adolescence but increases in weight more rapidly than does body weight after adolescence. The combined weight of liver, heart, and kidneys increases more slowly than brain weight and is about 50% of that of the brain during early childhood. Liver, heart, and kidney weight reaches 50% of its adult value before adolescence and exceeds brain weight by early adolescence. B, birth. SOURCE: Based on data from Himwich, 1973. Age at brain maturity: monkey, 4 yrs.; human, 12.5 yrs.; mouse, 90 days; rat, 120 days; hamster, 60 days; dog, 120 wks.

is never the same from moment to moment until growth processes are complete.

Normal Human Growth

After birth of the the full-term human infant, growth occurs at an average rate of 800 g/month or at an incremental rate of 25% of total body weight per month. Because this rate soon slows, the actual doubling of birth weight takes 5 to 6 months. At 1 year, the infant reaches almost three times its birth weight (Table 2-1). After the first year, growth proceeds

TABLE 2-1 Average Charges in Various Body Constituents, Birth to 17 Years

Age[a] (years)	Weight (kg)	Height (cm)	Surface Area (m²)	Surface Area (kg)	Fat (% of body weight)	Fat (kg)	Lean Body (kg)	Total Body Water (liters)	Extra-cellular Water (% wt)	Extra-cellular Water (liters)	Intra-cellular Water (% wt)	Total Body Protein (kg)	Total Bone Mineral (kg)	Total Potassium (mEq)
Birth	3.5	50	0.21	0.060	14	0.5	3.0	2.4	42	1.5	27	0.46	0.09	147
0.5	8.0	67	0.37	0.045	25	2.0	6.0	4.7	33	2.6	26	0.96	0.18	324
1	10.0	75	0.44	0.044	23	2.3	7.7	6.1	32	3.2	28	1.3	0.23	439
2	12.5	86	0.54	0.043	20	2.5	10.0	7.9	32	4.0	31	1.8	0.31	590
4	16.0	102	0.66	0.041	17	2.7	13.3	10.4	31	5.0	34	2.4	0.43	825
8	25.0	127	0.94	0.038	15	3.8	21.2	16.3	29	7.3	36	4.0	0.75	1,380
13, female	46	157	1.43	0.031	25	11.5	34.5	27.6	25	11.5	35	5.0	1.6	2,105
13, male	45	156	1.40	0.031	17	7.6	37.4	28.8	27	12.1	37	7.2	1.5	2,430
17, female	57	163	1.60	0.028	31	17.9	39.1	28.5	23	13.1	27	8.0	2.1	2,424
17, male	66	175	1.80	0.027	13	8.3	57.7	41.6	25	16.5	38	9.9	2.5	3,866
R[b]	17.6	3.4	8.1	2.2	1.6	26.2	16.1	14.6	1.2	9.9	25.6	19.4	25.6	21.4

[a] Sexes are combined through age 8 years.
[b] Ratio of values for 17 years old (sexes combined) to values for newborns.

SOURCE: Federation of American Societies for Experimental Biology, 1974; Owen and Brozek, 1966; Fomon, 1966; Forbes, 1968; Fomon et al., 1982; Forbes, 1978.

at about 200 g/month. A peak gain is reached at adolescence: 500 to 600 g/month for boys and somewhat less for girls (Cheek, 1968a).

Growth in body proportions also changes dramatically during childhood as it does in utero. By 2 years of age, at four times birth weight, the toddler achieves about 20% of adult weight and about 50% of adult height; skull circumference and brain size will already be near their adult values. At birth, the length of the head is 25% of body length; by adulthood the head is only about 14% of the length of the body. Thus, growth proceeds from the head downward. The midpoint of the newborn's body is at midabdomen, but by adulthood it is at the junction of the legs with the trunk.

During infancy and adolescence, children are growing and adding new tissue more rapidly than during any other period in their postuterine life, but their various organs, tissues, and metabolic processes are maturing at different rates. For example, the neuronal cell population is relatively complete by 2 years of age, but full myelination of neuronal tissue is not complete until their 18th year. The brain achieves 50% of its adult weight by 6 months of age, whereas approximately 50% of adult stature is not reached until 2 years of age. In contrast, 50% of the adult weight of the liver, heart, and kidneys is not reached until the children are about 9 years old, and the same point in growth of skeletal muscle and total body weight is not attained until approximately the 11th year of age. Thus, as the child grows, his or her body consists of differing proportions of various tissues and organs that comprise the body. Various tissues—brain, skeletal muscle, liver, heart, and kidney—have different metabolic rates and biochemical pathways, and their changing physical proportions will alter the disposition of xenobiotic compounds over time.

The major periods of rapid growth include infancy and puberty. Growth is most rapid in the combined period of in utero development plus infancy and puberty. From birth to 4 years of age, the rate of growth decreases from 50 cm/year to the childhood value of about 10 cm/year. During puberty, the rate of linear growth increases to approximately 12 cm/year. These periods of rapid growth are believed to be susceptible to adverse influence of toxicants (Karlberg, 1989).

The hypothalamus, pituitary glands, gonads, and other tissues are involved in the control and expression of growth during puberty. Growth hormone (GH), both directly and through stimulation of the somatomedins, plays a major role in growth through endocrine, autocrine, and paracrine mechanisms. GH deficiency is associated with impaired growth, and GH excess with increased growth. The androgens and estrogens, through anabolic and other endocrine mechanisms, also modulate growth at puberty. High levels of estrogens (or estrogen-like molecules such as some pesticides) can, through effects on the growing bone, decrease the optimum height attained in adulthood or rate of linear growth during puberty.

Human Compared to Animal Infants

The newborn rabbit, rat, mouse, and hamster can double their birth weights in less than 1 week, much faster than the human infant can (Altman and Dittmer, 1962). These different growth velocities may alter the toxicity of pesticides and other chemicals among different species of infant animals. In various theories of carcinogenesis, it has been postulated that the rapidity of DNA synthesis and cell proliferation affects carcinogenicity or other toxic manifestations of chemicals. Therefore one might expect the biological effect (either positive or negative) to be more pronounced in the more rapidly growing animal—e.g., the dog, rabbit, or rat—than in the human infant. The effects on the mouse or hamster should be less pronounced than those on the dog, rabbit, or rat but still more pronounced than those on the human infant. However, if toxicity is related to absolute rates of growth, the order of increased sensitivity would be the hamster (36 days for a 10-fold increase in birth weight), the mouse (54 days), the rabbit (80 days), the rat (160 days), the dog (165 days), the monkey (1,000 days), and humans (5,000 days). This issue is complex, but in the absence of other factors, direct carcinogens are more potent in rapidly growing animals (Cohen and Ellwein, 1991; Weinstein, 1991).

Toxicologic Implications of Growth in Cell Numbers and Size

The rapid growth from conception through infancy in both animals and humans is achieved primarily by an absolute increase in the number of cells in the body (hyperplasia). After infancy, the bulk of growth is the result of increase in cell size (hypertrophy). The major exception is in the growth related to increased secretion of a series of hormones in adolescence. The change from hyperplasia to hypertrophy has major implications for adult form and size. Because of the possibility of a permanent reduction in total cell number, factors that alter growth prior to 2 years of age, even if only transient, are much more likely to result in diminished adult size than similar transient effects that occur after 2 years of age. An example of this phenomenon is the permanent reduction in body size that results from congenital rubella.

During the period of hyperplasia, which varies in duration from one organ to another, the increased rate of DNA/RNA replication and of protein synthesis may have different implications for toxicity. In a study of the age-related sensitivity of 3-, 7-, 11-, and 29-week-old rats to the induction of anemia, phenylhydrazine (PHZ) produced more marked anemia in the older animals on a milligram-per-kilogram basis (Vesell, 1982). The author postulated that the "resistance" of the younger animals could have been secondary to their increased rate of red blood cell proliferation. When a dosage based on surface area was used, the results continued to demonstrate a proportionate age-related sensitivity to PHZ.

Increased rates of cell proliferation can be associated with negative as well as positive outcomes. In the mammary gland, the proliferation of cells is greatest at the time of menarche—the first menstrual period—and the greatest risk for carcinogenesis from radiation is when the radiation occurs immediately before and after that event (Norman, 1982; Miller et al., 1989). On the other hand, radiation during infancy for treatment of an enlarged thymus also increased the risk of breast cancer 30 to 40 years later (Hildreth et al., 1989). The role of cell proliferation in carcinogenesis remains unclear and may depend on a variety of other factors (Moolgavkar and Venzon, 1979; Moolgavkar and Knudson, 1981; Finhorn, 1982; Rajewsky, 1985; Moolgavkar, 1988; Moolgavkar et al., 1988, 1990a,b; Cohen and Ellwein, 1990, 1991; Moolgavkar and Luebeck, 1990; Cohen et al., 1991).

In one study, 3- and 14-month-old rats were given intravenous doses of N-nitrosomethylurea (NMU), a known carcinogen (Anisimov, 1981). The 3-month-old rats received doses of 200 or 100 mg/kg of body weight (bw) (30 or 15 mg per rat) (groups 1 and 2, respectively), and the 14-month-old animals received 100 mg/kg bw (30 mg per rat) (group 3). The animals were observed throughout their life span, and tumor incidence was compared with that in a nontreated control group (group 4). Mammary adenocarcinomas occurred only in the rats treated at 3 months of age: 71% of group 1 and 32% of group 2 animals. Cervicovaginal sarcomas and carcinomas were found only in the group 3 animals injected at 14 months of age. Kidney tumors occurred in groups 1, 2, and 3 but were most common (54%) in the group 2 animals (the 3-month-old rats receiving the lower dose of NMU). The incidence of kidney tumors in group 1 was 10% and in group 3, 17%; there were none in the control group. Neoplasia of the hematopoietic system was found equally in groups 1, 2, and 3 (33%, 32%, and 25%, respectively, compared with 9% in the controls). The author postulated that the younger rats developed mammary and renal tumors more readily than the older animals because the proliferation of mammary gland cells and renal epithelium had decreased with age. The lack of mammary tumors in the animals treated at 14 months may reflect the insufficient number of remaining months of life to allow for development of the tumors. However, the growth of mammary tissue generally precedes the growth of vaginal/uterine tissue in most species. The older animals developed more uterine, cervical, and vaginal tumors. The higher level of estrogens at their age (14 months) was believed to be associated with more rapid proliferative activity of the target tissues (Anisimov, 1981).

In a study by Drew et al. (1983) of the effect of age and duration of exposure on the carcinogenicity of vinyl chloride (VC) in female rats, hamsters, and mice, younger animals were generally more affected than older ones. Beginning at about 2 months of age, animals were exposed to VC for 6, 12, 18, and 24 months. Other animals were exposed for 6 or 12 months beginning 6, 12, or 18 months later than the original groups. For all animals, the life span was shortened when exposure began at 2 months of age, regardless of duration of exposure (except for rats exposed for 6 months). Exposures beginning 6, 12, or 18 months later did not shorten the life span except in B6C3F$_1$ mice, which survived about 315 days from the onset of exposure regardless of the age at which exposure was initiated.

The incidence of tumors was also directly related to age of onset of exposure and duration of exposure, but age of onset appeared to be the dominant factor. One exception to this finding was that hepatocellular carcinomas were more prevalent in rats whose 6-month exposure began 6 months later than that of the group treated at 8 weeks. The incidences of hemangiosarcomas and mammary gland adenocarcinomas, as well as neoplastic nodules, were all related to the earlier age of exposure. The greatest incidence of hemangiosarcomas and stomach adenomas was observed in hamsters whose 6-month exposures began at the earliest ages. Longer exposures did not change the incidence. Similar findings were noted in the mice. However, the incidence of hemangiosarcomas was not affected by age in the B6C3F$_1$ mice, but was related to age in the Swiss mice.

In general, exposures beyond 12 months did not change the incidence of tumors in rats, and 6-month exposures of hamsters and mice were sufficient to achieve maximum tumor incidence with only two exceptions. Of note is the finding that the incidence of hemangiosarcomas and stomach adenomas in hamsters was reduced by exposures greater than 6 months. However, Drew et al. (1983) postulated that the shortened life span resulting from the longer exposures could have caused the animals whose exposure began at 2 months to die before their tumors could be expressed. The authors pointed out that animals exposed early in life must be allowed to live out their full life span if appropriate tumor incidence is to be ascertained. When this was done, it was quite apparent that the animals exposed at the youngest age had, with an occasional exception, the highest incidence of a variety of tumors. These studies lend support to the concept that the rate of cell proliferation (greatest in the youngest animals) is a contributing factor to the carcinogenic effect of VC. Similar conclusions were drawn by Swenberg et al. (1992).

Toxicologic Implications of the Growth and Development of Organs

The pattern of the rate of growth of the various organs in the human infant varies from the pattern of the overall body growth rate. Figure 2-2A illustrates the growth rate for several organs in humans. The thymus grows most rapidly initially, and throughout most of childhood exceeds its adult size. The brain approaches adult size early in childhood, although behavioral development continues for many years. The uterus and testes grow slowly until adolescence, and the growth of the ovaries is similar to that of the kidney and spleen and to increase in total body weight. Figures 2-2B and 2-2C present growth curves for some of the same organs in the rat and mouse. Each of these species has a different pattern of specific organ growth, and these differences complicate comparative stud-

A. Human

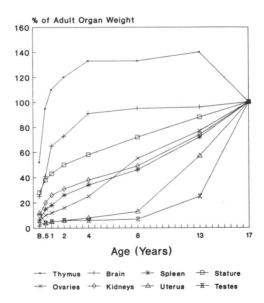

FIGURE 2-2 Organ development and stature or body weight as percentage of adult values by age in the human (A), rat (B), and mouse (C). SOURCE: Based on data from Altman and Dittmer, 1962.

B. Rat

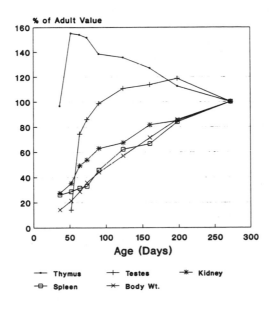

C. Mouse

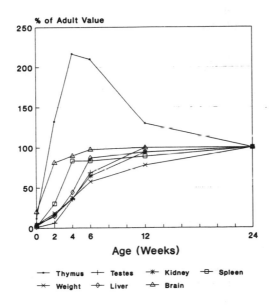

ies between species. Much less is known about the rate of functional development of various organs. The age period in which specific organs or tissues undergo their most rapid rate of development and the age at which development is completed have major implications for studies of toxicity to those organs in growing animals. Toxicity that is dependent on rates of cell proliferation (DNA / RNA replication and protein synthesis) might affect different tissues or organs at various stages of animal growth. Furthermore, whether the effect generated permanently alters the pattern of adult function, depresses adult function, or has no permanent effect may depend on whether the particular organ or tissue is still evolving or has reached its adult capacity at the time of exposure. Knowledge of the developmental pattern for various tissues and organs in animals used for studying pesticide toxicity is important in determining potential target sites and toxic end points in humans.

On another level, cellular functions within the body and within organs change with the period of overall growth. From early gestation to early infancy, blood cells are formed in the liver and spleen, but by late infancy only the bone marrow functions in the initial development of blood cells. However, "the volume of bone marrow in newborns is in fact so large that it is nearly equal to the marrow space occupied by hematopoietic cells in adults" (Nathan, 1989). For compounds that are potentially toxic to the hematopoietic system, the anatomic location and the relative size of the system may determine the degree of toxicity. Substances that accumulate in the liver or spleen in the young infant could have a direct effect on hematopoiesis—an effect that might be absent in the older child or adult. Similarly, the relatively large volume of hematopoietic tissue in the young could contribute to a more profound effect for substances that are widely distributed in the body but could reduce toxicity through dilution of compounds that concentrate in such tissue.

In several portions of the brain, the nerve cells (or neurons) that will ultimately be near the surface of the brain (in the cortex) must migrate from more central locations to the cortex during late fetal and early infant life (Rakic, 1970, 1972). This is only one of many developmental processes taking place in the brain that could be disrupted by substances toxic to the central nervous system (Langman et al., 1972). Although the developmental sequences of tissue organization and cellular maturation are similar in various animal species, the specific rate of each sequence relative to other sequences and across species differs significantly (Rodier, 1980). The myelination of nerve tracts in the spinal cord and peripheral nerves is a process that continues throughout childhood. Incomplete myelination of nerve fibers could alter their response to xenobiotic agents. The cumulative dose of cisplatin "that causes peripheral neuropathy tends to be higher in children and younger patients than in the elderly" (Legha, 1990).

Whether the decreased sensitivity in the child is related to incomplete myelination or to other metabolic differences is unknown. Thus, the impact of toxic products can produce quite different outcomes that vary both with time and with species.

Toxicologic Implications of Changes in Body Composition

The human infant is compositionally immature at birth. The infant's body is unlike the adult's in terms of body water and the relation of skeletal mass to other lean body tissue. Table 2-1 provides values for various body constituents at ages that represent significant developmental periods in the life of a child. The bottom row presents the value ratios for 17 year olds compared with newborns. From birth to age 17, body weight increases about 18-fold, stature slightly more than 3-fold, and surface area about 8-fold. Total body fat increases about 36-fold in the female and about 17-fold in the male. Body water increases about 15-fold, but extracellular water only 10-fold. Total body protein increases almost 20-fold, whereas bone mineral increases by 25-fold and total body potassium by about the same amount as total body protein because protein is located primarily in cells and potassium is primarily an intracellular ion.

The rate at which various body constituents increase and their relative proportions in the body shift during childhood (see Table 2-1). For example, total body fat accumulates most rapidly in infancy and again in adolescence, especially in the female. Bone mineral also increases markedly in adolescence—especially during the male's major growth spurt in late adolescence. Another relationship that has been known for many years is the correspondence between the extracellular fluid volume and the surface area of the body. Both of these sets of values also have a linear relationship to metabolic rate for most of childhood.

The implications of these changes in body composition for pesticide toxicity are not well defined experimentally, but one can hypothesize a number of possibilities. The relatively larger extracellular fluid volume of the infant would result in a somewhat greater dilution if the infant were to ingest equivalent amounts (on the basis of body weight) of water-soluble substances. Alternatively, lipid-soluble substances given in equivalent amounts on a body weight basis would be more concentrated in the fat of the young child because of the lower amount of fat per kilogram of body weight. Similarly, compounds that have an impact on bone growth would be more concentrated per unit of bone mass because of the bone's relatively smaller size per unit of body weight in the infant and child.

Premature and full-term infants differ from older children and adults in their body composition. Whereas the premature infant may be 85%

water, the lean body mass of the full-term infant is 82% water, compared with 72% water in the lean (fat-free or non-fat-containing) body mass of the adult. Most of the "excess" water in the infant is extracellular (Forbes, 1968). Thus, the overall amount of organ tissue per unit of whole body mass is less in the infant. One exception is the liver, which in the child is relatively larger per unit of body weight than in the adult. Several investigators have postulated that this relatively larger size of the liver could play a role in the capacity of the young child to metabolize drugs such as phenylbutazone and antipyrine more rapidly than the adult (Coppoletta and Wolbach, 1933; Alvares et al., 1975; Vesell, 1982). In addition, the organs themselves contain more water than do the organs of adults (Dickerson and Widdowson, 1960; Widdowson and Dickerson, 1960, 1964; Dickerson, 1962). Brain water decreases from 90% in the infant to 77% in the adult (Altman and Dittmer, 1973). Liver water decreases from 78% to 71% and kidney water from 84% to 81% (Widdowson, 1968). Although the total water content of skeletal muscle is relatively unchanged, the extracellular water of muscle decreases from 35% to 18%, but the intracellular water increases from 45% to 61%. The change in intracellular water is to a large extent related to an increase in cell size.

The changes in body water compartment sizes with increasing age may bear some relationship to changes in pesticide toxicity, depending on the distribution of the toxic compound. Those agents that are water soluble and distributed extracellularly will be more diluted in the youngest animals, or in human infants and small children, for comparable exposures on a milligram-per-kilogram basis. Thus one might expect toxicity to vary with age for water-soluble compounds. If the compound or its toxic metabolite is distributed intracellularly, the relatively decreased volume of intracellular water in the young would lead to toxicity varying indirectly with age. These age-related differences in toxicity emphasize the importance of ascertaining the distribution of water-soluble toxic materials within the specific tissue water compartments, extracellular or intracellular.

Cell size in infancy is relatively small. Thus, there is proportionately more cell membrane per unit of cell mass in the infant than in the adult (Cheek, 1968b). The effect of the relatively greater cell membrane and, therefore, of cell surface area compared to cell mass in the infant and young child has not been explored, but one might postulate that these proportions could increase the sensitivity of the cells of the young to compounds that act primarily at the level of the cell membrane.

DEVELOPMENT

Development, as used in this context, refers to the functional maturation of various cells, tissues, organs, organ systems, and the organism as a whole. As noted previously, the changes in function follow patterns similar to but not identical with physical growth. Functional development, like physical growth, is established to a large extent by genetic mechanisms. However, alterations in the patterns of functional development may be more readily modified by external (environmental) factors than the patterns of physical growth. In addition, there may well be external factors that modify growth but do not affect development and vice versa.

Genetics, Development, and the Environment

Understanding infant and child development in relation to the toxicity of pesticides requires an understanding of the constantly evolving interaction between a person's genetic endowment, developmental processes, and the environment. The infant or small child's phenotype (characteristics produced through interaction between genetic properties and environment) more closely resembles its genotype (genetic properties) than is true for the adult. At conception, each individual is genetically unique. From the time of conception, environmental factors may alter the genotype to produce a different phenotype. As the phenotype changes, it may alter the environment and be further altered by the environment. Specific enzyme systems may be enhanced, delayed, or altered permanently in their development by environmental factors. In addition, environmental exposures may impair, alter, or delay the development of some biochemical or physiological systems. Thus, not only will the programmed development of enzyme function alter the responses to xenobiotic compounds in immature, compared with mature, organisms, but environmental exposure to such substances before maturity may further modify response at a later time, or in some cases, throughout the life span. The specific stage in tissue or organ development when environmental factors can modify the cells to produce an effect apparent only in later life are termed critical periods of development. For example, giving insulin or glucosamine to newborn animals may permanently alter the mature animal's insulin levels and blood glucose values (Csaba and Dobozy, 1977; Csaba et al., 1979). Another example is the elevation of serum bilirubin in the human neonate, which produces altered brain function that becomes evident as the child matures. Similar elevations of serum bilirubin after infancy fail to produce these changes. Some of the damage to the central nervous system resulting from exposure to low levels of lead may not be apparent until the development of more mature functions in test animals (Csaba et al., 1979) or of such skills as reading and arithmetic in children. Similarly, neonatal exposure to diethylstilbestrol may produce effects later in life in the reproductive system (adenocarcinoma of the vagina and impaired function of the reproductive and immune systems) (Kalland, 1982).

As a general rule, compounds that interact with some genetic component of an individual are likely to be more active and cause greater impact on the young, whereas those compounds that depend for their activity on the development of acquired characteristics—such as elevated blood pressure, atherosclerosis, and loss of renal function—will be more active in the elderly. For example, the red blood cells of older mice are more easily damaged by oxidants, possibly because of acquired changes in their cell membrane (Tyan, 1982) and impaired ability to withstand oxidant stress. Similarly, the increasing presence of chromosome aberrations in the cells of the older person, perhaps because of longer exposure to toxic substances in the environment, may contribute to the increased carcinogenic susceptibility of the elderly (Singh et al., 1986).

Metabolism

The metabolic rate, a measure of the total energy expenditure of an organism, increases as the size of the organism increases. However, when energy expenditure or energy intake is evaluated per unit of mass, it decreases with increasing body size. Thus, the metabolism of exogenous substances administered in terms of body size will generally be more rapid in the immature than in the mature animal. This may have important implications for toxicity of pesticides—decreasing toxicity when the parent compound is the toxic agent and increasing toxicity when a breakdown product is the toxic substance. However, there is a variety of other aspects of changes in metabolism that can also act to modify toxicity.

The infant's body surface area per unit of body mass is two to three times greater than the adult's. This relatively large surface area per unit of mass is associated with a metabolic rate that is more than twice the adult's on a weight basis but closer to the adult value when compared per unit of surface area. Because food and water intake is related to metabolic rate, the infant's consumption will be greater than the adult's on a weight basis, but total energy consumption will be more nearly comparable to the adult's if surface area is the basis of comparison. Thus, infants differ from adults, not only in specific metabolic and pharmacokinetic measures, but also in overall metabolic rate.

Because of differences in the types of food consumed by the young (see Chapter 5), the amount of a food consumed, with its related additives, preservatives, and contaminants, may be 1 order of magnitude higher or lower for children than for adults per unit of body weight. The percent absorption varies with the development of the gastrointestinal system. The volume of distribution also depends on the degree of maturity as well as on the solubility characteristics of the substance. Because of developmental changes in enzyme activity, rates of deactivation or activation are related to the stages of maturation. The number of cell membrane receptors and the degree of protein binding are similarly variable with age. Rates of excretion by the liver and kidneys depend on both the overall organ development and the differential rates of maturation of specific organ components. For example, the development of the filtration function of the kidney is slower than the development of the reabsorption and secretory functions. Finally, the cellular response itself is dependent on intracellular maturation and differentiation. Thus, the human infant or infant animal may respond quite differently from the adult of its species to many xenobiotic substances. There is no reason to believe that these differences in response would not be equally applicable to pesticides.

Developmental Toxicity Studies

Because so many bodily functions are at various stages of development throughout infancy and early childhood, toxic effects of chemical agents during these age periods not only produce the same sorts of direct injuries to established organ tissues and functions seen in adults, but also have the potential to affect the later development of anatomic, physiologic, and metabolic processes.

During organogenesis, functional integrity does not necessarily coincide with morphological maturity, and relative organ size varies as development proceeds. Because of the evolving development of various organs and tissues, the effect of exposure to toxic substances will vary in a complex way with the age at exposure. Substances that are toxic to adults may have minimal effects at one stage of development, but at another stage these same substances may produce permanent damage to the organism or may be lethal. Such effects are particularly prominent for the central nervous system. For example, radiation treatment for medulloblastoma resulted in major cognitive problems at a later age for children less than 4 years old at the time of radiation, minimal problems later if treatment occurred at 5 to 7 years of age, and no residual cognitive difficulties when radiation was administered at 8 years and older (Chin and Maruyama, 1984).

The infant kidney is immature at birth, and the relatively poor glomerular filtration rate leads to delayed drug excretion and, therefore, an increased likelihood of toxicity (Kleinman, 1982). In the liver, the capacity to detoxify drugs by the process of conjugation develops slowly and is a major factor contributing to the toxicity of chloramphenicol in infants (Vesell, 1982).

The complexity of age-related effects of exposure for a specific tissue was demonstrated by Fleisch (1980), who identified the following factors that could change with age and thus alter the age-related sensitivity of blood vessels to various drugs:

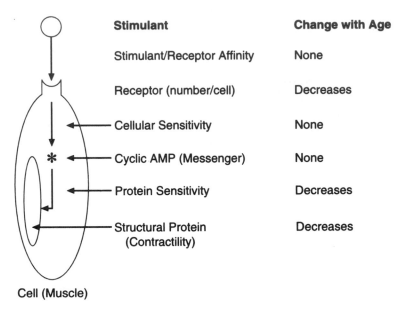

Stimulant	Change with Age
Stimulant/Receptor Affinity	None
Receptor (number/cell)	Decreases
Cellular Sensitivity	None
Cyclic AMP (Messenger)	None
Protein Sensitivity	Decreases
Structural Protein (Contractility)	Decreases

Cell (Muscle)

FIGURE 2-3   Changes in the arterial smooth muscle β-adrenergic system with age. The overall decrease in arterial smooth muscle relaxation with isoproterenol stimulation as the organism ages can be related to three separate changes in the model. SOURCE: Based on data from Fleisch, 1980.

- the number of impinging nerve cells,
- the number of drug receptors per cell,
- the relative proportion of one type of receptor to others,
- the cell membrane composition,
- the second messenger systems, and
- the concentration of reactive proteins.

The complexity of the age-related changes in vascular responsiveness is illustrated by analysis of the changes in contractile function of the arterial smooth muscle in the rat. The responsiveness of the venous smooth muscle to the stimulant isoproterenol does not change with age, whereas the responsiveness of the arterial smooth muscle increases in the postnatal period and then declines with increasing maturity. In the simplified model used to exemplify the response, isoproterenol combines with its receptor in the cell membrane and then stimulates a chemical messenger such as cyclic adenosine monophosphate (AMP), which in turn reduces the contractility of the structural protein (see Figure 2-3). In this model the affinity of isoproterenol for the receptor does not change with age, but the number of receptors per cell does decrease with age. The sensitivity of the chemical messenger to receptor stimulation does not change with maturity, but the sensitivity of the contractile protein decreases in the older rats. In addition, the overall contractile property of the structural protein shows a decline with maturation. Incidentally, there are no comparable unambiguous data on the changes in alternative stimulants with age, although age-related changes in function have been demonstrated.

Without detailed knowledge of all such age-related physiological changes and their potential interactions, it is impossible to extrapolate the impact of xenobiotic effects from mature to young animals. It is therefore necessary to examine such effects in immature animals of various ages.

A further complication is that developmental aspects of organs, tissues, and cells in the young may increase sensitivity to some drugs and decrease sensitivity to others. In fact, for the same compound one aspect of the immature organism may increase sensitivity and another aspect may decrease sensitivity. For example, the immaturity of some metabolic functions may lead to prolongation of the half-life of a toxic substance, but the larger volume of extracellular water may reduce its concentration. Reduced numbers of binding sites on target cells could also reduce the toxicity of the substance.

After reviewing a variety of studies of developmental toxicity, Calabrese (1986) concluded that there is no systematic way to predict differences in toxicity related to age. Because of differences in mechanisms of toxicity, rates of metabolism, enzyme development, stages of organ system development (some of which are more resistant to damage than others), and relative proportions of organ mass, it is unlikely that a predictive model of toxicity and development will evolve in the near future (Calabrese, 1986).

In 1983, Anisimov examined the age of animals as it relates to their sensitivity to carcinogens. He discussed four factors that may alter sensitivity on the basis of age:

- changing activity of carcinogen-metabolizing enzymes,
- changes in binding affinity,
- age-related alteration in the accuracy of DNA repair, and
- variation in proliferative activity of target tissues as a function of age

and of secondary modifiers of proliferation such as hormones and tissue-specific growth factor.

Because epidemiological data provide contradictory relationships between age at exposure and the development of cancer, and because such relationships vary with specific tumors, Anisimov examined 52 studies of chemical-induced carcinogenesis in animals of different age groups. In a tabulation of these studies for the effect of aging on latency, incidence, and tumor size, he noted that aging increased chemical carcinogenesis in 28 studies, decreased it in 19 studies, and had no effect in 5 studies. Analysis by animal species, target tissue, or chemical agent did not reveal any specific pattern (Anisimov, 1983). In his own studies of N-nitrosomethylurea in rats of various ages, Anisimov (1983) made several observations associated with increasing age at exposure: a decrease in tissue sensitivity to the action of the carcinogen in the mammary gland and kidney; an increased sensitivity in the cervix, uterus, and vagina; and no influence in the hematopoietic system (Anisimov, 1981).

If one were to examine the entire life span at many points, however, it is likely that sensitivity to pesticides and other toxic agents would not follow a regular progression in toxicity from infancy to senescence. One example is a study of the toxicity of 14 pesticides in mallard ducks at 1.5, 7, 30, and 180 days of age. In no case was there a regular progression of toxicity relative to age (Hudson et al., 1972). In all cases but two, there was either an initial decrease in toxicity followed by an increase (four cases) or an initial increase in toxicity in the earliest ages followed by a decrease (eight cases).

CONCLUSIONS AND RECOMMENDATIONS

Infants and children from the beginning of the third trimester of pregnancy until 18 years of age are the subject of this report. This age span was selected to cover the period of development from the onset of possible extrauterine existence to the age of maturation of essentially all biological systems. Reproductive toxicology and teratology are not a focus of this study.

The chapter discusses maturation in terms of physical growth, physiological and behavioral development, and the impact of the changing patterns of anatomical structure and function on the toxicity of chemicals in the immature individual. The changes do not begin at similar times or proceed at similar rates. At present, toxicity data on mature animals are insufficient for extrapolation to immature animals.

Conclusions

• Human infants and children differ from human adults not only in size but also, and more importantly, in the relative immaturity of biochemical and physiological functions in major body systems; body composition in terms of proportions of water, fat, protein, and mineral mass, as well as the chemical constituents of these body components; the anatomic structure of organs; and the relative proportions of muscle, bone, solid organs, and brain. These structural and functional differences between neonates and adults can potentially influence the toxicity of pesticides, due to qualitative and quantitative alterations in the magnitude of systemic absorption, distribution, binding, metabolism, interaction of the chemical with cellular components of target organs, and excretion. Without detailed knowledge of all age-related physiological changes and their potential interactions, it is not possible to extrapolate the impact of xenobiotic effects from mature to young animals.

• Whatever the target tissue or organ may be, some responses to pesticides may alter physiological processes that influence the uptake and metabolism of food constituents needed for optimal growth and development. Interference with such processes could result in disorders of overall growth or functional disorders of specific organ systems. It is often not known which developmental stage of individual biochemical systems, tissues, or organs will enhance, diminish, or not alter the infant's or child's sensitivity to the toxic effects of specific pesticides. In the experimental setting, therefore, the stage of growth and development of laboratory animals is a critical variable in evaluating the toxicity of pesticides.

• There are, however, specific periods in development when toxicity can permanently alter the function of a system at maturity. These special windows of vulnerability (critical periods) are often found in the early months of human pregnancies; however, some systems (e.g., the central nervous, immunologic, reproductive, and endocrine systems) continue to mature and may demonstrate particular sensitivity during the postnatal period.

• If a compound's toxicity is age related in one species, it is reasonable to assume that there will be an age relationship in other species in which the compound is toxic.

Recommendations

• Care must be taken when selecting an appropriate animal model for investigating the toxic effects of pesticides in infants and children, interpreting the data, and extrapolating the data from young animals to young humans. Toxicity in young rodents can vary substantially over a period of days, since maturation occurs so rapidly in these animals.

• In the evaluation of pesticide toxicity for immature animals, overall growth should be evaluated by measurement of growth rates and adult size at maturity.

• Studies should be conducted to examine age-related physiological changes and their potential interactions in immature animals of various ages.

• Because of the variable rates of organ development within and between species, specific organ system development should be evaluated by functional measures specific to each organ system. For example, if a chemical interfered with glomerular development in the kidney to the extent that the glomerular filtration rate were reduced by 50%, such an effect would not be likely to become apparent until maturity, and only then would the effect be documented by actual measurement of that specific function in the mature animal.

REFERENCES

Altman, P.L., and D.S. Dittmer, eds. 1962. Growth—Including Reproduction and Morphological Development. Washington, D.C.: Federation of American Societies for Experimental Biology.

Altman, P.L., and D.S. Dittmer, eds. 1973. Chemical composition of nervous tissue. Pp. 1206-1230 in Biology Data Book, 2nd edition, Vol. 2. Bethesda, Md.: Federation of American Societies for Experimental Biology.

Alvares, A.P., S. Kapelner, S. Sassa, and A. Kappas. 1975. Drug metabolism in normal children, lead-poisoned children, and normal adults. Clin. Pharmacol. Ther. 17:179-183.

Anisimov, V.N. 1981. Modifying effects of aging on N-methyl-N-nitrosourea-induced carcinogenesis in female rats. Exp. Pathol. 19:81-90.

Anisimov, V.N. 1983. Role of age in host sensitivity to carcinogens. International Agency for Research on Cancer. Sci. Publ. 51:99-112.

Calabrese, E.J. 1986. Age and Susceptibility to Toxic Substances. New York, Chichester, Brisbane, Toronto, Singapore: Wiley-Interscience Publication, John Wiley and Sons, Inc.

Cheek, D.B., ed. 1968a. A new look at growth. Pp. 3-18 in Human Growth: Body Composition, Cell Growth, Energy and Intelligence. Philadelphia: Lea and Febiger.

Cheek, D.B., ed. 1968b. Muscle cell growth in normal children. Pp. 337-351 in Human Growth: Body Composition, Cell Growth, Energy and Intelligence. Philadelphia: Lea and Febiger.

Chin, H.W., and Y. Maruyama. 1984. Age at treatment and long-term performance results in medulloblastoma. Cancer 53:1952-1958.

Cohen, S.M., and L.B. Ellwein. 1990. Proliferative and genotoxic cellular effects in 2-acetylaminofluorene bladder and liver carcinogenesis: Biological modeling of the ED01 study. Toxicol. Appl. Pharmacol. 104:79-93.

Cohen, S.M., and L.B. Ellwein. 1991. Genetic errors, cell proliferation, and carcinogenesis. Cancer Res. 51:6493-6505.

Cohen, S.M., D.T. Purtilo, and L.B. Ellwein. 1991. Pivotal role of increased cell proliferation in human carcinogenesis. Mod. Pathol. 4:371-382.

Coppoletta, J.M., and S.B. Wolbach. 1933. Body length and organ weights of infants and children. Am. J. Pathol. 9:55-59.

Csaba, G., and O. Dobozy. 1977. The sensitivity of sugar receptors: Analysis in adult animals of influences exerted at neonatal age. Endokrinologie 69:227-232.

Csaba, G., O. Dobozy, G. Lazary, and G. Kaizer. 1979. Possibility of long-lasting amplification of insulin receptors by a single treatment at newborn age. Acta Physiol. Acad. Sci. Hung. 53:487-492.

Dickerson, J.W.T. 1962. Changes in the composition of the human femur during growth. Biochem. J. 82:56-61.

Dickerson, J.W.T., and E.M. Widdowson. 1960. Chemical changes in skeletal muscle during development. Biochem. J. 74:247-257.

Drew, R.T., G.A. Boorman, J.K. Haseman, E.E. McConnell, W.M. Busey, and J.A. Moore. 1983. The effect of age and exposure duration on cancer induction by a known carcinogen in rats, mice and hamsters. Toxicol. Appl. Pharmacol. 68:120-130.

Federation of American Societies for Experimental Biology. 1974. Blood and other body fluids. Pp. 1751-2041 in Biological Handbooks: Biology Data Book, Second Edition. Bethesda, Md.: Federation of American Societies for Experimental Biology.

Finhorn, L. 1982. Oncodevelopmental biology and medicine. J. Int. Soc. Oncodevel. Biol. Med. 4:219-229.

Fleisch, J.H. 1980. Age-related changes in the sensitivity of blood vessels to drugs. Pharmacol. Ther. 8:477-487.

Fomon, S.J. 1966. Body composition of the infant. Part I: The male reference infant. Pp. 239-246 in Human Development, F. Falkner, ed. Philadelphia: W.B. Saunders.

Fomon, S.J., F. Brachka, E.E. Ziegler, and S.E. Nelson. 1982. Body composition of reference children from birth to age 10 years. Am. J. Clin. Nutr. 35:1169-1175.

Forbes, G.B. 1968. Changes in body water and electrolyte during growth and development. Pp. 80-86 in Body Composition in Animals and Man, Proceedings of a Symposium held May 4, 5, and 6, 1987 at the University of Missouri, Columbia, Publication 1958. Washington, D.C.: National Academy of Sciences.

Forbes, G.B. 1978. Body composition in adolescence. Pp. 239-272 in Human Growth, Vol. 2: Postnatal Growth, F. Falkner and J.M. Tanner, eds. New York: Plenum.

Hildreth, N.G., R.E. Shore, and P.M. Dvoretsky. 1989. Risk of breast cancer after irradiation of the thymus in infancy. New Engl. J. Med. 321:1281-1284.

Himwich, W.A. 1973. Problems in interpreting neurochemical changes occurring in developing and aging animals. Prog. Brain Res. 40:13-23.

Hudson, R.H., R.K. Tucker, and M.A. Haegele. 1972. Effect of age on sensitivity: Acute oral toxicity of 14 pesticides to mallard ducks of several ages. Toxicol. Appl. Pharmacol. 22:556-561.

Kalland, T. 1982. Long-term effects on the immune system of an early life exposure to diethylstilbestrol. Pp. 217-241 in Banbury Report 11: Environmental Factors in Human Growth and Development, V.R. Hunt, M.K. Smith, and D. Worth, eds. New York: Cold Spring Harbor Laboratory.

Karlberg, J. 1989. On the construction of the infancy-childhood-puberty growth standard. Acta Paediatr. Scand. Suppl 356:26-37.

Kleinman, L.I. 1982. The effect of lead on the maturing kidney. Pp. 153-171 in Banbury Report 11: Environmental Factors in Human Growth and Development, V.R. Hunt, M.K. Smith, and D. Worth, eds. New York: Cold Spring Harbor Laboratory.

Langman, J., M. Shimada, and P. Rodier. 1972. Floxuridine and its influence on postnatal cerebellar development. Pediatr. Res. 6:758-764.

Legha, S.S. 1990. Letter to the editor. New Engl. J. Med. 323:65.

Miller, A.B., G.R. Howe, G.J. Sherman, J.P. Lindsay, M.J. Yaffe, P.J. Dinner, H.A. Risch, and D.L. Preston. 1989. Mortality from breast cancer after irradiation during fluoroscopic examinations in patients being treated for tuberculosis. New Engl. J. Med. 321:1285-1289.

Moolgavkar, S.H. 1988. Biologically motivated two-stage model for cancer risk assessment. Toxicol. Lett. 43(1-3):139-150.

Moolgavkar, S.H., and D.J. Venzon. 1979. Two-event models for carcinogenesis: Incidence curves for childhood and adult tumors. Math. Biosci. 47:55-77.

Moolgavkar, S.H., and A.G. Knudson. 1981. Mutation and cancer: A model for human carcinogenesis. J. Natl. Cancer Inst. 66(6):1037-1052.

Moolgavkar, S.H., and G. Luebeck. 1990. Two-event model for carcinogenesis: Biological, mathematical, and statistical considerations. Risk Anal. 10:323-341.

Moolgavkar, S.H., A. Dewanji, and D.J. Venson. 1988. A stochastic two-stage model for cancer risk assessment. I. The hazard function and the probability of tumor. Risk. Anal. 8:383-392.

Moolgavkar, S.H., E.G. Luebeck, M. de Gunst, R.E. Port, and M. Schwarz. 1990a. Quantitative analysis of enzyme-altered foci in rat hepatocarcinogenesis experiments—I: Single agent regimen. Carcinogenesis 11:1271-1278.

Moolgavkar, S.H., F.T. Cross, G. Luebeck, and G.E. Dagle. 1990b. A two-mutation model for radon-induced lung tumors in rats. Radiat. Res. 121:28-37.

Nathan, D.G. 1989. The beneficence of neonatal hematopoiesis. New Engl. J. Med. 321:1190-1191.

Norman, J.E., Jr. 1982. Breast cancer in women irradiated early in life. Pp. 433-450 in Banbury Report No. 11: Environmental Factors in Human Growth and Development, V.R. Hunt, M.K. Smith and D. Worth, eds. New York: Cold Spring Harbor Laboratory.

Owen, G.M., and J. Brozek. 1966. Influence of age, sex and nutrition on body composition during childhood and adolescence. Pp. 222-238 in Human Development, F. Falkner, ed. Philadelphia, Pa.: Saunders.

Rajewsky, M.P. 1985. Tumorigenesis by exogenous carcinogens: Role of target-cell proliferation and state of differentiation (development). Pp. 215-224 in Age-Related Factors in Carcinogenesis, Proceedings of Symposium, International Agency for Research on Cancer and N.N. Petrov Research Institute of Oncology, IARC Scientific Publications No. 58, A. Likhachev, V. Anisomov, and R. Montanaro, eds. Lyon, France: International Agency for Research on Cancer.

Rakic, P. 1970. Neuron-glia relationship during granule cell migration in developing cerebellar cortex. A Golgi and electromicroscopic study in macacus rhesus. J. Comp. Neurol. 141:283-312.

Rakic, P. 1972. Mode of cell migration to the superficial layers of fetal monkey neocortex. J. Comp. Neurol. 145:61-83.

Rodier, P.M. 1980. Chronology of neuron development: Animal studies and their clinical implications. Dev. Med. Child. Neurol. 22:525-545.

Singh, S.M., J.F. Toles, and J. Reaume. 1986. Genotype- and age-associated in vivo cytogenetic alterations following mutagenic exposures in mice. Can. J. Genet. Cytol. 28:286-293.

Swenberg, J.A., N. Fedtke, and L. Fishbein. 1992. Age-related differences in DNA adduct formation and carcinogenesis of vinyl chloride in rats. Pp. 163-171 in Similarities and Differences Between Children and Adults, P.S. Guzelian, C.J. Henry, and S.S. Olin, eds. Washington, D.C.: ILSI Press.

Tyan, M.L. 1982. Age-related increase in erythrocyte oxidant sensitivity. Mech. Ageing Dev. 20:25-32.

Vesell, E.S. 1982. Dynamically interacting genetic and environmental factors that affect the response of developing individuals to toxicants. Pp. 107-124 in Banbury Report No. 11: Environmental Factors in Human Growth and Development, V.R. Hunt, M.K. Smith, and D. Worth, eds. New York: Cold Spring Harbor Laboratory.

Weinstein, I.B. 1991. Mitogenesis is only one factor in carcinogenesis. Science 251:387-388.

Widdowson, E.M. 1968. Biological implications of body composition. Pp. 71-79 in Body Composition in Animals and Man, Proceedings of Symposium, Publication 1598. Washington, D.C.: National Academy of Sciences.

Widdowson, E.M., and J.W.T. Dickerson. 1960. The effect of growth and function on the chemical composition of soft tissues. Biochem. J. 77:30-43.

Widdowson, E.M., and J.W.T. Dickerson. 1964. Chemical composition of the body. Pp. 1-247 in Mineral Metabolism: An Advanced Treatise, C.L. Comar and F. Bronner, eds. New York: Academic Press.

3

Perinatal and Pediatric Toxicity

BOTH ACUTE AND CHRONIC toxic reactions in the young are often considered together under the title of developmental toxicity. Such toxicity can be further subdivided by the organ system involved or by whether the toxic effect occurred before or after birth. The developmental purview of the committee extends from the beginning of the third trimester through 18 years of age; however, no single theoretical framework or unifying set of principles readily applies to so broad a developmental span. *Teratology*, the study of congenital malformations, has traditionally focused on the process of organogenesis, the sensitive period in prenatal development when birth defects can be induced by exposure to either endogenous (e.g., endocrine) or exogenous (e.g., xenobiotic) agents. One view of teratogenesis is that this type of abnormal development represents a special form of embryotoxicity.

Developmental toxicology includes the study of chemically induced alterations of the normal sequence of developmental processes. It both encompasses and expands the domain of abnormal development beyond that implied by teratology. Although the term denotes adverse chemical effects on development, its end points are not restricted to gross anatomical defects but encompass multiple expressions of abnormal outcome. This research specialty combines basic principles, concepts, and working assumptions from several disciplines, including developmental and cellular biology, pharmacology, and toxicology. A major objective is to understand how exogenous agents interfere with the normal progression of developmental events to produce phenotypically abnormal cells, tissues, organs, and function. Since this report's focus begins with the third trimester, the committee does not directly consider the teratogenicity of pesticides, i.e., their potential to produce gross structural malformations. Rather, the focus is on processes that occur after the completion of organogenesis and continue well into the postnatal period. However, the origins of this broader concern with peri- and postnatal toxicology are inextricably rooted in experimental teratology.

Studies of the toxicity of xenobiotic compounds in children have demonstrated the potential for either acute or chronic exposure to result in serious malfunctions at a later age. This potential exists because of the developmental character of the physiologic/biochemical/molecular function of the young individual. While a biologic system is developing, a toxic event can alter one aspect of that development so that all subsequent reactions are altered or modified. For example, transient elevations of serum bilirubin during the newborn period may produce changes in the basal ganglia of the brain that may not become apparent until several years later but are then permanent in nature.

ACUTE TOXICITY

In this section, the committee discusses and summarizes the relative sensitivity of infants, children, and adults to the acute toxicity of chemicals. Acute toxicity here is defined as toxicity resulting from a single exposure to a chemical. The injury may be immediate or delayed in onset. Both lethality and target organ injury will be considered as toxic end points. A limited number of findings from studies of laboratory animals are summarized where data on humans are inadequate. Because of the meager data base on age-dependent acute toxicity of pesticides, some examples of pharmacologic effects and adverse effects of therapeutic agents in pediatric and adult populations are described. Attention is focused, in turn, on age-related differences in the lethality of pesticides and other chemicals, differential effects of cholinesterase inhibitors in immature and mature subjects, and age-related effects of toxic and pharmacologic actions of selected therapeutic agents.

Data on age-related susceptibility to the lethal effects of chemicals are largely limited to acute LD_{50} studies in laboratory animals. Done (1964) was one of the first investigators to compile the results of LD_{50}s and other measures of lethality of a variety of chemicals in immature and mature animals. Immature animals were more sensitive to 34 chemicals, whereas mature animals were more sensitive to 24 compounds. Thiourea was 50 to 400 times more toxic (i.e., lethal) in adult than in infant rats. Conversely, chloramphenicol was 5 to 16 times more toxic in 1- to 3-day-old rats. Thus, Done (1964) concluded that immaturity does not necessarily entail greater sensitivity and that age-dependent toxicity is chemical dependent. Goldenthal (1971) tabulated LD_{50} values for newborn and neonatal animals as compared to adult animals primarily from data submitted by pharmaceutical firms in drug applications. Approximately 225 of these compounds were more acutely toxic (lethal) to neonates, whereas about 45 were more toxic to adult animals. Almost all the age-related differences in LD_{50}s in the reports of Goldenthal (1971) and Done (1964) were less than 1 order of magnitude; indeed, most varied no more than two- to threefold.

As discussed in Chapter 2, there are important differences between immature laboratory animals and humans. Nonprimate species are generally less mature at birth than are humans. Newborn mice and rats are among the most immature of commonly used test species, so it is not surprising that they often differ markedly from adult animals in sensitivity to chemicals. This phenomenon is particularly evident in the paper by Goldenthal (1971), who reported five times as many chemicals to be more acutely toxic to newborn than to adult animals. Since full-term human newborns are more mature, such pronounced age-dependent differences in toxicity would not be anticipated. Maturation in rodents is very rapid, so that even a few days of age can result in a marked disparity in test results (Done, 1964). Furthermore, organs and their associated functions mature at different rates in different species. Uncertainty in extrapolating findings among different species of mature animals is appreciable. When the additional variable of interspecies maturation patterns is introduced, the choice of an appropriate animal model for pesticide toxicity of neonates, infants, and children becomes even more complex.

The relative acute lethality of pesticides to immature and mature animals has been the subject of a number of studies. Goldenthal (1971), in his extensive compilation of LD_{50} values for newborn and adult animals, included several fungicides, herbicides, and the insecticide heptachlor. Each of these compounds was more toxic to newborn than to adult rats. Gaines and Linder (1986) more recently contrasted the acute toxicity of 36 pesticides given orally to weanling (4 to 6 weeks old) and to young adult Sherman rats. Age-related differences, where they existed, were usually no more than two- to threefold. Weanlings were more sensitive than adults to only 4 of the 36 compounds. Lu et al. (1965) observed that 14- to 16-day-old rats were intermediate between newborns (most sensitive) and adults (least sensitive) in their susceptibility to malathion poisoning. Such findings are in agreement with the observation that physiological and biochemical processes, which govern the pharmacodynamics of pesticides, mature quite rapidly in rodents. Indeed, metabolism and renal clearance of xenobiotic compounds and their metabolites soon approach and may exceed adult capacities in rodents within 2 to 3 weeks. This same phenomenon occurs in humans, albeit at a somewhat slower pace (i.e., within the first weeks to months of life). Higher metabolism may confer protection against pesticides or increased susceptibility to injury, depending on the relative toxicity (and rate of elimination) of the parent compound compared to its metabolites. The findings of Lu et al. (1965) are a good case in point. These investigators contrasted acute oral LD_{50} values for newborn, 14- to 16-day-old, and young adult Wistar rats. The adult animals were the most resistant to malathion, as would be anticipated, since adult rats most efficiently metabolize organophosphates and organophosphates are metabolically inactivated (Benke and Murphy, 1975). Conversely, the older rats of Lu et al. (1965) were the most sensitive to the acute toxicity of dieldrin. Thus, susceptibility to acute pesticide toxicity appears to be a function of age, species, and chemical.

Limitations of acute lethality data should be recognized. Acute doses of chemicals high enough to cause death may damage organ systems by mechanisms that are quite different from those that produce biological effects from chronic exposures to lower levels. MacPhail et al. (1987) examined age-related effects of a number of pesticides on lethality, serum chemistry, and motor activity in weanling and adult male rats. Although age was generally not an important determinant of toxicity for most of the pesticides, there were age-related differences in the effects of carbaryl and diazinon on motor activity. These results could not have been predicted on the basis of LD_{50} values for the two groups, leading MacPhail et al. (1987) to conclude that mortality may be a poor predictor of morbidity and that nonlethal end points should be used to assess the age-dependency of the neurobehavioral toxicity of pesticides. More sensitive indices should also be used to monitor other potentially vulnerable systems in infants and children, including the hormonal and reproductive systems, the immune system, the nervous system, developmental effects, and carcinogenesis/ mutagenesis. Unfortunately, relatively few well-controlled studies have been conducted, particularly in humans, in which sensitive end points are used to assess the relative toxicity of comparable doses of pesticides or other chemicals in pediatric and adult populations.

Cholinesterase inhibition, a mechanism by which organophosphate and carbamate insecticides produce excessive cholinergic effects, is a sensitive end point that can be monitored in humans and other mammals. Brodeur and DuBois (1963) reported that weanling (23-day-old) rats were more susceptible than adults to the acute toxicity of 14 of 15 organophosphates tested. The greater toxicity of parathion in weanling rats was tentatively attributed to deficient hepatic detoxification of parathion and its bioactive oxygen analogue, paraoxon (Gagne and Brodeur, 1972). A comprehensive investigation was reported by Benke and Murphy (1975) in five age groups of male and female Holtzman rats: 1, 12 to 13, 23 to 24, 35 to 40, and 56 to 63 days old. There was a progressive decrease in susceptibility to poisoning by parathion and parathion-methyl with increasing age up to

35 to 40 days for both sexes. Detailed experiments were conducted to determine the influence of aging on metabolic activation of the two compounds, as well as on detoxification systems (e.g., aryl esterase-catalyzed hydrolysis, glutathione-dependent dearylation and dealkylation, and binding in the liver and plasma). Benke and Murphy (1975) concluded that increased detoxification of the active oxygen analogues of parathion and parathion-methyl was largely responsible for the lower acute toxicity of the two insecticides in adult animals. Murphy (1982) subsequently pointed to two other factors that contributed to the lower sensitivity of adult rats to organophosphates: greater binding to noncritical tissue constituents and more rapid catabolism of the parent compounds.

The limited information available suggests that immature humans also experience greater susceptibility to organophosphate- and carbamate-induced cholinesterase inhibition and related effects. In 1976 in Jamaica, 79 people were acutely poisoned as a result of eating parathion-contaminated flour (Diggory et al., 1977). Seventeen of the patients died. Case-fatality ratios were highest (i.e., 40%) among children ranging from newborns to 4 years of age. Zwiener and Ginsburg (1988) presented the clinical histories of 37 infants and children exhibiting moderate to severe organophosphate and carbamate toxicity. Although most of these patients ingested the pesticides, six became intoxicated after playing on sprayed surfaces. Zwiener and Ginsburg (1988) noted that 76% of their subjects were younger than 3 years old. The investigators found there was a paucity of information in the literature on the toxicity of cholinesterase inhibitors in infants and children.

Parathion contamination of stored foodstuffs (Diggory et al., 1977) and aldicarb contamination of crops (Goldman et al., 1990) have resulted in the most widespread outbreaks of foodborne pesticide toxicity in North America. Goldman and co-workers investigated more than 1,000 cases of illness caused by consumption of aldicarb-contaminated watermelons and cucumbers. Unfortunately, infants and children were not studied as a subpopulation at risk. The investigators did calculate doses of aldicarb sulfoxide that produced illness in the general population and estimated that a 10-kg child could readily consume enough of the pesticide on watermelons to experience toxicity. The U.S. Environmental Protection Agency (EPA, 1988) concluded that infants and children are at the greatest risk of acute aldicarb toxicity. This conclusion was based on dietary consumption and contamination patterns, however, rather than on the greater sensitivity of infants and children to this potent cholinesterase inhibitor.

Although immature humans appear to be more susceptible than adults to the acute effects of cholinesterase inhibitors, the age-dependency of this phenomenon is not entirely clear. Some of the most applicable information has been provided by a study of the perinatal development of human blood esterases (Ecobichon and Stephens, 1973). Erythrocyte ace-tylcholinesterase and plasma pseudocholinesterase and arylesterase activities were measured in premature newborns of varying gestational age as well as in full-term newborns, children of different ages, and adults. Apparent K_m values for the three enzymes did not vary significantly with age for a variety of substrates, indicating that the enzyme properties were similar in all age groups. Enzymatic activity, however, did vary significantly with age. Levels of all three enzymes progressively increased during gestation, then rose markedly during the first year of life. Thereafter, erythrocyte cholinesterase and pseudocholinesterase activities increased gradually to adult levels. If one were to assume that one of these peripheral enzymes (e.g., erythrocyte cholinesterase) reflects brain acetylcholinesterase levels, then the most pronounced effects of cholinesterase inhibitors may be expected to occur in newborns, neonates, and infants, since a chemically induced depression of enzymatic activity may be more apparent when baseline cholinesterase levels are relatively low. Ecobichon and Stephens (1973) provided evidence of another mechanism of increased susceptibility of newborns—namely, diminished detoxification capacity (i.e., significantly lower plasma arylesterase and paraoxon hydrolysis activities). Children 2 to 8 years old had slightly lower activities than adults, suggesting that younger children may be somewhat more susceptible to cholinesterase inhibitors. The consequences of brain acetylcholinesterase inhibition on nervous system development and postnatal function remain largely unexplored.

Because of the paucity of data on the age-dependency of acute toxicity of pesticides in humans, the remainder of this section focuses on relative effects of therapeutic agents in pediatric and adult populations. Substantially more information should be available on drugs, due to their common use in all age groups and stringent requirements by the Food and Drug Administration (FDA) for demonstration of safety and efficacy. Data from well-controlled, parallel studies in infants, children, and adults, however, are quite limited for most drugs.

Done et al. (1977) reported what was termed a *therapeutic orphan* problem—namely, that safety and efficacy for children had not been proved for 78% of new drugs then marketed in the United States. A 1990 survey by the American Academy of Pediatrics revealed that the labeling of 80% of new drugs approved by the FDA between 1984 and 1989 did not include information on pediatric use. The FDA's policy has allowed the marketing of drugs that have been approved for adults but not studied in children, as long as labeling included disclaimers and no instructions about pediatric use. Without adequate information, physicians commonly prescribe such medications for children, possibly placing pediatric populations at increased risk of uncertain efficacy or adverse reactions.

The FDA (1992) proposed to amend labeling requirements for prescription drugs to promote their safe and effective use in children. Misunderstandings and concern about legal and ethical implications have limited clinical research in pediatric populations. The newly proposed guidelines provide alternative ways to assess effectiveness and safety in children without necessarily having to conduct comprehensive studies. Results from well-controlled studies in adults can be extrapolated to children under some circumstances, although separate pharmacokinetic studies are needed to establish appropriate pediatric dosage regimens. The intent of the proposed amendment is to provide more complete information on labeling of prescription drugs concerning use and possible hazards for children.

Several instances of severe adverse effects from pharmaceutical agents in pediatric populations have attracted widespread attention. During the 1950s, chloramphenicol produced a pallid cyanosis, which progressed to circulatory collapse and death in some newborns (Sutherland, 1959). This so-called *gray baby syndrome* has been attributed to the diminished hepatic glucuronide conjugation and renal secretory capacities of newborns. Weiss et al. (1960) reported blood half-lives of 26, 10, and 4 hours for chloramphenicol at birth, at 10 to 16 days of age, and in children 4 to 5 years old, respectively. Thus, there is a substantial increase in chloramphenicol metabolism and excretion capacity during the first days and weeks of life. Decreased metabolic and excretory capacities of newborns and neonates have been associated with exaggerated toxicity of a number of other chemicals, including benzyl alcohol (Gershanik et al., 1982), hexachlorophene (Tyrala et al., 1977), and diazepam (Nau et al., 1984). The hexachlorophene poisonings appeared to be associated with increased percutaneous absorption as well as deficient metabolism in newborns. *Floppy infant syndrome* in babies born to mothers given diazepam is apparently the result of a number of age-dependent factors, including a smaller volume of distribution and thus greater target organ concentrations of the lipophilic drug due to a smaller adipose tissue volume in newborns, increased amounts of free diazepam due to displacement of the drug from plasma protein binding sites by elevated free fatty acid levels, and a prolonged half-life as a result of diminished oxidative and conjugative metabolism (Warner, 1986). As discussed in Chapter 2, most physiological processes that govern the kinetics of drugs and other chemicals mature during the first year after birth. Indeed, profound changes in some processes (e.g., phase I and II metabolism) occur during the first days and weeks of life (Morselli, 1989). Thus, the most pronounced differences from adults in susceptibility to drug toxicity would be expected in newborns, neonates, and infants; the youngest are most likely to experience the most aberrant responses.

The net effect of immature physiological and biochemical processes on drug efficacy and toxicity is difficult to predict. The various processes mature at different rates and may enhance or offset one another. Local anesthetics provide a good illustration. These drugs are commonly administered to the mother during labor and delivery and may readily enter the maternal circulation and cross the placenta (Tucker and Mather, 1979). Cardiovascular depression and respiratory depression in newborns have occasionally been reported, although subtle neurophysiological impairment and behavioral changes are probably more common consequences (Dodson, 1976; Ostheimer, 1979). Premature and full-term newborns exhibit lower plasma protein binding of local anesthetics. This should result in increased amounts of free drug and a more pronounced pharmacologic response, but the greater volume of distribution in newborns reduces the concentration of drug at sites of action. Rates of hepatic microsomal metabolism and plasma pseudocholinesterase-catalyzed hydrolysis of anesthetics such as procaine are quite low in newborns. This deficit in metabolism, coupled with the larger distribution volume that must be cleared of drug, accounts for the prolonged half-life and long duration of action of lidocaine and its analogues in neonates (Morselli et al., 1980).

Hepatic metabolism and renal clearance of xenobiotic compounds change dramatically during the first year of life. Phase I metabolic reactions (e.g., oxidation) may rise from one-fifth to one-third of the adult rate during the first 2 to 3 postnatal weeks to two to six times the adult rate (Neims et al., 1976; Morselli, 1989). Different isozymes and enzymes mature at different ages. Certain phase II (e.g., glucuronidation) reactions do not reach adult levels for months, while maturation of alcohol dehydrogenase activity may take as long as 5 years (Kearns and Reed, 1989). The majority of xenobiotics, however, are metabolized most rapidly by individuals between 2 to 4 months and about 3 years of age. Thereafter, drug metabolism gradually declines to adult levels (Warner, 1986). Development of renal function displays a similar age-dependency. Glomerular filtration increases dramatically during the first week of life, approaching and exceeding adult values within 3 to 5 months. Renal tubular secretory and absorptive processes mature more slowly (Kearns and Reed, 1989). Older infants and children, therefore, may be less susceptible than adults to drugs that are metabolized to less toxic, more readily excretable metabolites. Spielberg (1992) noted that clearance of nearly all anticonvulsant drugs is quite limited in newborns, especially premature newborns. Conversely, clearance of such drugs (e.g., phenytoin, phenobarbital, carbamazepine, and diazepam) in infants and children, when calculated on a milligram-per-kilogram-of-body-weight basis, was well above that in adults until around puberty. Thus, children are less likely than adults to exhibit toxicity and require higher doses (on a milligram-per-kilogram-of-body-weight basis) of anticonvulsants to achieve therapeutic levels. In contrast, infants and children may be at greater risk from other drugs and chemicals that undergo metabolic activation (i.e., conversion to bioactive or cytotoxic metabolites). Unfortunately, there is a lack of information on such agents in humans in the published literature.

There was concern that acetaminophen (Tylenol), a drug that undergoes metabolic activation to hepatocytotoxic metabolite(s) via a P-450-mediated mixed-function oxidase (MFO) pathway, would cause increased morbidity and mortality in young children. This concern was never realized, however, since hepatotoxicity in young children was found to be less severe than in adults, and has rarely resulted in deaths of children (Rumack, 1984). Acetaminophen is metabolized by several parallel pathways. The two major detoxification pathways involve conjugation of the parent compound with sulfate or glucuronide. Thus only a small fraction of the drug remains to be oxidized by the P-450-mediated pathway to a reactive intermediate (*N*-acetyl-*p*-benzoquinonimine). This metabolite is conjugated with glutathione to produce nontoxic products or can bind covalently to cell proteins and nucleic acids, causing cellular injury (Hinson et al., 1990). Although prepubescent children have relatively high hepatic MFO activity, they also exhibit a greater capacity than adults to detoxify acetaminophen by phase II metabolic reactions, primarily sulfate conjugation (Miller et al., 1977). Also, higher glutathione levels in the young may contribute to protection from hepatotoxicity. Thus, the lower susceptibility of children to acetaminophen poisoning is due to their greater capacity to eliminate the drug by nontoxic pathways (Kauffman, 1992).

Clinical trials in infants and children are relatively infrequent for most classes of drugs, but this is not the case for many antineoplastic agents. Although some types of childhood cancer are refractory to chemotherapy, others have excellent cure rates (Petros and Evans, 1992). Therefore, phase I clinical trials are frequently conducted in both adult and pediatric populations to define the maximum tolerated dose (MTD) for appropriate dosage schedules in phase II trials. Antineoplastic agents include a wide variety of different types of chemicals that act by diverse mechanisms. Thus, results of phase I studies of anticancer drugs afford scientists some of the most comprehensive data sets for contrasting toxic effects of chemicals in children and adults. The investigations typically involve repetitive dosage regimens lasting days or weeks, however, rather than single, acute exposures.

Comparable clinical trials of antineoplastic agents in pediatric and adult patient populations have revealed toxic effects that are often similar qualitatively but different quantitatively (Glaubiger et al., 1982; Marsoni et al., 1985; Evans et al., 1989). In a compilation of data on 16 compounds for which there had been comparable phase I trials in adults and children,

TABLE 3-1   Maximum Tolerated Dose (MTD) of Some Anticancer
Drugs in Children and Adults

Drug	MTD (mg/m^2) Children	Adults	Ratio, MTD for Children/ MTD for Adults
Dianhydrogalactitiol	25	30	0.83
5-Azacytidine	200	225	0.89
TIC mustard	900	1,000	0.90
Piperazinedione	3	3	1.0
VP16-213	150	125	1.20
Diglycoaldehyde	≥7,500	6,000	1.25
m-AMSA	50	40	1.25
Daunomycin (mg/kg)	1.0	0.8	1.25
Adriamycin (mg/kg)	0.8	0.6	1.33
VM-26 (mg/kg)	4.0	3.0	1.33
3-Deazauridine (leukemia patients)	8.2	6.0	1.40
Azaserine (mg/kg) (total dose)	156	108	1.44
Anhydro-5-fluoro-cyclocytidine	≥300	≥200	1.50
Dihydroxyanthracenedione	18	12	1.5
3-Deazauridine (solid tumors)	2.8	1.5	1.85
Cyclocytidine	600	300	2.00
ICRF-187	>2,750	1,250	>2.20

SOURCE: Glaubiger et al., 1982.

the types of toxic effects that limited further dosage escalation were generally the same (Glaubiger et al., 1982). As shown in Table 3-1, the MTD
for children was higher than that for adults for 13 of the compounds.
Similar findings were reported by Marsoni et al. (1985). These investigators
compared the MTDs and recommended phase II doses in children and
adults for 14 drugs in patients with solid tumors and 8 drugs in patients
with acute leukemia. Children with solid tumors exhibited a greater dose
tolerance for 12 of the 14 drugs. Children with leukemia appeared to have
tolerances similar to those of adults.

Data on daunomycin in relation to the incidence of congestive heart
failure in children and adults have been compiled. Children seem to be
more sensitive than adults to this complication at comparable doses, even
though the MTD is approximately 20% higher in children than in adults.

The greater tolerance of children to many anticancer drugs may be
attributable to higher rates of metabolic or renal clearance. Both Glaubiger
et al. (1982) and Marsoni et al. (1985) expressed MTDs on a milligram-
per-square-meter rather than a milligram-per-kilogram-of-body-weight
basis. Had the relative doses been calculated as milligram per kilogram,
the interage differences should have been even more pronounced. Pinkel
(1958) observed that pediatric patients tolerated more methotrexate on a

milligram-per-kilogram basis than did adults, but the MTDs were similar when calculated on the basis of body surface area. Methotrexate is eliminated primarily by glomerular filtration and active renal tubular secretion of the parent compound. It is not surprising, therefore, that children with relatively high renal function exhibit greater rates of plasma elimination than do adults (Wang et al., 1979). In a study of 47 patients (3 to 39 years old) receiving methotrexate, Bleyer (1977) found a significantly higher incidence of neurotoxicity in the adults. Conversely, young infants have diminished renal function and exhibit lower systemic clearance and a greater potential for injury than do children (McLeod et al., 1992).

As maturation of xenobiotic metabolism and renal function generally parallel one another during the first year of life, it is not surprising that neonates and young infants may be at increased risk of injury from anticancer drugs that undergo metabolic inactivation. Vincristine is one such drug. It is detoxified in the liver and eliminated primarily via biliary excretion. Woods et al. (1981) reported a significantly higher incidence of neurotoxicity and hepatotoxicity in small infants than in children receiving vincristine. On the other hand, compounds that undergo metabolic activation may place children at greater risk than neonates or adults, since children have a higher metabolic capacity. Marsoni et al. (1985) observed that indicine N-oxide was one of the few anticancer drugs tested to have a lower MTD in children than in adults. Indicine N-oxide is believed to be converted to the toxic metabolite dehydroindicine by the liver. Cyclophosphamide is another drug that undergoes metabolic activation to cytotoxic metabolites. Certain of its metabolic pathways, however, also involve inactivation/detoxification. The half-life of cyclophosphamide is shorter in children (1 to 6.5 hours) than in adults (4 to 10 hours) (Crom et al., 1987). Although metabolic activation of drugs such as cyclophosphamide may be highest in children, the operability of concurrent detoxification pathways and inactivation of the reactive metabolites, coupled with rapid urinary excretion of the metabolites, apparently combine to hasten the elimination and thereby to negate expression of greater toxicity in children.

Because of the rapid increase in human immunodeficiency virus (HIV) positive children and the significant morbidity and mortality of the resultant disease, drugs for HIV treatment are being tested in both pediatric and adult populations. One of the most widely tested anti-HIV drugs is azidothymidine (AZT, Retrovir). McKinney et al. (1991) studied the effects of AZT in 88 children (mean age, 3.9 years; range, 4 months to 11 years). Maha (1992) reported that the efficacy and incidence of side effects (e.g., hematological abnormalities, primarily neutropenia) were similar in both adults and children but noted that the mean duration of therapy was much longer in the cohort of children, suggesting that they tolerated AZT somewhat better than did adults.

NEUROTOXICITY

Postnatal Effects of Neurotoxicants

Studies in animals suggest that the nature of an injury is determined by the stage of brain development at the time of exposure rather than by the relationship of the insult to the time of the birth event. Measures of brain development (e.g., gross brain weight and measures of biochemical change, physiologic function, and microanatomic structure) indicate that the processes and timing of brain development relative to birth differ among species (Himwich, 1973). These considerations are important in evaluating and comparing neurodevelopmental toxicology data from laboratory animals and human epidemiologic studies, especially when exposures occurred during the prenatal and weanling stages, reflecting different stages of brain development in different species. In humans, significant brain development and structural alteration occur until at least 4 to 6 years of age. It is plausible, therefore, that effects could result from exposures occurring several years after birth.

Studies evaluating microanatomic development of the brain indicate that the numerous brain structures have differing peak periods of growth. Therefore, toxic exposures at a particular time would differentially affect the structures undergoing peak development. Studies in animals indicate that exposures at different stages of brain development have differing effects on brain and behavioral function (Rodier, 1980). These critical periods or windows of vulnerability must be seriously considered when evaluating neurotoxic effects.

Because human brain development continues for years after birth, it can be hypothesized that postnatal exposure to xenobiotic compounds would alter the structure or function of the human nervous system. If this hypothesis is correct, there should be evidence of children suffering measurable effects from neurotoxic exposures at levels that do not affect adults. An alternative hypothesis suggests that children are less vulnerable because of the increased plasticity of the developing brain. In this case, children could be less vulnerable to insult. Unfortunately, the epidemiologic literature on childhood effects of neurotoxins is extremely difficult to evaluate because of the complex nature of human brain function and because of the multiple factors that affect brain development and confound evaluation.

The data on prenatal and early childhood exposure to lead indicate that effects occur at levels well below those that are toxic to adults (Bellinger et al. 1987). Irradiation studies also suggest vulnerability of the developing brain. Studies on fetal alcohol syndrome (FAS) and on neonatal drug addiction are based on less accurate dose data than are the lead studies, but the occurrence of permanent changes in brain capacity from fetal exposure is strongly suggestive of special vulnerability of the fetus. Damage from a given level of oxygen deprivation (anoxia) is generally more severe for the developing brain than for the mature brain (Menkes, 1981). In certain cases, vulnerability of the infant to neurotoxins may be related not only to the stage of neurologic development but also to the immaturity or failure of various other protective barriers. For example, the vulnerability of the neonatal brain to bilirubin exposure resulting in kernicterus may be related to the immaturity of the so-called blood-brain barrier. Bilirubin concentrations in the 40s (mg/dl) appear to cause no adverse effects in adults, but are not tolerated in children.

The data strongly suggest that exposure to neurotoxic compounds at levels believed to be safe for adults could result in permanent loss of brain function if it occurred during the prenatal and early childhood period of brain development. This information is of particular relevance to dietary exposure to pesticides, since policies that established safe levels of exposures to neurotoxic pesticides for adults could not be assumed to adequately protect a child less than 4 years of age. Knowledge of the degree of variations in neurotoxic dose levels between children and adults is necessary for establishing risk of exposure to the developing brain. Unfortunately, only minimal data are available on the effects of exposure at levels likely to occur in the food supply. The expansion of the knowledge base, particularly the refinement of animal models, is an important first step.

Measuring Neurotoxic Effects in Humans

Techniques for measuring neurotoxic effects attempt to match the various types of neurologic functions (Bondy, 1986; Triebig et al., 1987; Weiss, 1988). Acute severe clinical effects such as seizure, coma, or death are clear, measurable end points, whereas more subtle effects that occur at low exposures must be measured with more sensitive techniques.

Effects involving the peripheral nervous system can be assessed with the use of nerve conduction tests. Stimulus-response times are measured in animals to evaluate more complete reflex arcs. Specific sensory function may be quantified by using vibratory sensitivity measures (Singer et al., 1982; Wu et al., 1985). Specific sensory pathways may be measured by using evoked brain responses for auditory or visual signals (Otto, 1986; Weiss, 1988). Neurotoxic effects can be measured with electroencephalogram (EEG) technology (Dyer and Boyes, 1983; Dyer, 1985) and with biochemical measurements of neurotransmitter and neuroendocrine levels (Healy et al., 1984; Rosecrans et al., 1982; Finkelstein et al., 1988).

Cognitive and behavioral processes can be measured by testing a multiplicity of pathways and functions with methods that evaluate altered behavior in animals or psychological testing in humans. Unfortunately, testing is complicated by the fact that cognitive and behavioral outcomes can be influenced by many factors other than exposure to neurotoxins. Rigorous experimental or statistical designs are necessary to control for such confounding variables (see, for example, Weiss, 1983; Tilson and Mitchell, 1984; Weiss, 1988; Annau, 1990). Behavioral and developmental assessments have been conducted in children and in adults to identify age-related vulnerabilities to neurotoxins (Pearson and Dietrich, 1985). Animal models for behavioral and developmental studies are being evaluated (Buelke-Sam and Mactutus, 1990; Stanton and Spear, 1990; Tyl and Sette, 1990).

Vorhees (1986) attempted to define the areas of behavioral dysfunction that could be affected by prenatal brain damage, stating that behavorial teratogenosis could be expressed as impairment of several categories of neurobehavioral functions (e.g., sensory, cognitive, motor), delayed behavioral maturation of these functions, or other indices of compromised behavioral competence. He further noted that "the behavioral effects of some teratogens, even if concomitant with physical defects, may be the most significant devastating and noncorrectable of all the effects observed within the syndrome (associated with the teratogen)" (Vorhees, 1986, p. 43).

The Lead Model

The most extensive body of data describing the effects of a neurotoxin on the postnatal developing brain pertains to childhood exposure to lead. Progress in this area is reviewed below to illustrate the issues involved in evaluating the developmental effects of neurotoxins.

In the 1970s, researchers found that lead had measurable effects on the behavioral and cognitive function of children (Perino and Erinhart, 1974; Needleman et al., 1979; Graef, 1980) at blood-lead levels (20 to 40 $\mu g/dl$) considerably lower than the threshold previously considered to cause clinical lead disease or biochemical effects in adults.

During the 1980s, neurobehavioral and neurotoxic effects of lead exposure were found in children and in the human fetus at progressively lower levels of exposure (Moore et al., 1982; Needleman, 1983; Winneke et al., 1985; Bellinger et al., 1986; Mayer-Popken et al., 1986; Dietrich et al., 1987; Ruff and Bijur, 1989). Exposure levels resulting in blood-lead levels less than 20 $\mu g/dl$ were implicated. Measures of neurophysiologic and neurochemical disturbances (Otto and Reiter, 1984; Alfano and Petit, 1985; Otto et al., 1985; Moore et al., 1986) have supported the findings of toxic effects from exposure to low lead levels in humans (<30 µg/dl) and in animals. By testing for subtle neurologic, cognitive, and behavioral effects in children, these investigators elucidated neurodevelopmental toxic effects of lead.

The research that produced these findings was characterized by a variety of methodological approaches:

• Multifaceted approaches included a range of methods for biochemical and neurophysiologic measurements in animals and (where ethically possible) humans.

• Studies to evaluate subtle neurologic and developmental effects of lead included innovative methods that combined extensive batteries of different psychologic tests and rigorous statistical design focusing on various constructs to control confounding variables.

• Investigators looking for lead toxicity did not assume safe levels of exposure or protective effects. When biochemical alteration in function was found at what was considered to be subclinical levels of exposure, researchers looked for methods that would measure subtle functional changes.

Pesticides as Neurotoxicants

Many classes of compounds are used as pesticides. Some of them are known neurotoxicants. Important subclasses of the substances in use are known to have neurotoxic effects. Organophosphates and carbamates are used for demonstration purposes in this section because of the extensive data—not because they present greater potential risk than other compounds.

Data suggest that in addition to short-term effects, there are other neurologic effects of a long-term nature in adult humans. For example, symptoms of organophosphate-induced delayed neurotoxicity have been found several weeks after acute exposure and have continued for many months (Whorton and Obrinsky, 1983; Vasilesque et al., 1984; Cherniak, 1988). An intermediate syndrome starts several days after acute exposure and involves paralytic symptoms for many days (Senanayake and Karalliedde, 1987). Abnormal nerve conduction velocities have also been observed in some settings involving low-level, long-term exposure (Misra et al., 1988). Neurobehavioral and psychiatric effects have been reported in some epidemiologic studies of adult populations (Maizlish et al., 1987) and in studies of adult animals (Overstreet, 1984).

The evidence on chronic effects, particularly neurobehavioral effects of organophosphate and carbamate exposure, is less well established, but is strongly suggestive. Similar to the data on lead, there is strong evidence that acute, high-level exposure results in severe systemic disease caused by biochemical mechanisms that affect the nervous system directly. In addition, the data suggest that more long-term effects may result from exposure and that low-level exposure may have subtle, but measurable, effects on neurologic function.

The emerging data suggest that neurotoxic and behavioral effects may result from low-level chronic exposure to some organophosphate and carbamate pesticides. Sophisticated methods will be required to pursue this line of research. For many other pesticides, the data are far less complete. However, when animal studies have shown that a pesticide functions by disrupting neurologic cellular function and when systemic toxic effects are known to occur after high-level acute exposures, the possibility of low-level chronic neurotoxic and behavioral effects must be considered.

Effects of Pesticides in Children

In reviewing the data on the effects of pesticides, two questions must be addressed: Is there evidence that pesticides cause neurotoxic effects in children after acute exposure to high doses? Is there reason to suspect low-level, long-term developmental effects different from effects in adults?

Acute exposure of children to pesticides and resultant disease similar to neurotoxic effects in adults has been described for a range of pesticides, including organophosphates, carbamates, and organochlorines (Hayes, 1970; Mortenson, 1986). Pediatric cases involving neurotoxic effects due to acute exposure continue to be reported for other pesticides (e.g., Roland et al., 1985, who reported on exposure to insect repellents and encephalopathy). Data on children as segments of larger exposed populations have also been reported (e.g., CDC, 1986).

Very few pesticides have been well studied for effects on neurologic development in humans and animals. Studies on polybrominated biphenyls (PBBs) and polychlorinated biphenyls (PCBs) strongly suggest developmental effects from low-level exposures similar to the effects found for lead.

Data on exposure of humans were generated following a 1973-1974 exposure to PBBs in Michigan. Neuropsychological and developmental data were collected on children who were exposed in utero and during infancy. Physiological testing showed significant differences that were related to measures of body dose (Weil et al., 1981; Seagull, 1983).

In Taiwan, children exposed in utero to PCBs in contaminated cooking oil experienced deficits in developmental testing and abnormalities in behavioral assessment (Rogan et al., 1988). This study did not include good body burden measures, but sample sizes were large, permitting elucidation of more subtle effects.

Data on most compounds are not as extensive as those on PBBs, PCBs, and lead. Nevertheless, the pattern shown in the data on those compounds generate concern about the vulnerability of the developing human brain to any neurotoxic pesticides.

Levels of Pesticides Affecting Children

Although the vulnerability of the developing brain to neurotoxic exposure is of serious concern, it is entirely unclear from the data available whether exposures at levels consistent with usual dietary exposures would pose a substantial risk to the long-term neurologic development of children in general or to particular subgroups of children that are neurologically vulnerable.

It is theoretically possible that certain children with preexisting neurologic conditions such as hyperactivity might be more vulnerable to certain low-level neurotoxic exposures. There has been a scientific controversy surrounding the effects of "food additives" (i.e., dyes, flavors, and sugar) on children diagnosed as hyperactive. Responses vary with the study methodology, but even studies that do show effects do not show that all children in the hyperactive subpopulation are affected. These studies do not quantify effects of trace pesticide exposures, but they do raise the question, What would the dose curve for neurodevelopmental toxicants look like, and would all children be similarly vulnerable?

Comparability of Neurotoxicity Effects in Laboratory Animals

An evaluation of the accuracy with which adverse effects are detected across species (Stanton and Spear, 1990) was included in the proceedings of a workshop on "Similarities and Differences Between Children and Adults: Implications for Risk Assessment," sponsored by the International Life Sciences Institute (Kimmel et al., 1990). Species were subdivided into rodents, nonhuman primates, and humans and compared across several categories of neurobehavioral function (sensory, motivational/arousal, cognitive, motor, social). Such an analysis is extremely complex, and required a meticulously detailed comparison of hundreds of research reports for the seven toxicants considered. Overall, the investigators concluded that despite wide species differences in neurobehavioral functional categories, there was close agreement across species for the neurotoxic agents reviewed. Agents that produced cognitive, motor, and sensory deficits in humans generally resulted in corresponding deficits in laboratory animals. Although this relationship held up well at higher doses, comparability across species at lower doses was more difficult to assess. When the outcome measures were operationally similar, however, effects across species were observed with a high degree of reliability. This observation provides an essential basis for adequately predicting and formulating risk assessment guidelines for agents with potential developmental neurotoxicity.

IMMUNOTOXICITY

The primary function of the immune system is to provide resistance to pathogenic agents and surveillance against neoplastic cells. These functions are accomplished by both specific antibodies and cellular components of the immune system. Environmental agents may exert an influence on the immune system by altering cellular function or communication or by serving as a foreign structure and inducing a specific immune response. Altered immune function can result in impaired health by predisposing individuals to infectious disease, malignancy, or autoimmune disease. Because the immune system is not fully developed until adolescence, immunotoxic effects of environmental exposure in children and adults may differ.

Effects of Environmental Agents on the Immune System

Environmental agents may affect the immune system in a variety of ways. The potential outcomes can be summarized as follows:

- immunosuppression, or depressed function of the immune system;
- altered host resistance against infections or neoplastic agents;
- hypersensitivity, or autoimmune reactivity; and
- uncontrolled proliferation of immune components, such as lymphoma or leukemia (see section, "Carcinogenesis and Mutagenesis," below).

Animal Studies

Most of the studies investigating the effects of pesticides on the immune system have been conducted in animals and have focused on immunosuppression or impaired host resistance following subchronic exposure. For example, host resistance was evaluated in adult Swiss-Webster and B6C3F$_1$ mice following exposure to aldicarb (0.1 to 1,000 ppb) in drinking water (Thomas and Ratajczak, 1988; Thomas et al., 1990). After daily consumption for 34 days, no effect was noted on host resistance to infectious viral challenge, the functional ability of interferon-induced splenic NK cells to lyse YAC-1 lymphoma target cells, or cytotoxic T-cell function. In addition, there was no change in production of splenic antibody resulting from immunization with sheep erythrocytes, no effect on spleen lymphocyte blastogenesis to B- and T-cell mitogens, and no effect on the mixed lymphocyte culture response, blood counts, differential leukocyte counts, body weight, or relative lymphoid organ weights. The studies concluded that no exposure-related immunologic effects resulted from environmentally relevant concentrations of aldicarb.

The immunotoxic effect of sublethal exposure to dieldrin and aminocarb has also been examined (Fournier et al., 1988). Mice were exposed to the pesticides by gavage or intraperitoneal injection of sublethal ($<LD_{50}$) doses in corn oil or dimethyl sulfoxide on two occasions, then subsequently infected with mouse hepatitis virus (MHV3). Resistance to the viral infection indicated the status of cell-mediated immunity. Dieldrin increased the cumulative mortality of animals, whereas aminocarb did not. In addition, splenic lymphocytes from the dieldrin-treated mice were found to be functionally suppressed, as evidenced by their reduced ability to respond in a mixed lymphocyte culture. Aminocarb-exposed lymphocytes were not affected. These data indicate that cell-mediated immunity may be affected by pesticide exposure.

The immunotoxicity of captan was evaluated in rats and mice following oral administration (LaFarge-Frayssinet and Declöitre, 1982). Animals were fed a diet with or without 0.3% (wt/wt) captan [*cis*-N-(trichloromethylthio)-4-cyclohexene-1,2-dicarboximide] for 7, 14, 21, and 42 days. After 14 days of treatment, antibody formation was found to be depressed by about 70% in both species. The effect waned by day 42. Other effects noted on day 14 were reduced splenic T- and B-cell proliferation to mitogens. These responses also improved by day 42.

The effects of lindane, malathion, and dichlorophos on the immunocompetence of rabbits were assessed (Dési et al., 1978). Doses of 1/2.5 to 1/40 of the LD_{50} were given orally, in capsules, five times per week for 5 to 6 weeks. Animals were intravenously immunized weekly with *Salmonella typhi*, and antibody titers were assessed. Each of the pesticides caused a decreased antibody titer. Depression of red blood cell cholinesterase activity correlated with the immune suppression to show dose response.

Oral ingestion of lindane- and carbaryl-containing food *increased* antibody production in response to the antigenic stimulus, sheep red blood cells, in mice. However, *decreased* resistance to infection was noted following feeding of lindane. Duration of giardiasis was increased in mice, although nonreaginic antibody levels to the parasite were elevated (André et al., 1983).

Studies in mice with the organophosphorus pesticide O,O,S-trimethyl phosphorothioate (an impurity in malathion) demonstrated the ability of this chemical to block both generation of cytotoxic T lymphocytes and antibody responses at doses that did not affect body weight or splenic lymphocyte number (Rodgers et al., 1986). The macrophage appeared to be the affected splenic cell type. The suppression was reversible. Recovery time was dependent on the dose administered. A dose of 1 mg/kg was immunotoxic.

By contrast, another malathion impurity, O,S,S-trimethyl phosphoro-dithioate, was immunostimulatory (Rodgers et al., 1987). At nontoxic doses, mice demonstrated elevated cytotoxic T lymphocyte responses and heightened humoral immune responses.

The immunotoxic effects of the herbicide 2,3,7,8-tetrachlorodibenzo-p-dioxin (TCDD) have been studied extensively. In laboratory animals, the immune system appears to be a sensitive target organ. Immunosuppression is characterized by depressed cell-mediated immunity, which is most evident after perinatal exposure during the period of thymic organogenesis. The mechanism of immunosuppression in mice appears to be a defect in T-cell regulation, because nude mice (which lack T-cell populations) were more resistant than their normal littermates (Kerkvliet and Brauner, 1987). Exposure of adult animals to a TCDD concentration of 2.7 µg/kg resulted in depressed humoral immunity (Exon, 1984). In animals, the response is dependent on *Ah* locus, suggesting a genetic basis for susceptibility.

In the rat, the developing immune system has been shown to be more susceptible than the immune system of the adult to the immunotoxic effects of TCDD (Vos and Moore, 1974; Faith and Moore, 1977). Fetal and neonatal rats were exposed to TCDD through maternal dosing (5 µg/kg). The doses were administered by gavage on day 18 of gestation and on days 0, 7, and 14 of postnatal life. At this concentration, TCDD suppressed the developing immune system but not the immune system of the adult (Faith and Moore, 1977). In mice treated only at 1 month of age (not during the fetal or neonatal periods), there was reduced spleen cell response to phytohemagglutinin (PHA), which was not observed in mice treated at 4 months (Kerkvliet and Brauner, 1987). However, this effect was noted only at a toxic level of TCDD.

Few studies have examined the development of hypersensitivity following exposure to pesticides in laboratory animals. Localized dermal sensitivity has been reported for some pesticides such as naled, malathion, captan, Difulatan, DDT, and Omite (Ercegovich, 1973).

Studies in Humans

No studies have been conducted to examine the immunotoxic effects of pesticides on infants or children. Immunologic effects of chronic exposure to aldicarb in adults were investigated as a result of groundwater contamination by this carbamate pesticide in Wisconsin from 1981 to 1985 (Thomas et al., 1990). Levels of >1 to <61 ppb had been measured (enforcement standard for groundwater is 10 ppb). The average aldicarb level in the groundwater was 16.1 ppb. Adult women from 18 to 70 years of age were examined for immune status in 1985. The 23 women who consumed the contaminated groundwater were compared for health status, immune function, and fluid intake with 27 who consumed water with no known contamination. Aldicarb levels in the groundwater samples averaged 16.1 ppb. Results suggested an association between consumption of aldicarb and T-cell subset abnormalities, elevated response to *Candida* stimulation, increased number of T_8 cells, and increased percentage of T_8 to T_4 cells. The T-cell analyses were repeated on three more occasions and gave reproducible results. Dose-response data indicated a statistically significant association between aldicarb levels (using well-water values from individual households) and T_4:T_8 abnormalities as well as *Candida* stimulation results. However, although the stimulation results differed between groups, values for both groups were within normal limits. In addition, there was no self-reported clinical evidence of adverse health effects in the study groups (Thomas et al., 1990).

Health effects in humans from TCDD exposure were examined. In 1971 TCDD-contaminated sludge waste was mixed with waste oil and sprayed for dust control on residential, recreational, and commercial areas in eastern Missouri (Hoffman et al., 1986). Some reduction in activities in these areas was recommended in 1982. As a consequence, the longest period of exposure was 11 years. Individuals were exposed at nine residential sites. At least 1 ppb TCDD was found in all soil samples. Levels as high as 2,200 ppb were found in some samples.

The study involved 155 unexposed persons and 154 people exposed for 6 or more months. The exposed group had increased frequencies of abnormal T-cell subsets (10.4% compared with 6.8%). The T_4:T_8 ratio was less than 1 (8.1% compared with 6.4%). The exposed group had an increased frequency of anergy (11.8% compared with 1.1%) and relative anergy (35.3% compared with 11.8%). Anergy was correlated with the length of time the individual lived in the area. Chloracne was not observed. These results suggest an effect of TCDD exposure on the T-cell component of the immune system; however, the effect did not produce any clinical illness (Hoffman et al., 1986).

Hypersensitivity to pesticides has been examined. Few problems of dermatitis were noted after exposure to DDT and lindane, which were applied to the skin and clothing of individuals to control disease vectors (Ercegovich, 1973). Furthermore, there are no documented reports of sensitization to pesticides as a result of food or environmental exposure, nor are there reports of antibodies in sera from individuals exposed to pesticides, as would be expected if pesticides functioned as haptens and induced allergic responses.

CARCINOGENESIS AND MUTAGENESIS

Carcinogenesis is a multistage or multistep process by which a normal cell loses its ability to control its rate of proliferation and differentiation and becomes a cell from which a tumor may arise. These alterations may occur as a result of mutagenesis, which involves direct alteration of the structure of DNA, or as a result of nongenotoxic mechanisms that alter the expression of DNA or indirectly lead to mutagenesis. An increased rate of cell proliferation is an example of an indirect mechanism that can lead to carcinogenesis by increasing the likelihood that spontaneous mutation will occur or by decreasing the time available to repair DNA damage. Children may be more susceptible than adults to carcinogenesis or mutagenesis because as developing organisms, their rates of growth and thus of cell proliferation are much greater. Experimental and epidemiologic observations do not always support this, however.

Carcinogenesis in the Developing Organism

Animal Studies

Comparisons of tumor incidence observed in rodents at the same age and at the same dose rate but after different exposure durations indicate that tumor incidence is not solely a function of total accumulated lifetime dose but may depend on age at first exposure as well (Gaylor, 1988). This conclusion is supported by the observations of Toth (1968) and Rice (1979), who reported that in comparison to older animals, newborn and young animals are generally more susceptible to chemically induced tumor induction at some sites (including lung and liver) but are often more resistant to tumors at other sites (such as skin and breast). For example, intraperitoneal injections of the solvent urethane in mice produced a sixfold higher rate of leukemia when treatment was begun shortly after birth than when it was begun at about 45 days of age (Berenblum et al., 1966). Sensitivity to the induction of preneoplastic cells in the pancreas by the antibiotic azaserine is maximal in postnatal rats when the level of pancreatic DNA synthesis is high, whereas treatment is less effective in weanlings and ineffective in adults (Longnecker et al., 1977). When perinatal administration of ethylenethiourea was combined with 2 years of dietary administration to rats and mice, the incidence of thyroid tumors was slightly enhanced as compared to that obtained in the absence of perinatal exposure (NTP, 1992). By contrast, a number of studies do not support the conclusion that younger animals are more susceptible to carcinogenesis or mutagenesis than older animals. For example, Greenman (1987) failed to demonstrate an effect of age on 2-acetylaminofluorene-induced bladder cancer in mice and found that younger animals were more resistant to histopathologic changes in both the bladder and the liver. Singh et al. (1986) treated both young and old mice with ethylnitrosourea and observed that genetic alterations in bone marrow cells occurred with a greater frequency among older animals. Methylcholanthrene did not produce skin tumors when applied to newborn mice but did produce tumors in 42% of the mice treated as adults (Toth, 1968).

Anisimov (1983) surveyed the literature to determine the effects of aging on tumor latency, incidence, and size at different sites in different species for a variety of chemicals. Although these are not pesticides, the studies provide further evidence of end organ changes with age that may be applicable in the study of pesticide toxicity (Table 3-2). It is apparent from the table that results are contradictory and generalizations are impossible. Skin painting experiments with the dermal carcinogen dimethylbenz-[a]anthracene, for example, showed that younger mice are both more susceptible (Lee and Peto, 1970) and less susceptible (Stenbäck et al., 1981) than older mice to skin tumors. Increasing the age at which the carcinogen diethylnitrosamine was administered to rodents both increased (mice; Clapp et al., 1977) and decreased (rats; Reuber, 1976) the number of esophageal and forestomach tumors observed. The experiments that have been performed in animals to evaluate the effects of aging on susceptibility to chemical carcinogenesis clearly demonstrate that age may be an important factor but do not support the conclusion that younger animals are always more susceptible than older animals.

Cancer risk can thus be a function of age at first exposure, although increasing the age at first exposure does not necessarily decrease susceptibility. One explanation for this inconsistency is that as the number of cells in a target tissue increases with age, the total number of cell divisions may also increase, even if the mitotic rate decreases. There are likely to be a multitude of factors in addition to age and rates of cell proliferation that modulate carcinogenesis.

Increased susceptibility to carcinogenesis at younger ages, when it occurs, may be attributable to two factors: increased rates of cell proliferation and differing metabolic capabilities. The many roles that cell proliferation

TABLE 3-2 Effect of Aging on Latency, Incidence, and Size of Tumors at Different Sites

Site	Animal Species	Carcinogenic Agent	Age Group (months)	Effect of Aging	Reference
Skin	Mouse	MC, BP, TC	2-4 and 12-13	No effect	Peto et al. (1975); Cowdry and Suntzeff (1944)
		MC, DMBA	1.5-4 and 12-13	Decrease	Lee and Peto (1970); Cowdry and Suntzeff (1944)
		DMBA	2 and 11	Increase	Stenbäck et al. (1981)
		DMBA	14-20 and 22-24	Increase	Ebbesen (1977)
		UV-light	2-3 and 10	Decrease	Blum et al. (1942)
		Fast neutrons	1-3 and 21	Decrease	Castanera et al. (1971)
		Electrons	1 and 13	Decrease	Burns et al. (1981)
Soft tissues	Mouse	BP, DBA	1-3 and 6	Increase	Dunning et al. (1936)
		MC	6 and 20	Increase	Franks and Carbonell (1974)
		MC	3-4 and 12	Decrease	Saxen (1954); Stutman (1979)
		DMBA	2-6 and 13	Increase	Stenbäck et al. (1981)
		Plastic films	1 and 15.5	Increase	Paulini et al. (1975)
		Moloney sarcoma virus	3 and 30	Increase	Pazmino and Yuhas (1973)
	Rat	BP, MNU	3-4 and 9-14	Increase	Maiski et al. (1978); Anisimov (1982); Ovsyannikov and Anisimov (1983)
Bone	Rat	Radionuclides	2-3 and 8-10	No effect	Sundaram (1963); Streltsova and Moskalev (1964)
Mammary gland	Rat	DMBA, MC	Maximal sensitivity at 50 to 75 days		Huggins et al. (1961); Russo and Russo (1978)
		DMBA, MNU	3-4 and 14-16	Decrease	Syn-mao (1962); Anisimov (1981)
		FBAA	1-6 and 12	Decrease	Stromberg and Reuber (1975)
		Estrogens	1 and 20	Increase	Geschickter (1939)
		$^{5}_{7}$Se-seleno-methionine	3 and 24-26	Increase	Dedov (1982)
Liver	Mouse	DMH	2-3 and 12-13	No effect	Turusov et al. (1979)
	Rat	CCl$_4$	1-6 and 12	Increase	Reuber and Glover (1967)

Site	Species	Agent	Timing	Effect	Reference
Esophagus and forestomach		FBAA, DENA, AFB$_1$	1-6 and 12	Decrease	Reuber and Lee (1968); Stromberg and Reuber (1975); Kroes et al. (1975)
	Frog	DMNA	1.5 and 18	Decrease	Savchenkov et al. (1980)
	Mouse	DMNA, DMN	2 and 12-18	Increase	Khudoley (1981)
		DENA	2.5 and 17	Increase	Clapp et al. (1977)
Stomach	Rat	DENA	1-6 and 12	Decrease	Reuber (1976)
	Rat	MNNG	1.5-4.5 and 9	Decrease	Kimura et al. (1979)
Colon	Mouse	DMH	3 and 12	Increase	Turusov et al. (1979, 1981); Zimmerman et al. (1982)
	Rat	DMH	8-10 and 18	Decrease	Pozharisski et al. (1980)
			2 and 7	Decrease	Moon and Fricks (1977)
Pancreas	Mouse	MNU	3, 12, and 24	Increase	Zimmerman et al. (1982)
Kidney	Rat	FBAA, MNU, DMNA	1-6 and 12-18	Decrease	Reuber (1975); Savchenkov et al. (1980); Anisimov (1981)
Bladder	Mouse	DMBA (in vitro)	1.5-2 and 28-30	Increase	Summerhayes and Franks (1979)
Lung	Mouse	DENA	2.5 and 12	Increase	Clapp et al. (1977)
		MNU	3 and 24	Increase	Zimmerman et al. (1982)
		DBA, urethane	2.4 and 11-12	Decrease	Dourson and Baxter (1981)
	Rat	Fast neutrons	3 and 21	Increase	Castanera et al. (1971)
Pleura	Rat	Asbestos	2 and 10	Increase	Berry and Wagner (1976)
Uterus	Mouse	DMH	2 and 12	Increase	Turusov et al. (1979, 1981)
	Rat	MNU	3 and 14	Increase	Anisimov (1981)
Vagina	Mouse	DMBA	3 and 18	Increase	Anisimov (1982)
Ovary	Mouse	X-rays	2 and 12	Decrease	Cosgrove et al. (1965)
Testis	Rat	Fast neutrons	3 and 21	Increase	Castenera et al. (1971)

TABLE 3-2 (Continued)

Site	Animal Species	Carcinogenic Agent	Age Group (months)	Effect of Aging	Reference
Vascular wall	Mouse	DENA	2.5 and 17	Increase	Clapp et al. (1977)
	Rat	Vinyl chloride	1.5-4 and 12	Increase	Groth et al. (1981)
Hematopoietic system	Mouse	X-rays	1-2 and 6	Decrease	Kaplan (1947); Lindop and Rotblat (1962)
		MNU	3, 12, and 24	Increase	Zimmerman et al. (1982)
		PMS	6 and 10	No effect	Menczer et al. (1977)
	Rat	MNU	3 and 14	No effect	Anisimov (1981)
		Radionuclides	3 and 8-10	Increase	Streltsova and Moskalev (1964)
		X-rays	4 and 12	Decrease	Cosgrove et al. (1965)
	Frog	DMNA, DMN	1.5-2 and 12-18	Decrease	Khudoley (1981)

NOTE: MC, 3-methylcholanthrene; BP, benzo[a]pyrene; TC, tobacco smoke condensate; DMBA, 7,12-dimethylbenz[a]anthracene; DBA, 1,2,5,6-dibenzanthracene; MNU, N-nitrosomethylurea; FBAA, N-4(fluorobiphenyl)acetamide; CCl₄ carbon tetrachloride; DMH, 1,2-dimeth-ylhydrazine; AFB₁, aflatoxin B₁; MNNG, N-methyl-N'-nitro-N-nitrosoguanidine; DENA, N-nitrosodiethylamine; DMNA, N-nitrosodimethy-lamine; DMN, dimethylnitramine; PMS, pregnant mare serum.

SOURCE: Anisimov, 1983.

may play in carcinogenesis are described above; overall, increased rates of cell proliferation can contribute to an increased likelihood of carcinogenesis. For example, polycyclic aromatic hydrocarbons and aflatoxin B_1 produce liver tumors when administered to newborn rodents but not when administered to older animals, presumably because the liver proliferates rapidly in the developing organism but slowly in older animals. Differing metabolic capabilities may contribute to greater susceptibility if the developing organism has less competent detoxifying or conjugating abilities than the adult. Conversely, less competent activating enzymes may protect the developing animal from chemicals that require metabolic activation to their reactive forms to elicit effects. Ethylnitrosourea, which does not require metabolic activation, is very effective as a carcinogen in neonatal rodents as compared to adults, whereas diethylnitrosamine, which requires activation, is not (Vesselinovitch et al., 1979). In addition, there may be age-related differences in DNA repair abilities and in the fidelity of DNA replication.

Human Studies

Epidemiologic studies of the effects of age on susceptibility to carcinogenesis are conflicting. The risk of bladder cancer associated with employment in a "hazardous occupation" (e.g., an industry believed to be associated with an increased risk of bladder cancer, such as the rubber or leather industries, or work with dyestuffs, paint, and other organic chemicals) was greater in younger people (Hoover and Cole, 1973), whereas the risk of nasal cancer among nickel workers increased in proportion to age at beginning of exposure (Doll et al., 1970). Tucker et al. (1987) demonstrated that chemotherapeutic treatment of children with cancer using alkylating agents, which can form adducts with DNA and induce mutations, resulted in a significantly elevated risk of secondary leukemia. No study has been performed to determine whether similar treatment of adults has the same outcome, however, so it is not possible to conclude that children are more susceptible to chemically induced carcinogenesis on the basis of these limited data. Evidence from epidemiologic studies is thus inadequate to demonstrate a consistent increased susceptibility to carcinogenesis among children, nor would one assume that children would regularly be more susceptible to toxic end points in pesticide toxicity. These data emphasize the need to evaluate each pesticide specifically for age-related toxicity. The incidence of most cancers in humans increases with age, with the exception of certain tumor types that are associated with childhood and that are suspected to result from inborn genetic alterations or prenatal genetic damage. An example of a childhood tumor is retinoblastoma, in which a mutation occurs in the retinoblast population resulting from genetic damage either before or after conception, creating a population of altered retinal cells that is very susceptible to malignant transformation.

From 1973 through 1989, the incidence of cancer among children of all races from 0 to 14 years old increased 7.6%. The greatest increases were observed for acute lympocytic leukemia (23.7%), brain and nervous system cancers (28.6%), and cancers of the kidney and renal pelvis (26.9%). The incidence of several other childhood cancers decreased (bones and joints, −15.1%; Hodgkins disease, −1.5%; non-Hodgkins lymphomas, −0.9%). During the same period, total cancer incidence for the entire U.S. population increased approximately 16.1% (Miller et al., 1992).

METABOLISM AND PHARMACOKINETICS

Data on pharmacokinetics are basic to considerations of the relative risks of toxic injury from pesticides in both children and adults. The fundamental goal of pharmacokinetic studies is to delineate the uptake and disposition of pesticides, drugs, and other chemicals in the body. A basic tenet of toxicology is that toxic responses are a function of the concentration of the active chemical in target tissues. Thus the degree and duration of a toxic effect depend on the quantity of the reactive form of a chemical that reaches its target site and the length of time the agent remains there. These factors in turn depend on the magnitude of systemic absorption, binding, distribution, metabolism, interaction with cellular components, and elimination of the chemical from the tissue and body. The important structural and functional differences between infants and adults can have an impact on one or more of these pharmacokinetic processes, which in turn may result in different effects of chemicals on the two age groups.

This section focuses on age-related factors that influence the pharmacokinetics of pesticides, drugs, and other chemicals in humans. The study subjects are grouped as follows: premature newborns, full-term newborns, neonates (birth to 4 weeks), infants (4 weeks to 1 year), young children (1 to 5 years), older children (6 to 12 years), and adolescents (13 to 18 years). Consideration in this section is largely limited to information from studies in humans, since there are major difficulties in extrapolating from immature animals to immature humans. Nonprimate species are less mature in many respects than humans at birth. Maturation in most lower animals, however, is quite rapid; some adult-like characteristics and functions are attained in as little as 14 to 21 days in rodents. A difference of only a few days in exposure age can thus have a marked effect on the handling of a chemical and its ensuing effects in such species (Done, 1964; Neims et al., 1976). Various body structures and associated functions mature at different rates in different species. Utmost care must be exercised in selecting an appropriate animal model for developmental pharmacokinetic and toxicology studies, in interpreting the data, and in extrapolating the data to humans. Animal studies are presented here when data on humans are inadequate or when findings in animals elucidate ontogenetic mechanisms.

For infants and children, exposure to pesticides occurs primarily through ingestion, inhalation, and through the skin. The newborn may have previously encountered chemical agents in utero, but an in-depth examination of in-utero exposure is beyond the prescribed scope of this report. The major emphasis in this section is the absorption and disposition of ingested chemicals. Because children put all kinds of things into their mouths, they are at risk of ingesting pesticides from nonfood sources, including contaminated household objects, ornamental plants, sod, and paint. In certain situations, significant exposure may result from inhalation of pesticides or skin contact with contaminated surfaces (see Chapter 7). Dermal and inhalation exposures are also addressed because they may contribute to the total systemic dose and need to be considered when establishing prudent levels of dietary intake for infants and children.

Dermal and Pulmonary Exposure

The skin area of the infant per unit of body weight is double that of the adult, whereas the permeability of the infant's skin, except for those born prematurely, appears to be similar to that of the adult. These are important factors to remember when considering dermal absorption or penetration of xenobiotic compounds. The stratum corneum (the outer layer of the skin, which serves as the barrier to penetration by chemicals) is fully developed in the human newborn. Studies of the bacteria-inhibiting agent hexachlorophene in premature and full-term infants, the hormone testosterone in infant and adult monkeys (Wester et al., 1977), and alcohols in premature and full-term infants and human adults have shown no differences in penetration, but differences in absorption have been shown for fatty acids (Wester and Maibach, 1982).

There is little evidence to suggest that percutaneous absorption of chemicals varies greatly with age during the preadolescent period, since the overall thickness of the stratum corneum remains relatively constant throughout postnatal development (Rasmussen, 1979). There is a paucity of information, however, from well-controlled studies on percutaneous absorption of chemicals in this age group. McCormack et al. (1982) observed no difference in the rate of penetration of a series of alcohols through premature, full-term newborn, and adult skin specimens in vitro. They did find differences in penetration of a series of fatty acids, which the investigators attributed to differences in solubilization of the fatty acids in epidermal lipids. Wester et al. (1977) reported that the percutaneous absorption of testosterone was similar in the newborn and adult rhesus monkey.

Some studies have been conducted to assess the age-dependency of dermal absorption of pesticides in rodents. Solomon et al. (1977) found no significant differences between newborn and adult guinea pigs in blood or brain concentrations of the insecticide γ-benzene hexachloride following its topical application. Knaak et al. (1984), however, reported that the fungicide triadimefon was more rapidly absorbed through the skin of young rats than through the skin of adult rats. Shah et al. (1987) contrasted the percutaneous absorption of 14 pesticides in young (33-day-old) versus adult (82-day-old) female rats. The pesticides studied were structurally diverse, in that they included organophosphates, carbamates, organometallics, chlorinated hydrocarbons, biological insecticides, and a triazine compound. No clear age-related pattern of absorption was found. At least four of the compounds (atrazine, carbaryl, chlordecone, and chlorpyrifos) were absorbed more extensively by the younger animals. Six compounds, however, were better absorbed by the adults, and the others seemed to be equally well absorbed by both age groups. No one class of compounds was better absorbed in one particular age group, other than the organometallics, which were more extensively absorbed by the adult rats. Skin penetration was not well correlated with the octanol-water partition coefficients of this diverse group of chemicals (Shah et al., 1987).

The premature human newborn may be a special case. Studies have shown that formation of the stratum corneum is incomplete until just before birth at full term (Singer et al., 1971). Greaves et al. (1975) reported that premature newborns bathed in an antibacterial hexachlorophene solution had considerably higher blood levels of hexachlorophene than did full-term newborns monitored by Curley et al. (1971). Tyrala et al. (1977) similarly observed an inverse relationship between body weight and post-conceptional age versus blood hexachlorophene concentrations in a group of 54 premature and full-term newborns. Subjects with large areas of abraded skin exhibited particularly high blood levels. Thus diminished effectiveness of the epidermal barrier to absorption was apparently a major determinant of the elevated blood levels of hexachlorophene. Another important factor was the reduced capacity of premature and full-term newborns to metabolize and eliminate hexachlorophene. Dermal hexachlorophene exposures have resulted in a number of cases of brain damage in newborns (Powell et al., 1973; Shuman et al., 1974), suggesting that this very early age group could be at increased risk of toxicity from direct skin contact with pesticides.

Several factors may contribute to increased percutaneous absorption and toxicity of pesticides in neonates and infants. Dermal absorption of a variety of chemicals is markedly increased under diapers and rubber pants. These materials retard the evaporation of volatile chemicals and enhance the hydration and temperature of the skin, thereby increasing penetration by water-soluble chemicals. Damage to the stratum corneum, as in diaper rash, circumvents this barrier layer. In addition, some skin surfaces such as the male scrotum and the face are more absorbent than skin in other areas of the body.

The ratio of surface area to body weight in newborns and infants is approximately 2.5-fold greater than that of adults. Thus, if the exposed area of skin and percutaneous absorption rate in a neonate and adult were equivalent, the neonate would receive almost three times the systemic dose, on a kilogram-of-body-weight basis (Wester and Maibach, 1982). Nevertheless, there are few data to indicate which types of chemicals may be more extensively absorbed through the skin of neonates and infants.

The alveolar epithelium is another potential portal of entry into the body for pesticides. Most xenobiotics are absorbed through the alveolar epithelium into the pulmonary (blood) circulation by simple passive diffusion, rather than specific active transport processes (Schanker, 1978). Therefore, changes in several parameters with age may be of consequence in pulmonary absorption of pesticides, including alveolar surface area, thickness of alveolar membranes, porosity and other properties of the membranes, pulmonary blood flow, and respiratory volume. The normal respiratory volume of the resting infant is approximately twice that of the resting adult, when expressed per unit of body weight. The structural development of the human lung is known to continue postnatally (Hislop and Reid, 1981; Langston, 1983). There is a marked increase in alveolar surface area for the first 18 to 24 months of life. Thereafter, pulmonary structures continue to increase in size, and alveolar surface area increases gradually throughout childhood (Thurlbeck, 1982). There is a progressive increase up to 18 years of age in collagenous elastic fiber bundles in the alveolar walls. Some of these factors may counteract others in terms of their influence on the pulmonary absorption of chemicals. For example, the effect of increased respiratory volume may be offset by the infant's smaller surface area for absorption. Unfortunately, little information is available on the pulmonary absorption and bioavailability of xenobiotic compounds in infants and children.

One group of investigators has studied the pulmonary absorption of nonvolatile drugs in neonatal and adult animals. Hemberger and Schanker (1978) injected measured doses of a series of drugs into the tracheas of neonatal (3 to 27 days of age) and adult rats. Systemic absorption was determined by assay of the quantity of drug remaining in the lungs and trachea after 2 hours. Lipid-soluble compounds (procainamide and sulfisoxazole) were absorbed at similar rates by neonates and adults, indicating that the properties of the alveolar epithelium do not change substantially with age. Lipid-insoluble compounds (p-aminohippuric acid, mannitol, and tetraethylammonium bromide), however, were absorbed about twice as readily by the 3- to 12-day-old rats as by rats 18 days old or older. The lipid-soluble drugs were absorbed much more rapidly than the lipid-insoluble drugs, but the amount of all drugs absorbed per unit of time was directly proportional to the administered concentration, leading Hemberger and Schanker (1978) to conclude that there was no evidence for an absorption mechanism other than simple diffusion in either the neonate or the adult rats. Lipid-insoluble substances are believed to be absorbed in the lung primarily by diffusion through aqueous channels, or pores (Schanker, 1978). Because the alveolar membrane is reported to become thinner in neonatal rats as they age (Burri et al., 1974), Hemberger and Schanker (1978) concluded that greater porosity must account for the greater absorption of lipid-insoluble drugs in neonatal rats.

Oral Exposure

The gastrointestinal tract is the major portal of entry of pesticides into the body. Absorption depends on the physical and chemical properties of the pesticide, as well as on conditions within the gastrointestinal tract itself. Some of the more important conditions, or factors, include gastric emptying and intestinal motility, gut flora, acid and enzyme secretory activity, mucosal structure and surface area, cellular transport systems, and gastrointestinal blood supply. All these change with postnatal development. A change in one may tend to enhance chemical absorption, whereas a change in another may have the opposite effect. The complexity of the system, with its multiple, interrelating factors, makes it difficult to predict the net effect of maturation on absorption of pesticides and other chemicals.

Absorption

The majority of information on gastrointestinal absorption of chemicals in infants and children has come from nutrition and drug studies. The musculature of the gastrointestinal tract is relatively thin and quiescent before birth but then must rapidly make the transition from placental to intestinal nutrition (Balistreri, 1988). Ingested nutrients entering the gut and regulatory hormones secreted in response to food are important in the adaptation to extrauterine life. For example, plasma levels of the peptide YY increase substantially after consumption of milk during the first 2 weeks of life (Adrian et al., 1986). Peptide YY reduces the rate of gastric emptying and slows intestinal transit, thereby increasing the efficiency of absorption of nutrients. The rate and extent of absorption of orally administered drugs are often quite variable in newborns, largely because of irregular, unpredictable gastric emptying and peristalsis (Warner, 1986). Some drugs (such as digoxin and diazepam) appear to be absorbed as efficiently by full-term newborns and neonates as by adults. Premature and full-term newborns, however, exhibit slow, incomplete oral absorption of other drugs, including phenobarbital, chloramphenicol, rifampin, valproate, and phenytoin (Morselli et al., 1980; Morselli, 1989). Although poor bioavailability of phenytoin is generally seen during the first month of life, infants absorb the drug well. Similarly, infants absorb valproate at a rate comparable to that of adults.

Absorption of Proteins and Related Large Molecules. The microvillus surface of the adult small intestine serves as a barrier to penetration by many substances, particularly large, charged molecules such as proteins. There is evidence from human and animal studies, however, that the immature intestine can allow passage of intact macromolecules. Antibodies, including immunoglobulin, bovine serum albumin, α-lactalbumin, and other antigens, are reported to be absorbed intact by newborns (Grand et al., 1976). Acquisition of both allergies and passive immunity has been attributed to gastrointestinal absorption of intact proteins by neonates (see section on "Immunotoxicity," above). The period of macromolecular uptake varies in laboratory animals, from as short as 1 to 2 days after birth in guinea pigs to as long as 23 to 24 days in rabbits (Hoffmann, 1982). Selective absorption of intact γ-globulins from colostrum occurs in ruminant species for only a few days, but absorption is retained by suckling rats for up to 20 days. Absorption of α-globulins and other proteins such as cholera toxin has been shown to involve high-affinity binding to specific receptors on the microvillus surface of intestinal mucosal cells in rodents. The binding is followed by invagination of the membrane to form protein-filled vesicles, which migrate through the cytoplasm to the basolateral surface of the cell, where the vesicle contents are extruded (Walker and Isselbacher, 1974). Because this energy-dependent, endocytotic process involves receptor binding of specific proteins, it is unlikely that pesticides would be absorbed in significant quantities in human neonates by this mechanism. And, indeed, most data indicate that uptake of proteins in humans is limited and nonselective, rather than a specific, receptor-dependent transport process as is found in the rat (Walker and Isselbacher, 1974). Because macromolecular uptake in human newborns appears to be nonselective, there could be a transitory period during which compounds that are normally excluded by the gastrointestinal tract of children and adults are absorbed.

Age-Related Changes in Absorption. Although there is a paucity of data on absorption of drugs as related to age in human neonates and infants, some definitive investigations have been conducted using laboratory animals. Hoffmann (1982) summarized the results of a number of animal studies. The rate of penetration of four model compounds (antipyrine, sodium salicylate, tetraethylammonium bromide, and phenosulfonphthaline) from everted intestinal sacs of rats was evaluated in an in vitro experiment. Penetration of all compounds was two to three times greater for 10- than for 30-day-old rats, which in turn exhibited somewhat greater penetration rates than did adult animals. In an in vivo study, disappearance of the compounds from the duodenum of anesthetized rats was examined. Antipyrine, sodium salicylate, and tetraethylammonium bromide were each absorbed more rapidly by 10-day-old rats than by adult rats. These differences in the rate of absorption were not reflected by differences in blood levels, apparently because of a greater volume of distribution in the neonatal rats. Large synthetic molecules such as polyvinylpyrrolidinone have also been found to be well absorbed in neonatal and suckling mammals, as have heavy metals (Hoffmann, 1982). Closure, or decrease to the low-level uptake characteristic of adult animals, generally has been found to occur at the time of weaning.

Closure has been associated with structural and functional maturation of intestinal epithelial cells. It has been suggested that a marked reduction in pinocytotic activity is largely responsible for this phenomenon. Pinocytosis is believed to be the primary mechanism for nonselective absorption of macromolecules by the neonatal intestinal epithelium in mammals (Lecce, 1972). Bierring et al. (1964) observed large numbers of pinocytotic vacuoles associated with phagosomes at the base of microvilli in the intestinal epithelial cells of human fetuses. Udall and Walker (1982) saw such vacuoles associated with an extensive apical tubular network system, as well as pseudopodlike cytoplasmic extensions projecting through the lamina propria in the intestine of 1-week-old rabbits. The intestinal epithelium of the adult rabbit showed a marked decrease in the tubular network and pseudopods, which accompanied cessation of systemic uptake of bovine serum albumin. Pang et al. (1983) found that the membrane of newborn rabbits had a significantly higher lipid-to-protein ratio than did that of adult animals. Electron spin resonance spectra revealed that the membranes from the newborn rabbits were more disorganized and fluid, which could account for the more efficient penetration and diffusion of macromolecules during the perinatal period.

Absorption and Retention of Lead and Other Heavy Metals   Although lead and other heavy metals are not commonly used now as pesticides by themselves, they have been used in the past. Lead, in particular, has been widely used and studied. Lead is still of major concern as a health hazard for a number of reasons, including its prevalence in the environment and the increased sensitivity of children to its adverse effects. Barltrop (1965) reported one of the earliest investigations indicating that blood lead concentrations were age related. In a study group of 470 London children, lead levels in blood increased to a maximum in 3 year olds and then progressively decreased during the next 4 to 5 years of life. Subsequent studies demonstrated that young children had higher lead blood levels than adults living with the children (Barltrop et al., 1974; McNeil et al., 1975).

Research findings for both laboratory animals and humans indicate that increased absorption and retention of lead are important factors in the relatively high incidence of lead toxicity in the young. Suckling rats were shown to absorb a greater percentage of ingested lead than did older rats (Kostial et al., 1971). The increased absorption of lead and other divalent cations, including iron, calcium, cadmium, mercury, and manganese, typically lasted until the animals were weaned (Kostial et al., 1978). Ziegler et al. (1978) conducted metabolic balance studies in 14- to 746-day-old human subjects consuming cow's milk, infant formula, fruit juice, and strained fruits or vegetables that contained known small amounts of lead. Absorption and retention of lead increased with increasing dietary lead intake. Net absorption and retention averaged approximately 42% and 32%, respectively, when intake exceeded 5 μg/kg bw/day in 61 balance studies conducted by Ziegler et al. (1978). Data of other investigators indicate that human adults absorb only about 10% of ingested lead (Hursh and Suomela, 1968; Rabinowitz et al., 1976). Thus, greater gastrointestinal absorption, in concert with the increased retention and target organ deposition, appears to contribute significantly to the higher risk of lead poisoning in infants and young children.

Increased lead absorption in the very young is commonly attributed to pronounced pinocytotic activity in the gastrointestinal epithelium, but there appear to be other determinants as well. Because absorption decreases substantially when animals are weaned, Kostial et al. (1978) studied the effect of cow's milk on lead uptake. Inclusion of milk in the diet resulted in a substantially greater absorption not only of lead but also of cadmium, mercury, and manganese in 6- or 18-week-old rats. Still, absorption of the metals by the milk-fed animals was not as great as in 1-week-old suckling rats. The mechanism of milk's effect is unknown, although it is possible that the metals may bind to some milk constituent, and this binding would facilitate their penetration of the gastrointestinal mucosal barrier. Barltrop (1982) reported that raising dietary fat content from 5% to 10% enhanced lead absorption by 80% in rats and proposed that certain protein deficiencies could result in enhanced absorption because protein diets with a high sulfhydryl content impaired gastrointestinal mucosal binding and uptake of lead. Trace element deficiencies may also play a role. Six and Goyer (1972) demonstrated that iron deficiency resulted in increased lead deposition and toxicity in young rats, possibly because of increased lead absorption. Iron deficiency has been commonly associated with lead poisoning in children. Low dietary intake of calcium has also been shown to lead to greater lead uptake in immature animals and humans (Ziegler et al., 1978).

Other Factors Affecting Oral Absorption

There are additional physiological and morphological processes undergoing continuous maturational changes after birth that can affect gastrointestinal absorption of metals and other chemicals. The microflora of the gut changes considerably during the neonatal and infancy period (Long and Swenson, 1977). The fecal flora in milk-fed infants exhibit negligible demethylating ability, in contrast to that of weaned children and adults. This difference could be important for a chemical such as methylmercury, which is absorbed much more readily than inorganic mercury. Infants would be expected to absorb more of an ingested dose of methylmercury because much of it would not be demethylated in the gut (Rowland et al., 1983). The same would be true for the methylmercury that reenters the gastrointestinal tract via the bile.

Gastric pH varies considerably, falling during the initial hours after birth but returning to neutrality for 10 to 15 days; thereafter, it declines gradually, not reaching adult levels until about 2 years of age (Morselli et al., 1980). Agunod et al. (1969) observed that chemically stimulated hydrochloric acid secretion was low in the gastric juice of neonates but increased until it approached the lower limit of the normal range for adults after 3 months.

Achlorhydria may result in diminished absorption and bioavailability of acidic compounds. The converse should be true for basic compounds. The relatively small surface area of the neonatal intestine proportionally reduces absorption of all chemicals. Although villi and microvilli are present in the intestinal epithelium of newborns, cell proliferation is apparently quite slow (Grand et al., 1976). Autoradiographic experiments in rats demonstrated clearly that the villi of the small intestine of sucklings were shorter and that epithelial cell migration proceeded at a rate of 20%, or less than that, of weaned animals (Koldovsky et al., 1966). Varga and Csaky (1976) found that the blood supply to the gastrointestinal tract of rats changed with age. Fractional blood flow to the total gastrointestinal tract decreased from 20% in 20-day-old rats to 8% to 12% in adult animals.

The net effect of the aforementioned factors on pesticide absorption in the immature individual is hard to predict because they sometimes oppose one another, change at different rates in the maturing organism, and often are ill-defined in humans.

Distribution and Uptake of Chemicals

Distribution of a chemical to sites of action in different tissues, following systemic absorption, is governed by a number of factors. These include plasma protein binding, extracellular fluid volume, adipose tissue mass, organ blood flow, tissue uptake, and tissue binding. The factors exert their influence concurrently and may compete with one another. They may change to varying degrees at different rates during postnatal development. Thus, the net effect of maturation on the quantity of a particular chemical reaching a target tissue is difficult to ascertain. There are few data on binding and distribution of pesticides in infants and children, but a variety of drugs have been relatively well studied (Kearns and Reed, 1989). These are used to illustrate how chemical distribution and uptake can vary with age in the developing individual.

Protein Binding. Numerous chemicals, including a variety of pesticides, bind reversibly to plasma proteins. As long as the compounds are bound, they are not able to leave the bloodstream and reach sites of action (i.e., produce biological effects) in extravascular tissues. Although increased plasma protein binding thereby generally reduces the maximum bioactivity of chemicals, binding can prolong their effects by slowly releasing them to sites of action as well as to sites of inactivation or elimination. Much of our current knowledge of how altered plasma protein binding affects the health of infants and children has been gained from clinical studies of therapeutic agents.

Many drugs exhibit significantly lower plasma protein binding in premature and full-term newborns than in adults (Table 3-3). Investigators typically take blood samples from the umbilical cord at the time of delivery and determine the percentage of free, or unbound, drug in the sample. A diverse group of therapeutic agents, including both acidic and basic compounds, exhibit considerably reduced plasma protein binding in perinatal subjects (Morselli et al., 1980; Morselli, 1989). This group includes such common drugs as phenobarbital, digoxin, theophylline, phenytoin, lidocaine, imipramine, and diazepam. Though data for certain age groups are lacking for a number of these drugs, it appears that the age at which binding reaches adult levels is compound-specific. Rane et al. (1971), in a comprehensive study of the age-dependency of the plasma protein binding of phenytoin (a weak acid), found that adult values were approached in infants and young children 3 months to 2 years old. The

TABLE 3-3   Protein Binding of Some Drugs in Cord Plasma in Relation to Adult Plasma

Lower Binding	Higher Binding
Acid Drugs	
Ampicillin	Valporic acid[a]
Benzylpenicillin	(indirect evidence)
Nafcillin	Salicyclic acid[a]
Naproxen	Sulfisoxazole[a]
Salicylates	Cloxacillin[a]
Phenytoin	Flucloxacillin[a]
(same or lower binding)	
Phenylbutazone	
Phenobarbitone	
Pentobarbitone	
Cloxacillin	
Flucloxacillin	
Sulfamethoxypyrazine	
Sulfaphenazole	
Sulfadimethoxine	
Sulfamethoxydiazine	
Neutral drugs	
Digoxin	
Dexamethasone	
Basic drugs	
Diazepam	Diazepam[a]
Imipramine	
Desmethylimipramine	
Bupivacaine	
Lidocaine	
Propranolol	
Metocurine[a]	
D-Tubocurarine[a]	

[a] As compared with maternal plasma.

SOURCE: Adapted from Rane, 1992.

percent of unbound phenytoin diminished little thereafter in succeeding age groups. Values comparable to those of adults for binding of acidic drugs are often reached during the second to third year of life, whereas γ-globulins, which are believed to be important in binding of nonacidic compounds, may not attain adult levels until ages 7 to 12 years (Morselli et al., 1980). Reduced binding of imipramine has been reported in children younger than 10 years (Windorfer et al., 1974).

Quantitative and qualitative differences in circulating plasma proteins have been well documented (Morselli, 1976), and it is known that during the perinatal period there is a decrease in plasma protein binding. There appear to be four primary reasons for this decrease. First, the concentration of albumin, the most important binding protein for most compounds, is reduced. Second, there is a persistence of fetal albumin, which has a lower affinity for many drugs. Third, levels of γ-globulins and lipoproteins are also low postnatally. Fourth, a transient hyperbilirubinemia is universally present during the first few days after birth (Done, 1964; Nau et al., 1984). Bilirubin competes with drugs, particularly acidic ones, for albumin-binding sites. Rane et al. (1971), for example, demonstrated a correlation between the unbound fraction of phenytoin in umbilical cord blood and the total concentration of bilirubin in plasma of neonates. Conversely, large doses of highly bound drugs can displace bilirubin from plasma proteins, resulting in jaundice. Shortly after birth, lipolysis occurs, resulting in an elevation in free fatty acids in the blood. Free fatty acids compete with and can displace some drugs from plasma protein-binding sites. Nau et al. (1984) conducted a comprehensive study in which they measured binding of diazepam and N-desmethyldiazepam, its major active metabolite, in the serum of adults, fetuses, and neonates 1 to 11 days old. The free fatty acid concentration and free fraction of diazepam and its metabolite were highest on day 1. During the next 10 days of life, there were progressive, parallel decreases in serum free fatty acids, albumin, and free fraction of drug and metabolite. Bilirubin seemed to play a less important role in diazepam binding because it was relatively low on day 1 (the day the free fraction was highest) and peaked on day 3 during the time the free fraction was diminishing. Nau et al. (1984) concluded that increased plasma free fatty acids and albumin and decreased bilirubin levels, coupled with deficient metabolism and elimination of diazepam, predisposed the newborn to excessive, potentially adverse effects of the drug.

The extent of plasma protein binding can have a major impact on the magnitude and duration of chemical action and toxicity. Diminished binding, as mentioned previously, results in higher concentrations of free drug available for diffusion from the blood to sites of action in target tissues and sites of metabolism and/or excretion. Thus, diminished binding typically results in an increased intensity of pharmacological effects and potential for toxicity but a shorter duration of action. Clearance is directly proportional to the free fraction of chemical for compounds for which elimination is dependent on diffusion across cell membranes (e.g., into metabolizing liver cells) or glomerular filtration (and urinary excretion). Decreased binding will therefore normally result in increased elimination of a chemical. The neonate, however, often exhibits compromised hepatic metabolic and renal clearance capacities, so the duration of biological effects may be longer than anticipated.

Distribution Volumes. The distribution of xenobiotics and many natural compounds in the body is known to change with age. As mentioned previously, infants have a higher percentage of water in lean body tissues than do adults. The additional water is primarily extracellular, so that the volume of extracellular fluid per unit of body weight in infants is about twice that of adults (Widdowson and Dickerson, 1964). As a result, water-soluble chemicals have a greater volume (per unit of body weight) in which to distribute. The newborn is exceptionally resistant to the skeletal muscle relaxants succinylcholine and decamethonium. This resistance can be attributed to the distribution of these small, highly ionized molecules in the relatively large extracellular fluid volume, which in effect reduces their concentration and their resulting pharmacologic action. Penicillins have a higher volume of distribution* in neonates because of lower plasma protein binding and higher extracellular water content. Although decreased plasma protein binding of a drug such as lidocaine would be expected to result in an increased amount of free drug and thus an exaggerated pharmacologic response in newborns, its greater volume of distribution reduces its concentration at the site of action. One age-related difference negates the other. Morselli et al. (1980) pointed out that newborns need twice as long as adults to eliminate lidocaine because of the large distribution volume that must be cleared of the drug. In the newborn, the decreased metabolism of lidocaine is apparently offset by increased renal clearance; a low glomerular filtration rate is offset by diminished tubular reabsorption of the drug. Therefore, despite a number of age-dependent differences that affect pharmacokinetics, in this instance the differences nullify one another such that lidocaine's total body clearance and pharmacologic potency are comparable in newborns and adults.

Most drugs have a larger volume of distribution during infancy and early childhood, although the reverse is true for some other compounds (Done et al., 1977). The decreased binding and increased extracellular fluid volume typically seen postnatally result in a greater volume of distribution for relatively polar chemicals. Many drugs, however, have volume of distribution values in excess of the extracellular fluid volume or total body water as a result of their solubility in body fat. Adipose tissue usually makes up a smaller percentage of body weight in newborns and infants than in adults. Thus diazepam, a lipophilic drug, has a somewhat smaller volume of distribution in neonates and infants than in adults (Morselli et al.,

---

* Volume of distribution is an apparent volume based on the dose administered divided by the concentration in the plasma water (e.g., dose = 100 mg/kg; plasma water concentrations = 0.5 mg/ml; volume of distribution = 100 mg per kg/0.5 mg per ml = 200 ml/kg).

1980). Distribution volume differences, which may be observed for the first 10 years of life, appear to disappear more slowly than most other age-dependent differences that alter pharmacokinetics.

Barriers to Distribution. Barriers to tissue uptake, as well as distribution within organs, may vary with age, in conjunction with morphological and functional maturation. In most regions of the body only the vascular endothelium serves as a barrier to diffusion of chemicals from the bloodstream into surrounding tissue. Generally, diffusion is limited to the un-ionized, more lipid-soluble form of chemicals. In some organs (such as the liver), there are pores in the endothelium and gaps between adjacent endothelial cells, which facilitate passage of large, charged molecules. The vascular endothelium of certain organs (such as the brain and testes), however, is devoid of pores and pinocytotic vesicles, has tight cell junctions, and is encased in specialized pericapillary cells. The blood-brain barrier limits entry into the central nervous system to un-ionized, lipophilic compounds (Benet et al., 1990).

There is speculation that neonates and infants may be more susceptible to chemically induced neurotoxicity, in part because of the immaturity of their blood-brain barrier. Watanabe et al. (1990) point out that the central nervous system in developing individuals is potentially vulnerable to chemicals for a protracted period because the central nervous system requires longer than most other organ systems for cellular differentiation, growth, and functional organization. Therefore, any increase in accessibility to cytotoxic agents because of delayed maturation of the blood-brain barrier could have serious consequences.

One of the most commonly cited examples of this phenomenon is lead poisoning in infants and children. It is argued that children exhibit neurological disturbances at lower blood lead concentrations than adults, suggesting that lead enters the central nervous system of children in larger amounts (Barltrop, 1982). Although data on humans are lacking, laboratory studies show that heavy metals accumulate in the brain of immature animals in much greater amounts than in adults (Jugo, 1977; Kostial et al., 1978). These investigators found substantially higher levels of lead, mercury, and manganese in the brains of 1- and 2-week-old rats than in older animals given the agents by intravenous or intraperitoneal injection. The greater toxicity of morphine in immature rats was associated with higher brain-to-blood ratios (Kupferberg and Way, 1963). Pylkko and Woodbury (1961) attributed changes in the convulsant effects of strychnine and brucine in rats to maturation of the blood-brain barrier. The time of maturation of the blood-brain barrier in the rat appears to vary, ranging from about 1 week for 5,7-dihydroxytryptamine (Sachs and Jonsson, 1975) to as long as 3 weeks for cadmium (Wong and Klaassen,

1980). It is not known when this barrier becomes fully functional in humans.

Retention

Immature humans and laboratory animals typically exhibit greater systemic retention of heavy metals than do adults. Ziegler et al. (1978) found that retention of lead in human infants (14 to 746 days old) was dose dependent. Although urinary and fecal excretion of lead increased with dose, they apparently could not compensate for lead intake at higher doses. Investigations have shown greater whole body retention, elevated blood levels, and higher target organ concentrations in young animals than in older animals given heavy metals by injection (Kostial et al., 1978; Wong and Klaassen, 1980). Possible explanations for these age-related differences include diminished excretory capacity, high growth rates and high rate of protein synthesis, altered binding to proteins and other ligands in tissues, higher extracellular fluid volume, and greater permeability of tissue barriers in the immature organism. The brain and testes, two organs believed to have effective barriers to ionized molecules in the adult, exhibited substantially higher levels of lead, mercury (Kostial et al., 1978), and cadmium (Wong and Klaassen, 1980) in young rats. Conversely, the kidneys, another target organ, contained smaller amounts of these compounds in the young animals. Such differences may have significant toxicological implications. Data on tissue distribution as it relates to age are generally lacking for most other classes of chemicals.

Metabolism of Xenobiotic Compounds

The human newborn exhibits decreased capacity to metabolize a variety of drugs and other xenobiotic compounds (Warner, 1986; Reed and Besunder, 1989). The premature newborn is usually more deficient than the full-term newborn, although metabolic functions increase rapidly in both during the initial days after birth. A deficiency in MFO activity does not necessarily entail greater susceptibility to toxicity; indeed, it may have the opposite effect (Done, 1964). Inefficient metabolism would make the newborn more susceptible to the action of compounds that are converted to less active, more readily excreted metabolites. Conversely, inefficient metabolism should confer protection against the compounds that are metabolically activated (i.e., converted to reactive, cytotoxic metabolites). Human fetuses and newborns have higher MFO activity, in relation to adult values, than do nonprimates (Neims et al., 1976). Thus, human newborns should be more sensitive to chemicals requiring metabolic activation, whereas most laboratory animals should be more sensitive to chemicals that are detoxified and eliminated via the MFO system.

TABLE 3-4 Risk Assessment for Infants and
Children: Pharmacokinetic Factors

Drug	Newborn $t_{1/2}$ (hrs)	Adult $t_{1/2}$ (hrs)
Drugs with Low Hepatic Clearance		
Aminophylline	24-36	3-9
Amylobarbitone	17-60	12-27
Caffeine	103	6
Carbamazepine	8-28	21-36
Diazepam	25-100	15-25
Mepivacaine	8.7	3.2
Phenobarbitone	21-100	52-120
Phenytoin	21	11-29
Tolbutamide	10-40	4.4-9
Drugs with Intermediate or High Hepatic Clearance		
Meperidine	22	3-4
Nortriptyline	56	18-22
Morphine	2.7	0.9-4.3
Lidocaine	2.9-3.3	1.0-2.2
Propoxyphene	1.7-7.7	1.9-4.3

SOURCE: Adapted from Rane, 1992.

It is difficult to generalize about age-dependent deficiencies in the metabolism of xenobiotic compounds because different enzymatic pathways
seem to exhibit dissimilar maturational patterns (Neims, 1982). There are
a number of forms of cytochrome P-450 in the human liver that appear to
have distinctive substrate specificity and unique developmental patterns.
Pelkonen et al. (1973) found that fetal liver was much more deficient in aryl
hydroxylase activity than in other hepatic microsomal monooxygenases.
Metabolism of caffeine and theophylline, which initially undergo N-de-
methylation, is particularly slow in the human newborn. Neonates exhibit
a plasma half-life for caffeine of approximately 4 days, as compared to 4
hours for adults (Aldridge et al., 1979). These investigators observed that
adult levels and patterns of caffeine metabolism were reached at 7 to 9
months of age. Glucuronidation is one of the most inefficient pathways
for metabolism during early development and may take the longest to
mature (Done et al., 1977). In contrast, mixed-function-catalyzed oxidation
of a number of other drugs increases rapidly during the first days of life,
soon approaching and exceeding adult values (Neims, 1982). Thus, the
ontogeny of metabolism of the xenobiotic compounds, and its implications
in toxicology, is quite compound-specific.

The time course of postnatal development of MFOs has been delineated
for several drugs (see Table 3-4). Loughnan et al. (1977) measured the
plasma half-life of phenytoin in 2-day- to 96-week-old subjects given the
drug intravenously. Although the half-life was prolonged and variable
during the first week of life for full-term newborns, the premature new-

born exhibited even longer and more inconsistent values. Adult values seemed to be reached by 7 weeks of age, but the number of study subjects more than 1 week old was too small to be definitive. Neims et al. (1976) used the data of several investigators to estimate phenytoin half-life values as a function of age. A marked increase in phenytoin metabolism (i.e., a decrease in half-life) was observed during the first few days postnatally, followed by a progressive increase to levels two- to threefold greater than those of adults in neonates and infants more than 2 weeks old. The large interindividual variability seen in newborns diminished with age. A similar, rapid increase in metabolism and a decrease in variability with age were seen when data for phenobarbital were evaluated (Neims et al., 1976). Both drugs are metabolized by aromatic hydroxylation. Clearance of many drugs, when normalized to unit body weight, is two- to fourfold greater in infants and young children than in older children and adults. As illustrated in Figure 3-1, plasma levels and the dose of theophylline required to maintain therapeutic levels vary substantially with age. Such increases in metabolic clearance, evident for many drugs from the age of 2 to 3 months to the age of 2 to 3 years, tend to decline gradually during childhood until adult values are reached (Morselli et al., 1980).

Metabolism of xenobiotic compounds in the newborn and neonate may be both qualitatively and quantitatively different from that in the adult. Although chloramphenicol is primarily metabolized by hydrolysis and by glucuronidation, a glycolic acid derivative not found in adults has been identified in newborns (Morselli et al., 1980). Premature newborns, unlike infants and adults, exhibit substantial N-methylase activity. This enzyme, acting in the presence of N-demethylase deficiency, can convert theophylline to caffeine. This process is the opposite of what occurs in adults. Aldridge et al. (1979) evaluated the pattern of caffeine metabolites as affected by age in neonates and infants. The researchers found that the proportion of individual metabolites varied until an adultlike metabolite pattern was reached at 7 to 9 months. The toxicological implications of age-dependent qualitative differences in the metabolism of xenobiotic compounds are for the most part unknown.

Enzyme Development. The ontogeny of metabolism of xenobiotic compounds in humans remains largely unexplored. The number of investigations undertaken during fetal and neonatal periods is largely dictated by the availability of tissue. Accordingly, there have been a considerable number of studies of fetal hepatic enzymic differentiation during the first and second trimesters, but relatively few studies during the third trimester and postnatally (Rane and Sjoqvist, 1972; Neims et al., 1976). The major components of the monooxygenase system are present in fetal liver during midgestation. Smooth endoplasmic reticulum, the principal site of local-

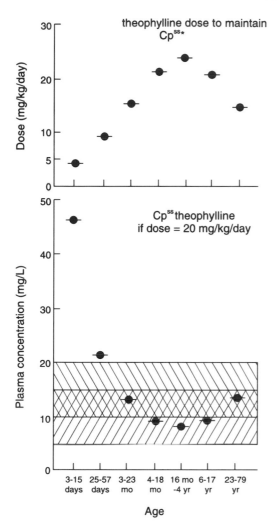

FIGURE 3-1 Theophylline dose requirements and plasma concentrations. (Top) Estimated dose requirements of theophylline (mg/kg/day) to maintain a steady-state plasma concentration Cp^{ss} of 10 mg/liter. (Lower) Estimated plasma concentrations of theophylline at steady state if dose is kept at 20 mg/kg/day. Shaded areas indicate tentative therapeutic level for bronchodilation and antiapneic activity. SOURCE: Aranda, 1984.

ization of the MFO system in the hepatocyte, is also present at this time (Gillette and Stripp, 1975). Levels of cytochrome P-450 and other monooxygenase components appear to remain relatively constant through parturition. Measurements at term have shown that P-450 levels and NADPH-cytochrome c reductase activity are each about 50% of adult values (Aranda et al., 1974). Postnatal increases in enzyme activity could conceivably result from any one or a combination of the following: increased synthesis of enzyme, conversion of inactive to active enzyme, decreased catabolism of enzyme, disappearance of an endogenous inhibitor, or appearance of an activating substance. Studies of the differentiation of a variety of liver enzymes during the late fetal and postnatal periods indicate that de novo synthesis is largely responsible for increased enzymatic activity (Greengard, 1977). This synthesis is attributed to gene expression as a result of natural (hormones and other endogenous substrates) and unnatural (drugs and other chemicals) stimuli encountered in the extrauterine environment. It is possible that relative proportions of different forms, or isozymes, of P-450 change with development of the individual. Warner (1986) pointed out that the period of most rapid MFO metabolism (from 2 or 3 months to 3 years) coincides with the period when concentrations of endogenous substrates (such as steroid hormones) are low. The gradual decline in metabolism to adult levels at puberty parallels the increase in sex steroids accompanying maturation.

Liver Development. The liver undergoes a series of integrated morphological and functional changes perinatally, including differentiation of hepatocytes and emergence of constitutive enzymes. One important function, which requires coordinated maturation of a series of processes, is enterohepatic circulation. Bile flow depends on the adequate synthesis, conjugation, secretion, and recirculation of bile acids (Balistreri et al., 1983). Suchy et al. (1981) found that serum levels of cholylglycine and chenodeoxycholate, the two major bile acids, became markedly elevated in normal neonates during the first 4 days of life. The levels then gradually declined over the next 4 to 6 months to values typical of children and adults. The initial increase, which the investigators termed *physiologic cholestasis*, was attributed to impaired intestinal reabsorption and hepatic transport processes. They speculated that the transport deficit could be caused by functional immaturity in hepatocellular uptake, binding, conjugation, or secretion of bile acids.

Two important possible consequences of the physiologic cholestasis during the first months of life are inefficient intestinal fat digestion and inhibition of biliary excretion. Impaired fat digestion could be of toxicologic importance when lipophilic chemicals are ingested with oils. Studies in rats have demonstrated that halogenated hydrocarbons such as carbon tetrachloride are more poorly absorbed and less acutely toxic when given in oils than when given undiluted or in aqueous vehicles (Kim et al., 1990). The oils serve as a reservoir in the gut to retard systemic absorption of lipophilic chemicals. Biliary excretion is one of the two major pathways for elimination of chemicals from the body. Bilirubin, a product of hemoglobin catabolism, can accumulate in the body if its excretion in the bile is unduly hindered. A transient physiological hyperbilirubinemia is commonly seen in newborns, but it usually lasts only a few days, rather than months as does physiological cholestasis (Suchy et al., 1981). The physiologic hyperbilirubinemia is generally attributed primarily to the newborn's diminished glucuronidation capacity, which in turn is believed to result from low hepatic microsomal uridine diphosphoglucuronyltransferase activity (Kawade and Onishi, 1981). Glucuronide conjugation is depressed to a greater extent and for a longer time during perinatal development than sulfate or glycine conjugation (Dutton, 1978). There have been a number of reports of toxicities associated with neonates' decreased ability to conjugate and eliminate chemicals in the bile and urine. Chemicals implicated in such cases include chloramphenicol (Sutherland, 1959; Weiss et al., 1960), hexachlorophene (Tyrala et al., 1977), benzyl alcohol (Gershanik et al., 1982), and diazepam (Nau et al., 1984).

Excretion

Renal excretion is the principal pathway for elimination of most chemicals from the body. Although volatile parent compounds and metabolites can be exhaled, this route of elimination is of little quantitative significance for most pesticides in current use. As described above, biliary excretion may play a role in elimination of some parent compounds and metabolites, notably conjugates formed in phase II-type reactions of liver metabolism. The conjugates may be eliminated concurrently in both bile and urine.

The kidneys are anatomically and functionally immature at birth. Although nephrogenesis is complete in the full-term human newborn, anatomic immaturity is still manifest for several weeks (Lorenz and Kleinman, 1988). The final phase of postnatal anatomical kidney development is that of increase in nephron size. Renal blood flow is relatively low in the neonate as a result of receiving a smaller percentage of total cardiac output and high intrarenal vascular resistance. Developmental changes in glomerular filtration rate parallel changes in renal blood flow (Lorenz and Kleinman, 1988). Glomerular filtration rate in premature newborns may be as low as 5.0% of that in adults, whereas in full-term newborns it is typically 30% to 40% of adult values. The glomerular filtration rate increases rapidly in the infant, becoming equivalent to that of the adult per unit of surface area within 10 to 20 weeks (West et al., 1948). Glomerular function appears to be more advanced at birth than renal tubular function (Weil, 1955). Tubular functions include both active and passive reabsorptive and secretory processes for specific agents. The deficient transport processes and reduced glomerular filtration rate produce smaller medullary solute gradients than are produced in adults, which in turn result in a diminished capacity of the neonate to concentrate urine. Maturation of tubular transport systems is relatively slow, in that maximum capacity may not be reached until about 8 months of age (West et al., 1948; Calcagno and Rubin, 1963).

Excretion of chemicals by the kidneys depends primarily on glomerular filtration and tubular secretion and reabsorption. A decrease in one or the other in neonates can result in delayed clearance of a chemical from the bloodstream and the body. Under such circumstances, the chemical may have a prolonged duration of action and there may be an increased risk of toxicity. Aminoglycoside antibiotics, such as kanamycin and gentamicin, are examples of compounds that are not metabolically degraded or bound to plasma proteins to a signficant extent. They are primarily eliminated by glomerular filtration. Total body clearance values for premature newborns are only about 5% of adult values, whereas full-term newborn values are about 10% to 30% of the values of adults. Increase in aminoglycoside clearance during the neonatal period is closely related to maturation of glomerular function and correlates well with increased creatinine clearance (McCracken et al., 1971). Penicillins, on the other hand, form a class of compounds eliminated primarily by the kidneys via tubular secretion. Newborns and neonates given penicillins typically exhibit prolonged blood half-lives. Morselli et al. (1980) note that glomerular filtration and tubular secretion mature more rapidly than does tubular reabsorption, so clearance values for some compounds may be quite high during the first 2 to 24 months of life.

SCALING AND REGRESSION ANALYSIS

Scaling is the mathematical process used to adjust the dosage of therapeutic drugs or toxic substances to achieve comparable effects between animals of different size in one species or between different species of animals of markedly different sizes. In this section, body weight, surface area, metabolic rate, and regression analysis will be used as variables to relate dosages of compounds and body size. The choice of an appropriate scaling variable is important because of the need to compare toxicities in newborn animals weighing only a few grams with their mature counterparts weighing hundreds of grams and then extrapolating these findings to human infants weighing several kilograms and to human adults weighing 50 to 100 kg.

Although there are sound theoretical bases for choosing any one of the variables that can bridge great differences in size, there is no current consensus on which is most appropriate. Part of the difficulty in using a single approach is that the rate of development of the processes involved in absorption, distribution, metabolism, and excretion of xenobiotic compounds vary with age and across species. Thus, although some agreement is necessary to establish one system of scaling so that one set of studies can be compared with another, no scaling method will resolve all the differences between animals of different ages and of different species.

Body Weight

Traditionally, the therapeutic effectiveness or the toxicity of xenobiotic compounds in animals of different sizes has been evaluated by comparing dosages based on weight and expressing these as, for example, milligrams of compound per kilogram of body weight or micrograms of compound per 100 g of body weight. In comparing animals that differ in size as much as mice and rats or rats and human adults do, or immature and mature animals, the values per unit of weight have often been found to be unpredictable for equivalent levels of therapeutic efficacy or toxicity. For example, the therapeutic dose of the anticancer drug methotrexate in the mouse is 1.5 mg/kg; in the rat, 0.5 mg/kg; and in humans, 0.07 mg/kg (Pinkel, 1958). Within one species, the vitamin K analog synkavite at a dosage of 0.16 mg/kg produces a bilirubin level of more than 30 mg/dl in the newborn rat, but only 3 mg/dl in the adult rat. In other words, a dose of 0.64 mg/kg in the adult rat is required to increase bilirubin levels by the same amount that is produced by a dose of slightly less than 0.04 mg/kg in the newborn rat (Wynn, 1963). In pesticide studies, the ratio of the LD_{50} of adult to newborn rats was 2.4 for parathion (3.6 mg/kg for adults and 1.5 mg/kg for 23-day-old rats), whereas for octamethyl pyrophosphate the LD_{50} ratio was reversed (0.2; the adult LD_{50} was 10 mg/kg and for the 23-day-old rat LD_{50} was 49 mg/kg) (Brodeur and DuBois, 1963).

Other Effects of Body Size

Another way to scale dosages of xenobiotic compounds between large and small animals is on the basis of some function of body size other than weight. The growth rate varies from organ to organ or component to component in relation to increases in body size. Examination of this phenomenon led to the use of the allometric expression in the form

$$y = aB^x,$$

in which y is the weight of an organ or component of the body, B is total body weight, a is a constant, and x is an exponent of body weight. Transforming this equation yields

$$\log y = x \log B + \log a,$$

and then a plot of log y versus log B yields a simple linear graph with x as the slope and log a as the intercept. This type of representation indicates that the growth of a particular organ bears some constant relationship to the growth of the animal, but that relationship may vary from structure to structure. In physiology, it became apparent that a variety of physiological functions also increased allometrically with body size. These relationships were extensively reviewed by Boxenbaum (1982) and Boxenbaum and D'Souza (1990).

When these allometric concepts were applied to metabolic rate, the equation would be

$$\text{metabolic rate} = a \cdot \text{wt}^x,$$

where a is a constant that depends on the units used for metabolic rate and weight (wt), and x would be a fraction that was less than 1.0. Brody (1945) recognized that the expression for metabolic rate was similar to the expression for the surface area of the body. He further made the assumption that since the rate of cooling of a solid body was proportioned to its surface area, by analogy the metabolic rate should be directly proportional to surface area. Thus if surface area = $a \cdot \text{wt}^{2/3}$, then metabolic rate = $a \cdot \text{wt}^{2/3}$. (For an extensive discussion of these relationships, see Calabrese [1986].)

When the potency or toxicity of a substance is related to the persistence of some concentration of the compound in the body, the effect will be related to the rate at which the original substance is metabolized to less potent or to more potent compounds. In many cases, the rate at which a xenobiotic compound is metabolized will be related to the overall metabolic rate of the animal. Because smaller animals, in general, have greater metabolic rates per unit of body mass than do larger animals, the rate of metabolism of many substances per unit of body weight will be more rapid in smaller animals than in larger ones.

Except for the energy required for growth (a small fraction of energy consumption), energy intake and expenditure are essentially equivalent. Total energy expenditure involves the energy required for resting metabolism, for diet-induced thermogenesis, and for physical activity. The total expenditure of energy is the overall metabolic rate, and this can be closely approximated by the total energy intake when body weight is relatively stable. The advantage of this is that energy intake is generally more easily measured than is metabolic rate.

Surface Area

A variety of methods related to metabolic rate have been developed to equalize the values for energy expenditure (and consumption) between individuals of different sizes. The two principal approaches have been to use a power function of weight, such as weight$^{2/3}$ or weight$^{3/4}$, or to use calculated surface area. Although surface area is reasonably approximated by weight$^{2/3}$, it is more precisely a power function of weight with an additional factor based on a power function of height or length. The surface area methodology has been the more widely used of these two approaches, although weight$^{2/3}$ and surface area have often been used interchangeably despite the fact that they are not actually equivalent (Brody, 1945; Lindstedt, 1987). Values for usual energy intakes by human individuals of different sizes and the relationship between kilocalories per kilogram of body weight and kilocalories per square meter of body surface area are shown in Table 3-5. Because of the comparability of energy requirements throughout infancy and childhood when calculated on a surface area basis ($\approx$2,000 kcal/m^2), it has become common to assume that the metabolism of therapeutic drugs would also be comparable on this basis. Therefore, many medication dosages for children are calculated in terms of milligrams per square meter.

Although these assumptions are reasonably valid for many drugs in individuals of a comparable age group, they are not necessarily valid across all age groups or for all drugs (Lamanna and Hart, 1968). Almost all the comparisons between animals of different sizes (within and across species) in which surface area has been used to compare physiological functions, metabolism of drugs, or other body parameters have used comparisons between mature animals. Because of differences in the rate of maturation of individual organs and their functions, within and between species, as the animal proceeds from birth to maturity, it is not likely that the simple surface area relationship will be true for the comparison of immature and mature animals. Infants, children, and adults, just as immature and mature animals of other species, differ from each other in stages of functional development of individual organ systems, the processes of growth, and maturation of enzyme systems involved in the metabolism of xenobiotic compounds, as well as in their metabolic rates expressed on a weight or surface-area basis (see Table 3-5). Compounds such as chloramphenicol, which is poorly detoxified in the newborn (Calabrese, 1978); acetaminophen, which is excreted primarily as a sulfate conjugate in the young and as a glucuronide in the adult (Sonawane, 1982); isoniazid, which is acetylated at a reduced rate in the newborn (Nyhan, 1961); and caffeine, which is poorly biotransformed in the infant (Neims, 1982), are examples of drugs that differ in metabolism and toxicity in relation to age, whether doses are compared in terms of weight or surface area.

Metabolic Rate

Another method for comparing children of all sizes with adults in order to establish equivalent dosages of drugs is to relate the dose directly to metabolic rate (Calabrese, 1986). Dosage is represented as the amount per unit of metabolic rate, e.g., milligrams per 1,000 kcal per 24 hours. This

TABLE 3-5 Representative Values of Weight, Length (Height), Surface Area, and Caloric Intake for Individuals at Various Ages

Age (years)	Weight (kg)	Length (Height) (cm)	Surface Area (m²)	Caloric Intake	Kilocalorie/ Kilogram[a]	Kilocalorie/Square Meter[b]
Premature birth	1.0	36	0.09	140	140	1,555
Full-term birth	3.5	50	0.21	400	115	1,905
2	13.0	88	0.56	1,200	92	2,140
8	25.0	127	0.93	2,000	80	2,150
13, female	46.0	157	1.43	2,200	48	1,540
13, male	45.0	156	1.40	2,800	62	2,000
Adult, female	57.0	163	1.60	2,300	40	1,440
Adult, male	70.0	178	1.85	3,000	43	1,620

[a] Kilocalories per kilogram are given to indicate the three- to fourfold differences in metabolic rate when calculated on a weight basis.
[b] Kilocalories per square meter are shown to demonstrate the relative stability of this measure throughout childhood and also to emphasize the differences between children and adults, even on this basis.

SOURCE: Based on data from Hamill et al., 1979; NRC, 1989.

TABLE 3-6   Comparison of Doses Across Age Groups, Based on
Weight, Surface Area, and Metabolic Rate

Age (years)	Dose Basis		
	100 mg/kg	3,784 mg/Square Meter[a]	2,333 mg/1,000 kcal[a]
Premature birth	100	340	327
Full-term birth	350	795	933
2	1,300	2,120	2,800
8	2,500	3,520	4,666
13, female	4,600	5,410	5,130
13, male	4,500	5,300	6,530
Adult, female	5,700	6,050	5,365
Adult, male	7,000	7,000	7,000

NOTE: To illustrate the relative differences in dosage on these three bases, the dosage for the adult male is held constant in absolute figures (7,000). By dividing that dose by the adult male's weight, surface area, and metabolic rate, one obtains the figures of 100 mg/kg (column 2), 3,784 mg/m² (column 3), and 2,333 mg/1,000 kcal (column 4). These figures are then used to calculate the absolute dosages for each individual on each basis. These data demonstrate that the three bases for equating drug dosage between individuals of different ages and weights will result in absolute dosages that may vary severalfold for the same age and size person.

[a] These values per square meter (m²) and per 1,000 kcal were used to provide equivalent adult dosages.

method has the theoretical advantage of directly relating dosage to the variable that is commonly responsible for differences in drug metabolism between individuals of different sizes and ages. An example of dosages based on weight, surface area, and metabolic rate is given in Table 3-6, in which the adult dose is made constant. With the exception of the very-low-birth-weight infant, dosages based on kilocalories consumed (or metabolized) yield higher values for children than the other approaches for equivalent adult dosage. Even this approach can yield erroneous results, as illustrated by Schmidt-Nielsen (1972), whose calculations of the dosage of lysergic acid needed to produce the condition called *musth* in the male elephant were as follows:

Based on body weight, elephant versus cat	=	297 mg
Based on metabolic rate, elephant versus cat	=	80 mg
Based on body weight, elephant versus human	=	8 mg
Based on metabolic rate, elephant versus human	=	3 mg
Based on brain weight, elephant versus human	=	0.4 mg

Another problem arises when one considers whether compounds are metabolized to less toxic or to more toxic substances in the liver. This difference is of particular importance in extrapolation from one animal species to another, especially to the human species. For example, the extent of the postnatal development of cytochrome P-450-dependent oxidations is contingent on both species and substrate as well as on the sex of the animal. In the hydroxylation of aromatic compounds, maximum activity is reached at about 30 days in male and female rats but decreases after that in the female and is constant or slowly increasing in males for another 30 days (Klinger, 1982). In humans, there appears to be a relatively consistent difference; humans have a significantly lower rate of hepatic metabolism than found in other mammalian species. Boxenbaum and D'Souza (1990) discussed this difference between human and other animal species in detail and used the term *neoteny* to explain these differences. Neoteny is the "retention of formerly juvenile characteristics by adult descendants . . . produced by retardation of somatic development" (Gould, 1977).

The significance of this concept for pesticide toxicology is that the hepatic metabolism of many xenobiotics in the mature human subject will occur at a reduced rate when compared to the rates in other mature animals even when corrected for scaling difference. It remains uncertain at this time how the problem of neoteny applies to immature animals. This is an area that deserves investigation although the studies will be complex because of the difficulty in identifying comparable stages of maturity in young animals. Further description of the maturation of enzyme systems is provided in the monograph by Calabrese (1986) entitled *Age and Susceptibility to Toxic Substances* and in the detailed review by Klinger (1982).

Regression Analysis

An alternative approach to scaling that has overcome some of the problems mentioned is that of regression equations that use the allometric concept without a predetermined power value. In this approach, the log of dose for a specific outcome is plotted against the log of body weight. For many substances and many animal species a linear relationship exists, but the slopes and intercepts are specific for each compound and each species. Krasovskii (1976) explained this process in detail, and after examining the action of several hundred compounds stated, "[I]t was shown that the regularities of the comparative sensitivity of the animals to 80-85% of the substances can be characterized by a straight line equation." When this approach was used to evaluate human toxicity of 107 substances based on data from four to six animal species, the results indicated there was a transfer error that did not exceed a value of 300% to 400% between calculated and observed values.

For risk assessment purposes, the problem becomes simpler as long as both exposure and toxicity are expressed on the same basis, e.g., micrograms per kilogram, micrograms per square meter, or micrograms per 1,000 kilocalories. To clarify this concept, assume that pesticide X is used on food A. The residue at the point of consumption is found to be 100 $\mu g/100$ g (1 $\mu g/g$) of food A. If the average adult consumption of food A is found to be 210 g with a total energy intake of 3,000 kcal/day (Table 3-5) and the average 2-year-old male infant consumes the same proportion of his total calories—1,200—as food A (Table 3-5), the following data apply:

Adult: 210 g of food A in a 3,000 kcal total intake
2 year old: (1,200 kcal ÷ 3,000 kcal) × 210 g food A = 84 g food A
Adult exposure = 210 g × 1 $\mu g/g$ = 210 μg of pesticide X
2-year-old exposure = 84 g × 1 $\mu g/g$ = 84 μg of pesticide X

If these intakes are then scaled by weight, surface area, or calories for both adult and child, the value of intake of pesticide X becomes 70 $\mu g/$ 1,000 kcal for both (see Table 3-7).

If toxicity is evaluated on the same basis as exposure is assessed (i.e., on the basis of weight, surface area, or calories metabolized) using the same ratio of dosages, the results will be identical whatever basis is used. For example, if the no-observed-effect level (NOEL) were 6 $\mu g/kg$ for the adult and 13 $\mu g/kg$ for the child, then the NOEL expressed on the basis of surface area would be 228 $\mu g/m^2$ in the adult and 300 $\mu g/m^2$ in the child. On a metabolic rate basis, the NOEL for adult and child would both be 140 $\mu g/1,000$ kcal. Thus, on whatever basis one calculated toxicity, both the infant and adult would be consuming X at half the NOELs for their age group. Similar calculations would be needed to translate these ratios from the human species to the species to be used for toxicity testing.

Because the diets of small children are limited in diversity, it would be reasonable for the 2 year old to consume twice as much of food A as the adult as a proportion of his total energy intake. In this situation, the amount of residue ingested by the infant would be 13 $\mu g/kg$, 300 $\mu g/$ m^2, and 140 $\mu g/1,000$ kcal. Because of the relative increase in intake by the child compared to the adult, the child would be at the child's NOEL whatever the basis of the calculation.

If the efficacy or toxicity of a compound is not related to its rate of metabolism, using energy consumption as a basis for relating dosages probably would not provide equivalent levels of toxicity. In acute toxicity, for example, if the toxicity depended on peak concentration, comparable dosage would be on a simple weight basis because volumes of distribution based on body water compartments are more closely related directly to weight than to surface area or metabolic rate. Even under such circumstances, differences in rates of absorption, extent of protein binding,

TABLE 3-7 Expression of Exposure Values for the Adult and Child

	Consumption and Physiological Parameters				Pesticide X Exposure Values		
Age	Pesticide X	Weight	Surface Area	Food A	Weight	Surface Area	Food A
Adult	210 µg	70 kg	1.85 m^2	3,000 kcal	3.0 µg/kg	114 µg/m^2	70 µg/1,000 kcal
Child	84 µg	13 kg	0.56 m^2	1,200 kcal	6.5 µg/kg	150 µg/m^2	70 µg/1,000 kcal

plasma half-life, or concentration at the receptor site could modify the dosage generated on a weight basis.

In chronic exposures, the differential rate of development of metabolizing enzymes such as glucuronyl transferase and the P-450-dependent mixed-function oxidases can have an impact on toxicity that is independent of overall metabolic rate. If the parent compound were the toxic substance, delayed enzymatic degradation would enhance toxicity. However, if a metabolite were the source of toxicity, the slower metabolism would result in reduced toxicity.

If the metabolism of a xenobiotic material is fully elucidated, it would be appropriate to use a reference base or denominator (weight, surface area, or metabolic rate) that best reflects the pharmacokinetics of that compound. When the developmental pharmacokinetics of a substance are not well delineated, as is often the case for pesticides, it has been demonstrated empirically that on a weight basis, the toxicity difference between immature and mature animals, based on comparative $LD_{50}s$, is usually less than a factor of 5 and has only occasionally been reported to exceed a factor of 10 for any pesticide studied to date (Calabrese, 1986).

CONCLUSIONS AND RECOMMENDATIONS

Conclusions

In this chapter, the committee summarized the mechanisms of toxicity and explored the potentially vulnerable organ systems or functional systems in the developing animal. In addition, it reviewed data on toxicants that provide the basis for these conclusions. Although the committee was interested in drawing conclusions from toxicity testing of pesticides, in many cases there were no data on developing animals. To illustrate the principles of toxicity as they pertain to developing animals, the committee therefore utilized information from testing of other toxicants, including drugs.

As described in Chapter 2, the nervous system, the immune system, and reproductive systems continue to develop after birth. This observation heightens concern that toxicity during these postnatal developmental stages or periods may have lasting consequences throughout adult life.

• Differences in toxicity between young and mature animals may be in either direction but are generally modest. The younger animal may be more sensitive or may be less sensitive than the older animal to comparable levels of exposure of toxic agents. The direction of these differences appears to be compound specific as well as age specific because toxicity may not vary linearly with age. In those instances where such measures as LD_{50}s are significantly different, the differences are usually less than 1 order of magnitude and often substantially less.

• Data on age-dependent pharmacokinetics of pesticides are lacking for most animals, and the data base on pharmacokinetics and metabolism of drugs in immature humans is also limited. Available information reveals that some functional immaturities offset or cancel one another, whereas others tend to be additive. For example, inefficient gastrointestinal absorption of the antiinflammatory drug indomethacin is offset by decreased biliary and renal clearance. Maturation of most biochemical and physiological processes occurs within the first 2 years; indeed, substantial changes occur within the first days and weeks of life. Compared to adults, therefore, neonates and infants can be anticipated to have the greatest differences in pharmacokinetics and susceptibility to pesticide toxicity— the youngest being the most likely to exhibit aberrant responses. Metabolic and renal clearance of many xenobiotics reaches and exceeds adult levels (when expressed on a body weight basis) during the first year. Therefore, older infants and young children may metabolize pesticides more extensively and eliminate them more rapidly than adults. This rapid elimination may confer increased resistance or susceptibility to toxicity, depending on the nature of the compound to which the individual is exposed.

• On the basis of our understanding of mechanisms of action of toxicants in mature animals, including the human adult, it is generally possible to predict that similar mechanisms of action will occur in immature animals, including the human infant, child, or adolescent (i.e., biochemical *mechanisms* of toxicity are similar across age and developmental stages). For example, if a toxicant is cytotoxic (causes cell death) in the adult, it should cause cell death in the immature animal by the same mechanism. This principle suggests that mechanisms of action should also be comparable across species.

• Studies of the toxicity of xenobiotic compounds in developing mammals—both laboratory animals and humans—demonstrate the potential for acute and chronic toxicity. Toxicity in the perinatal and pediatric periods is of special concern, since systems and structures under development at those times are important for survival over the lifetime of the mammal.

• Studies in laboratory animals have demonstrated an age-related difference in acute toxicity; however, the direction of the difference is dependent on the chemical, and the magnitude of the effect is usually no more than 1 order of magnitude and often is considerably less. A developing animal may be more, less, or equally sensitive to a given chemical than is an adult.

• Studies of the influence of age on the toxicity of xenobiotic compounds in laboratory animals may provide an incomplete or inaccurate picture of similar effects in humans. Rodents are less mature at birth than humans. Thus, more pronounced age-related differences would be anticipated in rodents than in humans during the neonatal period. Rodents and most other commonly used test animals mature very rapidly during this time, so that differences of even a few days in age can profoundly affect susceptibility to the toxicity of xenobiotic compounds.

• Very few data were found on the relative toxicity of pesticides in immature and mature humans. There is, however, a limited data base on pharmaceutical agents. As in animals, age dependency of humans to therapeutic and side effects is drug and age dependent. Since premature and full-term newborns are the most different anatomically and physiologically from adults, it follows that they typically exhibit the most pronounced differences from adults in sensitivity to drugs.

• Pharmacokinetic processes control the amount of bioactive chemicals in target organs and, in turn, the magnitude and duration of toxic effects. The net effect of immaturity on the various processes that affect chemical disposition is difficult to predict. The situation is complex in a number of respects. Most laboratory animals are less mature than humans at birth, although maturation in animals is more rapid. In addition, various body structures and associated functions mature at different rates in different species.

• Comparison of exposures between immature and mature animals and across species is complex, and no single mathematical expression is universally applicable. The toxicity of a pesticide varies with its rate-limiting pharmacokinetic processes.

• Data from studies in humans (e.g., in children and adults treated with cytotoxic chemotheraupetic agents) show that toxic effects are similar qualitatively but may differ quantitatively. That is, the types of toxic effects that limit treatment were similar in children and adults; however, the dose at which treatment limitations were reached was different. Indeed, for some drugs, the maximum tolerated dose is greater in developing than in human adults (i.e., such drugs are less toxic to infants and children).

• Studies of the ontogeny of plasma esterases in humans suggest that immature individuals may be at greater risk for acute toxic effects of pesticides that are cholinesterase inhibitors. The increase in sensitivity appears to be greatest during the first months of life, apparently because of relatively low levels of esterases and a diminished capacity to detoxify such pesticides.

• Although the principles of developmental toxicity following in utero exposure have received considerable attention, there has been little attention to the principles of developmental toxicity following exposure to pesticides in the postnatal and perinatal periods.

• The central nervous system (CNS) continues to develop during the second and third trimester and during postnatal life. These developmental processes include many—if not all—the developmental processes that occur during prenatal development. As expected, alteration of the development of the nervous system can be blatant or silent—until the function is needed.

• The immune system is responsible for modulating host defenses against a range of human diseases. The successful development and functioning of the immune system require recognition and response to a range of cellular and circulating signals acting by endocrine, autocrine, and paracrine mechanisms. These complex control systems offer multiple opportunities for disruption by environmental chemicals—such as agricultural pesticides.

• Assessment of the effects of pesticides on the developing human nervous system is difficult because the methodology for such assessment is complex and poorly delineated. Development of the CNS is characterized by exacting architectural complexity and localization of function occurring over a prolonged period postnatally. The effects of altered neurologic development may be measured either as anatomic or behavioral and cognitive outcomes.

• The scientific consensus on appropriate neurodevelopmental outcome measures for evaluation of exposures (in animal models and human epidemiologic studies) is still evolving (NRC, 1992). Measurement of these end points is complicated, not only because of the elaborate nature of the end points that are measured but also because the timing of the insult may change the outcome and the functional end points may not be manifested until long after the exposure.

• Despite the difficulties in measuring effects, exposure to xenobiotic compounds has been found to alter CNS development at the anatomic and functional level. The alteration in development can be irreversible, thus resulting in permanent loss of function. These damaging effects of xenobiotic compounds on the CNS of the developing organism can occur at exposure levels that are safe for the adult.

• Certain classes of pesticides, including organophosphates, carbamates, and organochlorines, are known to have neurotoxic effects, especially as the result of high-dose acute exposures. Generally, data are insufficient for evaluation and determination of the neurodevelopmental effects of low-dose exposure to these broad classes of agents, and risk assessment for low-level exposures is not possible using current data. Nevertheless, the biochemical (neurotoxic) mode of action of these classes of compounds and the distinctive qualitative vulnerability of the child's developing nervous system makes the evaluation of low-dose neurodevelopmental effects a concern.

Recommendations

• The establishment of a standard developmental assessment model or protocol would allow one to interpret toxicity studies in immature animals in a systematic manner. The committee recommends that a standard protocol for evaluation of immature animals be established and required as part of the basic assessment of pesticides for toxicity to immature animals.

• Given the potential for lifelong effects following perinatal or pediatric toxicity, it is essential to develop toxicity testing procedures that specifically evaluate in appropriate animal models vulnerability during the developmental period and the adverse effects, if any, over the life of the animal.

• Given that toxicity is generally age related, consideration of this phenomenon must be included in regulatory action, testing methodologies, and public health policies.

• In general, it is possible to extrapolate data on the end points of toxic concern from adults to developing humans, although the dose that produces the toxicity is likely to be different. The developing human frequently seems to be more resistant than the adult to cytotoxicity from the anticancer and anti-AIDS drugs tested.

• Extrapolation of toxicity data from laboratory animals to humans—especially those for developing organisms—may be inaccurate. Careful attention to species differences in disposition and metabolism as well as in stages of maturation of organ systems is essential for accurate policy development and public health protection.

• Greater attention is needed to develop a broader understanding of the principles guiding developmental toxicity of organisms, especially humans, following birth and during critical periods of postnatal development, including infancy and puberty.

• Neurodevelopmental effects must be part of the battery of end points evaluated for toxicants, including pesticides and agricultural chemicals.

• Analysis of the impact or toxicity of agricultural chemicals on the immune system is essential. Regulatory development of a battery of consensus tests is critical to protect the developing immune system. At present, there is a paucity of information on the effects of many chemicals on the developing and indeed on the mature immune system.

• Although it is extremely difficult to assess neurodevelopmental effects, the CNS may be peculiarly vulnerable during a prolonged period of development, even if the exposure is at a level known to be safe for adults. Thus, a feasible, streamlined, and publicly credible method of assessment must be developed. Effectiveness of animal model and epidemiologic evaluation must be considered. Regulatory development of a battery of consensus tests will be difficult but necessary to ensure public confidence. Agencies involved should actively support research and innovation in this area of assessment.

• Since the kinetics of a variety of chemicals can be profoundly different in immature and mature subjects, the influence of immaturity on pesticide kinetics and toxicity is complex and must therefore be assessed on a case-by-case and chemical-by-chemical basis.

• When the pharmacokinetics of a specific compound are understood well enough to indicate that the metabolism (and elimination) of the substance is, in fact, proportional to the animal's rate of metabolism, as is often the case, then comparisons on a metabolic rate basis, on a surface area basis, or weight$^{2/3}$ basis would be reasonable. If surface area is used for adjusting from animals to humans, it should also be used for adjusting from infants to adults. In the many situations when such data are not available, the use of a simple body weight relationship for toxicity testing may be used as long as potential exposure is calculated on the same basis. Since most dietary pesticide exposure data are based on body weight, this is an added reason to use body weight as the basis for examining toxicity. In any case, scaling methods should be consistent.

REFERENCES

Adrian, T.E., H.A. Smith, S.A. Calvert, A. Aynsley-Green, and S.R. Bloom. 1986. Elevated plasma peptide YY in human neonates and infants. Pediatr. Res. 20:1225-1227.
Agunod, M., N. Yamaguchi, R. Lopez, A.L. Luhby, and G.B. Glass. 1969. Correlative

study of hydrochloric acid, pepsin and intrinsic factor secretion in newborns and infants. Am. J. Dig. Dis. 14:400-414.

Aldridge, A., J.V. Aranda, and A.H. Neims. 1979. Caffeine metabolism in the newborn. Clin. Pharmacol. Ther. 25:447-453.

Alfano, D.P., and T.L. Petit. 1985. Postnatal lead exposure and the cholinergic system. Physiol. Behav. 34:449-455.

André, F., J. Gillon, C. André, S. LaFont, and G. Jourdan. 1983. Pesticide-containing diets augment anti-sheep red blood cell nonreaginic antibody responses in mice but may prolong murine infection with *Giardia muris*. Environ. Res. 32:145-150.

Anisimov, V.N. 1981. Carcinogenesis and ageing. I. Modifying effects of aging on *N*-methyl-*N*-nitrosourea-induced carcinogenesis in female rats. Exp. Pathol. 19:81-90.

Anisimov, V.N. 1982. Carcinogenesis and ageing. III. The role of age in initiation and promotion of carcinogenesis. Exp. Pathol. 22:131-147.

Anisimov, V.N. 1983. Role of age in host sensitivity to carcinogens. Pp. 99-112 in Modulators of Experimental Carcinogenesis: Proceedings of a Symposium Organized by IARC and the All-Union Cancer Research Centre of the USSR Academy of Medical Sciences, V. Turusov and R. Montesano, eds. IARC Scientific Publ. No. 51. Lyon, France: International Agency for Research on Cancer.

Annau, Z. 1990. Behavioral toxicology and risk assessment. Neurotoxicol. Teratol. 12:547-551.

Aranda, J.V. 1984. Maturational changes in theophylline and caffeine metabolism and disposition: Clinical implications. Pp. 868-877 in Proceedings of the Second World Conference on Clinical Pharmacology and Therapeutics, L. Lemberger, and M.M. Reidenberg, eds. July 31-August 5, 1983. Washington, D.C.: American Society for Pharmacology and Experimental Therapeutics.

Aranda, J.V., S.M. MacLeod, K.W. Renton, and N.R. Eade. 1974. Hepatic microsomal drug oxidation and electron transport in newborn infants. J. Pediatr. 85:534-542.

Balistreri, W.F. 1988. Anatomical and biochemical ontogeny of the gastrointestinal tract and liver. Pp. 33-57 in Nutrition During Infancy, R. C. Tsang and B. L. Nichols, eds. Philadelphia: Hanley and Belfus.

Balistreri, W.F., J.E. Heubi, and F.J. Suchy. 1983. Immaturity of the enterohepatic circulation in early life: Factors predisposing to "physiologic" maldigestion and cholestasis. J. Pediatr. Gastroenterol. Nutr. 2:346-354.

Barltrop, D. 1965. The relationship between some parameters employed in the diagnosis of lead poisoning in childhood with a special reference to the excretion of delta-aminolaevulinic acid. Thesis. University of London.

Barltrop, D. 1982. Nutritional and maturational factors modifying the absorption of inorganic lead from the gastrointestinal tract. Pp. 35-41 in Banbury Report, No. 11: Environmental Factors in Human Growth and Development, V.R. Hunt, M.K. Smith, and D. Worth, eds. New York: Cold Spring Harbor Laboratory.

Barltrop, D., C.D. Strehlow, I. Thorton, and J.S. Webb. 1974. Significance of high soil lead concentrations for childhood lead burdens. Environ. Health Perspect. 7:75-82

Bellinger, D., A. Leviton, H.L. Needleman, C. Waternaux, and M.B. Rabinowitz. 1986. Low-level lead exposure and infant development in the first year. Neurobehav. Toxicol. Teratol. 8:151-161.

Bellinger, D., A. Leviton, C. Waternaux, H. Needleman, and M. Rabinowitz. 1987. Longitudinal analyses of prenatal and postnatal lead exposure and early cognitive development. New Engl. J. Med. 316:1037-1043.

Benet, L.Z., J.R. Mitchell, and L.B. Sheiner. 1990. Pharmacokinetics: The dynamics of drug absorption, distribution, and elimination. Pp. 3-32 in The Pharmacological Basis of Therapeutics, 8th ed., A.G. Gilman, T.W. Rall, A.S. Nies, and P. Taylor, eds. New York: Pergamon Press.

Benke, G.M., and S.D. Murphy. 1975. The influence of age on the toxicity and metabolism of methyl parathion and parathion in male and female rats. Toxicol. Appl. Pharmacol. 31:254-269.

Berenblum, I., L. Boiato, and N. Trainin. 1966. On the mechanisms of urethane leukemogenesis in newborn C57BL mice. II. Influence of thymectomy and of subsequent thymus reimplantation. Cancer Res. 26:361-363.

Berry, G., and J.C. Wagner. 1976. Effect of age at inoculation of asbestos on occurrence of mesotheliomas in rats. Int. J. Cancer. 17:477-483.

Bierring, F., H. Andersen, J. Egberg, F. Bro-Rasmussen, and M. Matthiessen. 1964. On the nature of the meconium corpuscles in human fetal intestinal epithelium. I. Electron micorscopic sutdies. Acta Path. Microbiol. Scand. 61:365-376.

Bleyer, A.W. 1977. Clinical pharmacology of intrathecal methotrexate. II. An improved dosage regimen derived from age-related pharmacokinetics. Cancer Treat. Rep. 61:1419-1425.

Blum, H.F., H.G. Grady, and J.S. Kirby-Smith. 1942. Relationships between dosage and rate of tumor induction by ultraviolet radiation. J. Natl. Cancer Inst. 3:91-97.

Bondy, S.C. 1986. The biochemical evaluation of neurotoxic damage. Fundam. Appl. Toxicol. 6:208-216.

Boxenbaum, H. 1982. Interspecies scaling, allometry, physiological time, and the ground plan of pharmacokinetics. J. Pharmacokinet. Biopharm. 10:201-227.

Boxenbaum, H., and R.W. D'Souza. 1990. Interspecies pharmacokinetic scaling, biological design and neoteny. Adv. Drug Res. 19:139-196.

Brodeur, J., and K.P. DuBois. 1963. Comparison of acute toxicity of anticholinesterase insecticides to weanling and adult male rats. Proc. Soc. Exp. Biol. Med. 114:509-511.

Brody, S. 1945. Bioenergetics and Growth. New York: Reinhold.

Buelke-Sam, J., and C. Mactutus. 1990. Testing methods in developmental neurotoxicity for use in human risk assessment. Conference on the Qualitative and Quantitative Comparability of Human and Animal Developmental Neurotoxicity. Work Group II Report. Neurotoxicol. Teratol. 12:269-274.

Burns, F.J., E.V. Sargent, and R.E. Albert. 1981. Carcinogenicity and DNA break repair as a function of age in irradiated rat skin. Proc. Am. Assoc. Cancer Res. 22:91. (Abstract)

Burri, P.H., J. Dbaly, and E.R. Weibel. 1974. The postnatal growth of the rat lung. I. Morphometry. Anat. Rec. 178:711-730.

Calabrese, E.J. 1978. Pollutants and High Risk Groups. New York: Wiley and Sons. 266 pp.

Calabrese, E.J. 1986. Age and Susceptibility to Toxic Substances. New York, Chichester, Brisbane, Toronto, Singapore: Wiley-Interscience Publication, John Wiley and Sons, Inc.

Calcagno, P.L., and M.I. Rubin. 1963. Renal extraction of PAH in infants and children. J. Clin. Invest. 43:1632.

Castanera, T.J., D.C. Jones, D.J. Kimeldorf, and V.J. Rosen. 1971. The effect of age at exposure to a sublethal dose of fast neutrons on tumorigenesis in the male rat. Cancer Res. 31:1543-1549.

CDC (Centers for Disease Control). 1986. Aldicarb food poisoning from contaminated melons. MMWR 35(16):254-258.

Cherniak, M.C. 1988. Toxicological screening for organophosphorus-induced delayed neurotoxicity: Complications in toxicity testing. Neurotoxicology 9:249-272.

Clapp N.K., W.C. Klima, and L.H. Cacheiro. 1977. Temporal advancement of diethylnitrosamine carcinogenesis in aging mice. Proc. Am. Assoc. Cancer Res. 18:62. (Abstract)

Cosgrove, G.E., A.C. Upton, L.H. Smith, and J.N. Dent. 1965. Radiation glomerulosclerosis and other late effects: Influence of radiological factors and AET. Radiat. Res. 25:725-735.

Cowdry, E.V., and V. Suntzeff. 1944. Influence of age on epidermal carcinogenesis induced by methylcholanthrene in mice. Yale J. Biol. Med. 17:47-58.

Crom, W.R., A.M. Glynn-Barnhart, J.H. Rodman, M.E. Teresi, R.E. Kavanagh, M.L. Christensen, M.V. Relling, and W.E. Evans. 1987. Pharmacokinetics of anticancer drugs in children. Clin. Pharmacokin. 12:168-213.

Curley, A., R.E. Hawk, R.D. Kimbrough, G. Nathenson, and L. Finberg. 1971. Dermal absorption of hexachlorophene in infants. Lancet 2:296-297.

Dedov, V.I. 1982. Late sequelae of the internal irradiation of the endocrine system of female rats. Izv. Akad. Nauk SSSR. Ser. Biol. 3:454-458.

Dési, I., L. Varga, and D. Farkas. 1978. Studies on the immunosuppressive effect of organochlorine and organophosphoric pesticides in subacute experiments. J. Hyg. Epidemiol. Microbiol. Immunol. 22:115-122.

Dietrich, K., K. Krafft, R. Bornschein, P. Hammond, O. Berger, P. Succop, and M. Bier. 1987. Low-level fetal lead exposure effect on neurobehavioral development in early infancy. Pediatrics 80:721-730.

Diggory, H.J.P., P.J. Landrigan, K.P. Latimer, A.C. Ellington, R.D. Kimbrough, J.A. Liddle, R.E. Cline, and A.L. Smrek. 1977. Fatal parathion poisoning caused by contamination of flour for international commerce. Am. J. Epidemiol. 106:145-153.

Dodson, W.E. 1976. Neonatal drug intoxication: Local anesthetics. Pediatr. Clin. N. Am. 23:399-411.

Doll, R., L.G. Morgan, and F.L. Speiger. 1970. Cancer of the lung and nasal sinuses in nickel workers. Br. J. Cancer 24:623-632.

Done, A.K. 1964. Developmental pharmacology. Clin. Pharmacol. Therap. 5:432-479.

Done, A.K., S.N. Cohen, and L. Strebel. 1977. Pediatric clinical pharmacology and the "therapeutic orphan." Annu. Rev. Pharmacol. Toxicol. 17:561-573.

Dourson, M.L., and C.S. Baxter. 1981. Reduced prevalence and growth rate of urethane-induced lung adenomas in ageing adult strain A mice. Toxicology 20:165-172.

Dunning, W.F., M.R. Curtis, and F.D. Bullock. 1936. The respective roles of heredity and somatic mutation in the origin of malignancy. Am. J. Cancer 28:681-712.

Dutton, G.J. 1978. Developmental aspects of drug conjugation, with special reference to glucuronidation. Annu. Rev. Pharmacol. Toxicol. 18:17-35.

Dyer, R.S. 1985. The use of sensory evoked potentials in toxicology. Fundam. Appl. Toxicol. 5:24-40.

Dyer, R.S., and W.K. Boyes. 1983. Use of neurophysiological challenges for the detection of toxicity. Fed. Proc. 42(15):3201-3206.

Ebbesen, P. 1977. Effect of age on non-skin tissues on susceptibility of skin grafts to 7,12-dimethylbenz(a)anthracene (DMBA) carcinogenesis in BALB/c mice, and effect of age of skin graft on susceptibility of surrounding recipient skin to DMBA. J. Natl. Cancer Inst. 58:1057-1060.

Ecobichon, D.J., and D.S. Stephens. 1973. Perinatal development of human blood esterases. Clin. Pharmacol. Therap. 14:41-47.

EPA (U.S. Environmental Protection Agency). 1988. Aldicarb. Special Review Technical Support Document. Office of Pesticides and Toxic Substances. Washington, D.C.: U.S. Environmental Protection Agency.

Ercegovich, C.D. 1973. Relationship of pesticides to immune responses. Fed. Proc. 32:2010-2016.

Evans, W.E., W.P. Petros, M.V. Relling, W.R. Crom, T. Madden, J.H. Rodman, and M. Sunderland. 1989. Clinical pharmacology of cancer chemotherapy in children. Clin. Pharmacol. 36:1199-1230.

Exon, J.H. 1984. The immunotoxicity of selected environmental chemicals, pesticides and heavy metals. Pp. 355-368 in Chemical Regulation of Immunity in Veterinary Medicine. New York: Liss.

Faith, R.E., and J.A. Moore. 1977. Impairment of thymus-dependent immune functions by exposure of the developing immune system to 2,3,7,8-tetrachlorodibenzo-p-dioxin (TCDD). J. Toxicol. Environ. Health 3:451-464.

FDA (Food and Drug Administration). 1992. Specific requirements on content and format for labeling for human prescription drugs: Proposed revision of pediatric use subsection in the labeling. Fed. Regist. 57:47423-47427.

Finkelstein, Y., M. Wolf, and A. Biegon. 1988. Brain acetylcholinesterase after acute parathion poisoning: A comparative, quantitative, histochemical analysis - post mortem. Ann. Neurol. 24(2):552-557.

Fournier, M., G. Chevalier, D. Nadeau, B. Trottier, and K. Krzystyniak. 1988. Virus-pesticide interactions with murine cellular immunity after sublethal exposure to dieldrin and aminocarb. J. Toxicol. Environ. Health 25:103-118.

Franks, L.M., and A.W. Carbonell. 1974. Effect of age on tumor induction in C57BL mice. J. Natl. Cancer Inst. 52:565-566.

Gagne, J., and J. Brodeur. 1972. Metabolic studies on the increased susceptibility of weanling rats to parathion. Can. J. Physiol. Pharmacol. 50:902-915.

Gaines, T.B., and R.E. Linder. 1986. Acute toxicity of pesticides in adult and weanling rats. Fundam. Appl. Toxicol. 7:299-308.

Gaylor, D.W. 1988. Risk assessment: Short-term exposure at various ages. In Phenotypic Variation in Populations: Relevance to Risk Assessment, A.D. Woodhead, M.A. Bender, and R.C. Leonard, eds. New York: Plenum.

Gershanik, J., B. Boecler, H. Ensley, S. McCloskey, and W. George. 1982. The gasping syndrome and benzyl alcohol poisoning. New Engl. J. Med. 307:1384-1388.

Geschickter, C.F. 1939. Mammary carcinoma in the rat with metastasis induced by estrogen. Science 89:35-37.

Gillette, J.R., and B. Stripp. 1975. Pre- and postnatal enzyme capacity for drug metabolite production. Fed. Proc. 34:172-178.

Glaubiger, D.L., D.D. von Hoff, J.S. Holcenberg, B. Kamen, C. Pratt, and R.S. Ungerleider. 1982. The relative tolerance of children and adults to anticancer drugs. Front. Radiat. Therap. Oncol. 16:42-49.

Goldenthal, E.L. 1971. A compilation of LD_{50} values in newborn and adult animals. Toxicol. Appl. Pharmacol. 18:185-207.

Goldman, L.R., M. Beller, and R.J. Jackson. 1990. Aldicarb food poisonings in California, 1985—1988: Toxicity estimates for humans. Arch. Environ. Health 45:141-147.

Gould, S.J. 1977. Ontogeny and Phylogeny. Cambridge, Mass.: Belknap Press of Harvard Press. 520 pp.

Graef, J.W. 1980. Management of low level lead exposure. In Low Level Lead Exposure: The Chemical Duplication of Current Research, M.L. Needleman, ed. New York: Raven Press.

Grand, R.J., J.B. Watkins, and F.M. Torti. 1976. Development of the human gastrointestinal tract: A review. Gastroenterology 70:790-810.

Greaves, S.J., D.G. Ferry, E.G. McQueen, D.S. Malcom, and P.M. Buckfield. 1975. Serial hexachlorophene blood levels in the premature infant. New Zealand Med. J. 81:334-336.

Greengard, O. 1977. Enzymic differentiation of human liver: Comparison with the rat model. Pediatr. Res. 11:669-676.

Greenman, D.L., A. Boothe, and R. Kodell. 1987. Age-dependent responses to 2-acetyl-aminofluorene in BALB/c female mice. J. Toxicol. Environ. Health 22:113-129.

Groth, D.H., W.B. Coate, B.M. Ulland, and R.W. Hornung. 1981. Effects of aging on the induction of angiosarcoma. Environ. Health Perspect. 41:53-57.

Hamill, P.V.V., T.A. Drizd, C.L. Johnson, R.B. Reed, A.F. Roche, and W.M. Moore. 1979. Physical growth: National Center for Health Statistics Percentiles. Am. J. Clin. Nutr. 32:607-629.

Hayes, W.J. 1970. Epidemiology and general management of poisoning by pesticides. Pediatr. Clin. N. Am. 17:629.

Healy, M.P., M. Aslam, P.G. Harrison, and N.P. Fernando. 1984. Lead induced convulsion in young infants. J. Clin. Hosp. Pharm. 9(3):199-207.

Hemberger, J.A., and L.S. Schanker. 1978. Pulmonary absorption of drugs in the neonatal rat. Am. J. Physiol. 234:C191-C197.

Himwich, W. 1973. Problems in interpreting clinical changes in developing and aging animals. Prog. Brain Res. 40:13-23.

Hinson, J.A., D.W. Roberts, R.W. Benson, K. Dalhoff, S. Loft, and H.E. Poulsen. 1990. Mechanism of paracetamol toxicity (Letter). Lancet 335:732.

Hislop, A., and L. Reid. 1981. Growth and development of the respiratory system. Pp. 390-432 in Pediatrics, 2d ed., J.A. Davis and J. Dobbing, eds. London: Hinemann.

Hoffman, R.E., P.A. Stehr-Green, K.B. Webb, R.G. Evans, A.P. Knutsen, W.F. Schramm, J.L. Staake, B.B. Gibson, and K.K. Steinberg. 1986. Health effects of long-term exposure to 2,3,7,8-tetrachlorodibenzo-p-dioxin. J. Am. Med. Assoc. 255:2031-2038.

Hoffmann, H. 1982. Absorption of drugs and other xenobiotics during development in experimental animals. Pharmacol. Ther. 16:247-260.

Hoover, R., and P. Cole. 1973. Temporal aspects of occupational bladder carcinogenesis. New Engl. J. Med. 288:1041-1043.

Huggins, C., L.C. Grand, and F.P. Brillantes. 1961. Mammary cancer induced by a single feeding of polynuclear hydrocarbons, and its suppression. Nature 189:204-207.

Hursh, J.B., and J. Suomela. 1968. Absorption of ^{212}Pb from the gastrointestinal tract of man. Acta Radiol. Ther. Phys. Biol. 7:108-120.

Jugo, S. 1977. Metabolism of toxic heavy metals in growing organisms: A review. Environ. Res. 13:36-46.

Kaplan, H.S. 1947. Observation on radiation-induced lymphoid tumors of mice. Cancer Res. 7:141-147.

Kauffman, R.E. 1992. Acute acetaminophen overdose: An example of reduced toxicity related to developmental differences in drug metabolism. Pp. 97-103 in Similarities and Differences Between Children and Adults: Implications in Risk Assessment, P.S. Guzelian, C.J. Henry, and S.S. Olin, eds. Washington, D.C.: ILSI Press.

Kawade, N., and S. Onishi. 1981. The prenatal and postnatal development of UDP-glucuronyltransferase activity toward bilirubin and the effect of premature birth on this activity in the liver. Biochem. J. 196:257-260.

Kearns, G.L., and M.D. Reed. 1989. Clinical pharmacokinetics in infants and children: A reappraisal. Clin. Pharmacokinet. 17:29-67.

Kerkvliet, N.I., and J.A. Brauner. 1987. Mechanisms of 1,2,3,4,6,7,8-heptachlorodibenzo-p-dioxin (HpCDD)-induced humoral immune suppression: Evidence of primary defect in T-cell regulation. Toxicol. Appl. Pharmacol. 87:18-31.

Khudoley, V.V. 1981. The role of age in carcinogenesis induced by N-nitrosodimethylamine and N-dimethylnitramine in frog Rana temporaria L. Vopr. Onkol. 10:67-71.

Kim, H.J., J.V. Bruckner, C.E. Dallas, and J.M. Gallo. 1990. Effect of oral dosing vehicles on the pharmacokinetics of orally administered carbon tetrachloride in rats. Toxicol. Appl. Pharmacol. 102:50-60.

Kimmel, C.A., D.C. Rees, and E.Z. Francis, eds. 1990. Special issue on qualitative and quantitative comparability of human and animal developmental neurotoxicity. Neurotoxicol. Teratol. 12:173-292.

Kimura, M., T. Fukuda, and K. Sato. 1979. Effect of aging on the development of gastric cancer in rats induced by N-methyl-N-nitro-N-nitrosoguanidine. Gann 70:521-525.

Klinger, W. 1982. Biotransformation of drugs and other xenobiotics during postnatal development. Pharmacol. Ther. 16:377-429.

Knaak, J.B., K. Yee, C.R. Ackerman, G. Zweig, and B.W. Wilson. 1984. Percutaneous absorption of triadimefon in the adult and young male and female rat. Toxicol. Appl. Pharmacol. 72:406-416.

Koldovsky, O., P. Sunshine, and N. Kretchmer. 1966. Cellular migration of intestinal epithelia in suckling and weaned rats. Nature 212:1389-1390.

Kostial, K., I. Simonovic, and M. Pisonic. 1971. Lead absorption from the intestine in newborn rats. Nature 233:564.

Kostial, K., D. Kello, S. Jugo, I. Rabar, and T. Maljkovic. 1978. Influence of age on metal metabolism and toxicity. Environ. Health Perspect. 25:81-86.

Krasovskii, G.N. 1976. Extrapolation of experimental data from animals to man. Environ. Health Perspect. 13:51-58.

Kroes, R., J.W. Weiss, and J.H. Weisburger. 1975. Immune suppression and chemical carcinogenesis. Recent Results Cancer Res. 52:65-76.

Kupferberg, H.J., and E.L. Way. 1963. Pharmacological basis for increased sensitivity of the newborn rat to morphine. J. Pharmacol. Exper. Ther. 141:105-112.

LaFarge-Frayssinet, C., and F. Declöitre. 1982. Modulatory effect of the pesticide captan on the immune response in rats and mice. J. Immunopharmacol. 4:43-52.

Lamanna, C., and E.R. Hart. 1968. Relationship of lethal toxic dose to body weight of the mouse. Toxicol. Appl. Pharmacol. 13:307-315.

Langston, C. 1983. Normal and abnormal structural development of the human lung. Pp. 75-91 in Abnormal Functional Development of the Heart, Lungs, and Kidneys: Approaches to Functional Teratology, R.J. Kavlock and C.T. Grabowski, eds. New York: A.R. Liss.

Lecce, J.G. 1972. Selective absorption of macromolecules into intestinal epithelium and blood by neonatal mice. J. Nutr. 102:69-75.

Lee, P.N., and R. Peto. 1970. The effect of the age of mice on the incidence of skin cancer. Br. J. Cancer 24:849-852.

Lindop, P.J. and J. Rotblat. 1962. The age factor in the susceptibility of man and animals to radiation. I. The age factor in radiation sensitivity in mice. Br. J. Radiol. 35:23-31.

Lindstedt, S.L. 1987. Allometry: Body size constraints in animal design. Pp. 65-79 in Drinking Water and Health: Pharmacokinetics in Risk Assessment, Vol. 8. Washington, D.C.: National Academy Press.

Long, S.S., and R.M. Swenson. 1977. Development of anaerobic fecal flora in healthy newborn infants. J. Pediatr. 91:283-301.

Longnecker, D.S., J. French, E. Hyde, H.S. Lilja, and J.D. Yager, Jr. 1977. Effect of age on nodule induction by azaserine and DNA synthesis in rat pancreas. J. Natl. Cancer Inst. 58:1769-1775.

Lorenz, J.M., and L.I. Kleinman. 1988. Ontogeny of the kidney. Pp. 58-85 in Nutrition During Infancy, R.C. Tsang and B.L. Nichols, eds. St. Louis, Mo.: C.V. Mosby.

Loughnan, P.M., A. Greenwald, W.W. Purton, J.V. Aranda, G. Watters, and A.H. Neims. 1977. Pharmacokinetic observations of phenytoin disposition in the newborn and young infant. Arch. Dis. Child. 52:302-309.

Lu, F.C., D.C. Jessup, and A. Lavallée. 1965. Toxicity of pesticides in young versus adult rats. Food Cosmet. Toxicol. 3:591-596.

MacPhail, R.C., S. Padilla, and L.W. Reiter. 1987. Age-related effects of pesticides. Second International Symposium on the Performance of Protective Clothing. Tampa, Fla. (Abstract).

Maha, M. 1992. Retrovir® (AZT): A comparison of clinical trial results in adults and children. Pp. 107-112 in Similarities and Differences Between Children and Adults: Implications in Risk Assessment, P.S. Guzelian, C.J. Henry, and S.S. Olin, eds. Washington, D.C.: ILSI Press.

Maiski, I.N., B. Botev, G.P. Airapetian, T.A. Pokrovskaya, T. Vatseva, L. Markholev, I. Vatev, and I. Mladenov. 1978. Development of induced tumors in rats of different ages. Onkologiia Sofia. 15:83-87.

Maizlish, N., M. Schnenker, J. Seiber, and S. Samuels. 1987. A behavioral evaluation of pest control workers in short-term low-level exposure to the organophosphate diazinon. Am. J. Indust. Med. 12(2):153-172.

Marsoni, S., R.S. Ungerleider, S.B. Hurson, R.M. Simon, and L.D. Hammershaimb.

1985. Tolerance to antineoplastic agents in children and adults. Cancer Treat. Rep. 69:1263-1269.

Mayer-Popken, O., W. Denkhaus, and H. Konietzko. 1986. Lead content of fetal tissues after maternal intoxication. Arch. Toxicol. 58(3):203-204.

McCormack, J.J., E.K. Boisits, and L.B. Fisher. 1982. An in vitro comparison of the permeability of adult versus neonatal skin. Pp. 149-164 in Neonatal Skin, Structure and Function, H.I. Maibach and E.K. Boisits, eds. New York: Dekker.

McCracken, G.H., N.R. West, and L.J. Horton. 1971. Urinary excretion of gentamicin in the neonatal period. J. Infect. Dis. 123:257-262.

McKinney, R.E., Jr., M.A. Maha, E.M. Connor, J. Feinberg, G.B. Scott, M. Wulfsohn, K. McIntosh, W. Borkowski, J.F. Modlin, P. Weintrub, K. O'Donnell, R.D. Gelber, G.K. Rogers, S.N. Lehrman, and C.M. Wilfert. 1991. A multicenter trial of oral zidovudine in children with advanced human immunodeficiency virus disease. New Engl. J. Med. 324:1018-1025.

McLeod, H.L., M.V. Relling, W.R. Crom, K. Silverstein, S. Groom, J.H. Rodman, G.K. Rivera, W.M. Crist, and W.E. Evans. 1992. Distribution of antineoplastic agents in the very young child. Br. J. Cancer 66(Suppl):S23-S29.

McNeil, J.L., J.A. Ptasnik, and D.B. Croft. 1975. Evaluation of long-term effects of elevated blood lead concentrations in asymptomatic children. Arh. Hig. Toksikol. (Yugoslavia) 26(Suppl.):97-118.

Menczer, J., H. Komarov, M. Shenboum, V. Insler, and B. Czernobilsky. 1977. Attempted induction of granulosa cell tumor in BALB-c mice by gonadotropin administration. Gynecol. Invest. 8:314-322.

Menkes, J.H., ed. 1981. Metabolic disease of the nervous system. In Textbook of Child Neurology. Philadelphia: Lea & Febiger.

Miller, B.A., L.A.G. Ries, B.F. Hankey, C.L. Kosary, and B.K. Edwards, eds. 1992. Cancer Statistics Review: 1973-1989. NIH Pub. No. 92-2789. Bethesda, Md.: National Cancer Institute.

Miller, R.P., R.J. Roberts, and L.J. Fischer. 1977. Acetaminophen elimination kinetics in neonates, children and adults. Clin. Pharmacol. Therap. 19:284-294.

Misra, U.K., D. Nag, W.A. Khan, and P.K. Ray. 1988. A study of nerve conduction velocity, late responses and neuromuscular synapse functions in organophosphate workers in India. Arch. Toxicol. 61:496-500.

Moon, R.C., and C.M. Fricks. 1977. Influence of gonadal hormones and age on 1,2-dimethylhydrazine-induced colon carcinogenesis. Cancer 40:2502-2508.

Moore, M.R., A. Goldburg, I.W. Bushnell, R. Day, and W.M. Fyro. 1982. Prospective study of the neurological effects of lead in children. Neurobehav. Toxicol. Teratol. 4(6):739-743.

Moore, M.R., M.J. McIntosh, and I.W. Bushnell. 1986. The neurotoxicology of lead. Neurotoxicology 7(2):591-556.

Morselli, P.L. 1976. Clinical pharmacokinetics in neonates. Clin. Pharmacokin. 1:81-98.

Morselli, P.L. 1989. Clinical pharmacology of the perinatal period and early infancy. Clin. Pharmacokinet. 17:13-28.

Morselli, P.L., R. Franco-Morselli, and L. Bossi. 1980. Clinical pharmacokinetics in newborns and infants: Age-related differences and therapeutic implications. Clin. Pharmacokin. 5:485-527.

Mortenson, M.L. 1986. Management of acute childhood poisonings caused by selected insecticides and pesticides. Pediatr. Clin. N. Am. 33(2):421-445.

Murphy, S.D. 1982. Toxicity and hepatic metabolism of organophosphate insecticides in developing rats. Pp. 125-134 in Report No. 11: Environmental Factors in Human Growth and Development, V.R. Hunt, M.K. Smith, and D. Worth, eds. New York: Cold Spring Harbor Laboratory.

Nau, H., W. Luck, and W. Kuhnz. 1984. Decreased serum protein binding of diazepam and its major metabolite in the neonate during the first postnatal week relate to increased free fatty acid levels. Br. J. Clin. Pharmacol. 17:92-98.

Needleman, H.L. 1983. Lead at low dose and the behavior of children. Neurotoxicology 4(3):121-133.

Needleman, H.L., C. Gunnoe, A. Leviton, R. Reed, H. Peresie, C. Maher, and P. Barrett. 1979. Deficits in psychologic and classroom performance of children with elevated dentine lead levels. New Engl. J. Med. 300:689-695.

Neims, A.H. 1982. The effect of age on drug disposition and action. Pp. 529-537 in Banbury Report No. 11: Environmental Factors in Human Growth and Development, V.R. Hunt, M.K. Smith, and D. Worth, eds. New York: Cold Spring Harbor Laboratory.

Neims, A.H., M. Warner, P.M. Loughnan, and J.V. Aranda. 1976. Developmental aspects of the hepatic cytochrome P450 monooxygenase system. Annu. Rev. Pharmacol. Toxicol. 16:427-445.

NRC (National Research Council). 1989. Recommended Dietary Allowances: 10th Edition. Washington, D.C.: National Academy Press. 302 pp.

NRC (National Research Council). 1992. Environmental Neurotoxicology. Washington, D.C.: National Academy Press.

NTP (National Toxicology Program). 1992. NTP Technical Report on the Perinatal Toxicity and Carcinogenicity Studies of Ethylene Thiourea in F/344 rats and B6C3F$_1$ mice (feed studies). NTP Technical Report 388. Research Triangle Park, N.C.: National Toxicology Program.

Nyhan, W.L. 1961. Toxicity of drugs in the neonatal period. J. Pediatr. 59:1-20.

Ostheimer, G.W. 1979. Neurobehavioral effects of obstetric analgesia. Br. J. Anaesth. 51:355-405.

Otto, D.A. 1986. Electrophysical Assessment of Neurotoxicity in Children. Health Effects Research Laboratory. Research Triangle Park, N.C.: U.S. Environmental Protection Agency.

Otto, D., and L. Reiter. 1984. Developmental changes in slow cortical potentials of young children with elevated lead body burden. Neurophysiological considerations. Ann. N.Y. Acad. Sci. 425:377-383.

Otto, D., G. Robinson, G.S. Bauman, S. Schroeder, P. Mushak, D. Klienman, and L. Boon. 1985. 5-year follow-up study of children with low-to-moderate lead absorption: Electrophysiological evaluation. Environ. Res. 38(1):168-186.

Overstreet, D.M. 1984. Behavioral plasticity and the cholinergic system program. Neuropsychopharmacol. Biol. Psychiatry 8(1):135-151.

Ovsyannikov, A.I. and V.N. Anisimov. 1983. Effect of age on carcinogenesis induced by single subcutaneous administration of N-nitroso-N-methylurea in the rat. Vopr. Onkol.

Pang, K. Y., J.L. Bresson, and W.A. Walker. 1983. Development of the gastrointestinal mucosal barrier: Evidence for structural differences in microvillus membranes from newborn and adult rabbits. Biochim. Biophys. Acta 727:201-208.

Paulini, K., G. Beneke, B. Körner, and R. Enders. 1975. The relationship between the latent period and animal age in the development of foreign body sarcomas. Beitr. Pathol. 154:161-169.

Pazmino, N.H. and L.M. Yuhas. 1973. Senescent loss of resistance to murine sarcoma virus (Moloney) in the mouse. Cancer Res. 33:2668-2672.

Pearson, D.T., and K.N. Dietrich. 1985. The behavioral toxicology and teratology of childhood: Models, methods, and implications for intervention. Neurotoxicology 6(3):165-182.

Pelkonen, O., E.H. Kaltiala, T.K.I. Larmi, and N.T. Karki. 1973. Comparisons of activities of drug-metabolizing enzymes in human fetal and adult livers. Clin. Pharmacol. Ther. 14:840-846.

Perino, J., and C.B. Erinhart. 1974. The relation of subclinical lead level to cognitive and sensorimotor impairment in black preschoolers. J. Learn. Dis. 7:26-30.

Peto, R., F.J. Roe, P.N. Lee, L. Levy, and J. Clack. 1975. Cancer and ageing in mice and men. Br. J. Cancer. 32:411-426.

Petros, W.P., and W.E. Evans. 1992. Antineoplastic agents. Pp. 390-405 in Pediatric Pharmacology: Therapeutic Principles in Practice, 2d ed., S.J. Yaffe and J.V. Aranda, eds. Philadelphia: Saunders.

Pinkel, D. 1958. The use of body surface area as a criterion of drug dosage in cancer chemotherapy. Cancer Res. 18:853-856.

Powell, H., O. Swarner, L. Gluck, and P. Lampert. 1973. Hexachlorophene myelinopathy in premature infants. J. Pediatr. 82:976-981.

Pozharisski, K.M., L.G. Prvanova, V.N. Anisimov, and V.F. Klimashevski. 1980. Effect of age on carcinogenesis in intestines. Exp. Onkol. 2:20-21.

Pylkko, O.O., and P.M. Woodbury. 1961. The effect of maturation of chemically induced seizures in rats. J. Pharmacol. Exp. Ther. 131:185-190.

Rabinowitz, M.B., G.W. Wetherill, and J.D. Kopple. 1976. Kinetic analysis of lead metabolism in healthy humans. J. Clin. Invest. 58:260-270.

Rane, A. 1992. Drug disposition and action in infants and children. Pp. 10-21 in Pediatric Pharmacology: Therapeutic Principles in Practice, S.J. Yaffe, and J.V. Aranda, eds. Philadelphia: Saunders.

Rane, A., and F. Sjoqvist. 1972. Drug metabolism in the human fetus and newborn infant. Pediatr. Clin. North Am. 19:37-49.

Rane, A., P.K.M. Lunde, B. Jalling, S.J. Yaffe, and F. Sjoqvist. 1971. Plasma protein binding of diphenylhydantoin in normal and hyperbilirubinemic infants. Pediatr. Pharmacol. Ther. 78:877-882.

Rasmussen, J.E. 1979. Percutaneous absorption in children. Pp. 25-38 in Year Book of Dermatology, R.L. Dobson, ed. Chicago: Year Book Medical.

Reed, M.D., and J.B. Besunder. 1989. Developmental pharmacology: Ontogenetic basis of drug disposition. Pediatr. Clin. North Am. 36:1053-1074.

Reuber, M.D. 1975. Hyperplastic and neoplastic lesions of the kidney in Buffalo rats of varying ages ingesting N-4-(4-fluorobiphenyl)acetamide. J. Natl. Cancer Inst. 54:427-429.

Reuber, M.D. 1976. Effects of age and sex on lesions of the oesophagus in Buffalo strain rats ingesting diethylnitrosamine. Exp. Cell Biol. 44:65-72.

Reuber, M.D., and E.L. Glover. 1967. Hyperplastic and early neoplastic lesions of the liver in Buffalo strain rats of various ages given subcutaneous carbon tetrachloride. J. Natl. Cancer Inst. 38:891-899.

Reuber, M.D., and C.W. Lee. 1968. Effect of age and sex on hepatic lesions in

Buffalo strain rats ingesting diethylnitrosamine. J. Natl. Cancer Inst. 41:1133-1140.

Rice, J.M. 1979. Perinatal Carcinogenesis. NCI Monogragh No. 51. DHEW Publ. No. (NIH) 79-1622. Washington, D.C.: U.S. Government Printing Office.

Richman, D.D., M.A. Fischl, M.H. Grieco, M.S. Gottlieb, P.A. Volberding, O.L. Laskin, J.M. Leedom, J.E. Groopman, D. Mildvan, M.S. Hirsch, G.G. Jackson, D.T. Durack, S. Nusinoff-Lehrman, and the AZT Collaborative Working Group. 1987. The toxicity of azidothymidine (AZT) in the treatment of patients with AIDS and AIDS-related complex: A double-blind, placebo-controlled trial. New Engl. J. Med. 317:192-197.

Rodgers, K.E., T. Imamura, and B.H. Devens. 1986. Organophosphorus pesticide immunotoxicity: Effects of O,O,S-trimethyl phosphorothioate on cellular and humoral immune response systems. Immunopharmacology 12:193-202.

Rodgers, K.E., N. Leung, C.F. Ware, and T. Imamura. 1987. Effects of O,S,S-trimethyl phosphorodithioate on immune function. Toxicology 43:201-216.

Rodier, P.M. 1980. Chronology of neuron development: Animal studies and their clinical implications. Med. Child Neurol. 22:525-545.

Rogan, W.J., B.C. Gladan, K.-L. Hung, S.-L. Koong, L.-Y. Shih, J.S. Taylor, Y.-C. Wu, D. Yang, N.B. Ragan, and C.-C. Hsu. 1988. Congenital poisoning by polychlorinated biphenyls and their contaminants in Taiwan. Science 241:334-336.

Roland, E.H., J.E. Jan, and J.M. Riss. 1985. Toxic encephalopathy in a child after brief exposure to insect repellants. J. Can. Med. Assoc. 132(2):155-156.

Rosecrans, J.A., J.-S. Hong, R.E. Squibb, J.H. Johnson, W.E. Wilson, and H.A. Tilson. 1982. Effects of perinatal exposure to chlordecone (kepone) on neuroendocrine and neurochemical responsiveness of rats to environmental challenges. Neurotoxicology 3(2):131-142.

Rowland, I.R., R.D. Robinson, R.A. Doherty, and T.D. Landry. 1983. Are developmental changes in methylmercury metabolism and excretion mediated by the intestinal microflora? Pp. 745-758 in Reproductive and Developmental Toxicity of Metals, T.W. Clarkson, G.F. Nordberg, and P.R. Sager, eds. New York: Plenum.

Ruff, H.A., and P.E. Bijur. 1989. The effects of low to moderate lead levels on neurobehavioral functions of children. J. Dev. Behav. Pediatr. 10(2):103-109.

Rumack, B.H. 1984. Acetaminophen overdose in young children. Am. J. Dis. Child. 138:428-433.

Russo, J., and I.H. Russo. 1978. DNA labeling index and structure of the rat mammary gland as determinants of its susceptibility to carcinogenesis. J. Natl. Cancer Inst. 61:1451-1459.

Sachs, C., and G. Jonsson. 1975. 5,7-Dihydroxytryptamine induced changes in the postnatal development of central 5-hydroxytryptamine. Med. Biol. 53:156-168.

Savchenkov, M.F., V.V. Benemansky, and V.Y. Levina. 1980. Blastomogenic activity of dimethylnitrosamine in animals of different age. Hyg. Sanit. 4:17-20.

Saxen, E.A. 1954. On the factor of age in the production of subcutaneous sarcomas in mice by 20-methylcholanthrene. J. Natl. Cancer Inst. 14:547-569.

Schanker, L.S. 1978. Drug absorption from the lung. Biochem. Pharmacol. 27:381-385.

Schmidt-Nielsen, K. 1972. How Animals Work. London: Cambridge University Press.

Seagull, E.A.W. 1983. Developmental abilities of children exposed to polybrominated biphenyls (PBB). Am. J. Public Health 3:281-285.

Senanayake, N., and L. Karalliedde. 1987. Neurotoxic effects of organophosphorus insecticides: An intermediate syndrome. New Engl. J. Med. 316:761-763.

Shah, P.V., H.L. Fisher, M.R. Sumler, R.J. Monroe, N. Chernoff, and L.L. Hall. 1987. Comparison of the penetration of 14 pesticides through the skin of young and adult rats. J. Toxicol. Environ. Health 21:353-366.

Shuman, R.M., R.W. Leech, and E.C. Alvord, Jr. 1974. Neurotoxicity of hexachlorophene in the human: I. A clinicopathologic study of 248 children. Pediatrics 54:689-695.

Singer, E.J., P.C. Wegmann, M.D. Lehman, M.S. Christensen, and L.J. Vinson. 1971. Barrier development, ultrastructure, and sulfhydryl content of the fetal epidermis. J. Soc. Cosmet. Chem. 22:119-137.

Singer, R., M. Moses, J. Valciukas, R. Lilis, and I.J. Salikoff. 1982. Nerve conduction velocity studies of workers employed in the manufacture of phenoxy herbicides. Environ. Res. 29:297-311.

Singh, S.M., J.F. Toles, and J. Reaume. 1986. Genotype- and age-associated in vivo cytogenetic alterations following mutagenic exposures in mice. Can. J. Genet. Cytol. 28:286-293.

Six, K.M., and R.A. Goyer. 1972. The influence of iron deficiency on tissue content and toxicity of ingested lead in the rat. J. Lab. Clin. Med. 79:128-136.

Solomon, L.M., D.P. West, J.F. Fitzcoff, and A.M. Becker. 1977. Gamma benzene hexachloride in guinea pig brain after topical application. J. Invest. Dermatol. 68:310-312.

Sonawane, B.R. 1982. Developmental aspects of acetaminophen hepatotoxicity: Influence of age and acute starvation. Pp. 507-520 in Banbury Report No. 11: Environmental Factors in Human Growth and Development, V.R. Hung, M.K. Smith, and D. Worth, eds. New York: Cold Spring Harbor Laboratory.

Spielberg, S.P. 1992. Anticonvulsant adverse drug reactions: Age dependent and age independent. Pp. 104-106 in Similarities and Differences Between Children and Adults: Implications in Risk Assessment, P.S. Guzelian, C.J. Henry, and S.S. Olin, eds. Washington, D.C.: ILSI Press.

Stanton, M.E., and L.P. Spear. 1990. Conference on the qualitative and quantitative comparability of human and animal developmental neurotoxicity. Work group I report: Comparability of measures of developmental neurotoxicity in humans and laboratory animals. Neurotoxicol. Teratol. 12:261-267.

Stenbäck, F., R. Peto, and P. Shubik. 1981. Initiation and promotion at different ages and doses in 2200 mice: II. Decrease in promotion by TPA with ageing. Br. J. Cancer 44:15-23.

Streltsova, V.N., and Y.I. Moskalev. 1964. Blastomogenic Effect of Ionizing Radiation. Moscow: Meditsina Press.

Stromberg, K., and M.D. Reuber. 1975. Histopathology of breast lesions induced in BUF rats of varying ages by ingestion of N-4-(4'-fluorobiphenyl) acetamide. J. Natl. Cancer Inst. 54:1223-1230.

Stutman, O. 1979. Chemical carcinogenesis in nude mice: Comparison between nude mice from homozygous matings and heterozygous matings and effect of age and carcinogen dose. J. Natl. Cancer Inst. 62:353-358.

Suchy, F.J., W.F. Balistreri, J.E. Heubi, J.E. Searcy, and R.S. Levin. 1981. Physiologic cholestasis: Elevation of the primary serum bile acid concentrations in normal infants. Gastroenterology 80:1037-1041.

Summerhayes, I.C., and L.M. Franks. 1979. Effects of donor age on neoplastic transformation of adult mouse bladder epithelium *in vitro*. J. Natl. Cancer Inst. 62:1017-1023.

Sundaram, K. 1963. Longterm consequences of ^{90}Sr in rats and the problem of carcinogenesis. Pp. 139-144 in Cellular Basis and Aetiology of Late Somatic Effects of Ionizing Radiation, R.J.C. Harris, ed. London: Academic Press.

Sutherland, J.M. 1959. Fatal cardiovascular collapse of infants receiving large amounts of chloramphenicol. J. Dis. Child. 97:761-767.

Syn-mao, L. 1962. Role of Hormonal and Nervous Factors in Development of Induced Mammary Tumors in Rats. Thesis. University of Kiev.

Thomas, P.T., and H.V. Ratajczak. 1988. Assessment of carbamate pesticide immunotoxicity. Toxicol. Ind. Health 4:381-390.

Thomas, P., H. Ratajczak, D. Demetral, K. Hagen, and R. Baron. 1990. Aldicarb immunotoxicity: Functional analysis of cell-mediated immunity and quantitation of lymphocyte subpopulations. Fundam. Appl. Toxicol. 15:221-230.

Thurlbeck, W.M. 1982. Postnatal human lung growth. Thorax 37:564-571.

Tilson, H.A., and C.L. Mitchell. 1984. Neurobehavioral techniques to assess the effects of chemicals on the nervous system. Annu. Rev. Pharmacol. Toxicol. 24:425-450.

Toth, B. 1968. A critical review of experiments in chemical carcinogenesis using newborn animals. Cancer Res. 28:727-738.

Triebig, G., I. Cxuzda, H.J. Krekeler, and K.H. Schaller. 1987. Pentachlorophenol and the peripheral nervous system: A longitudinal study in exposed workers. Br. J. Ind. Med. 44:638-641.

Tucker, G.T., and L.E. Mather. 1979. Clinical pharmacokinetics of local anesthetics. Clin. Pharmacokinet. 4:241-278.

Tucker, M.A., A.T. Meadows, J.D. Boice, M. Stovall, O. Oberlin, B.J. Stone, J. Birch, P.A. Vo'te, R.N. Hoover, J.F. Fraumeni, and the Late Effects Study Group. 1987. Leukemia after therapy with alkylating agents for childhood cancer. J. Natl. Cancer Inst. 78:459-464.

Turusov, V.S., L.S. Baslova, and V.A. Krutovskikh. 1979. Effect of age, castration and pregnancy on 1,2-dimethylhydrazine-induced carcinogenesis in CBA mice. Bull. Exp. Biol. Med. 87:458-460.

Turusov, V.S., N.S. Lanko, and Y.D. Parfenov. 1981. Effect of age on induction by 1,2-dimethylhydrazine tumors of intestines in mice. Bull. Exp. Biol. Med. 91:705-707.

Tyl, R., and W.F. Sette. 1990. Weight of evidence and quantitative evaluation of developmental neurotoxicity data. Conference on the Qualitative and Quantitative Comparability of Human and Animal Developmental Neurotoxicity. Work Group III Report. Neurotoxicol. Teratol. 12:273-280.

Tyrala, E.E., L.S. Hillman, R.E. Hillman, and W.E. Dodson. 1977. Clinical pharmacology of hexachlorophene in newborn infants. J. Pediatr. 91:481-486.

Udall, J.N., and W.A. Walker. 1982. Macromolecular transport across the developing intestine. Pp. 187-198 in Banbury Report No. 11: Environmental Factors in Human Growth and Development, V.R. Hunt, M.K. Smith, and D. Worth, eds. New York: Cold Spring Harbor Laboratory.

Varga, F., and T.Z. Csaky. 1976. Changes in the blood supply of the gastrointestinal tract in rats with age. Pflugers Arch. 364:129-133.

Vasilesque, C., M. Alexander, and A. Dan. 1984. Delayed neuropathy after organophosphorus insecticide poisoning: A clinical, electrophysiological, and nerve biopsy study. J. Neurol. Neurosurg. Psych. 47:543-548.

Vesselinovitch, S.D., K.V.N. Rao, and N. Mihailovich. 1979. Neoplastic response of mouse tissues during perinatal age periods and its significance in chemical carcinogenesis. In Perinatal Carcinogenesis, NCI Monogragh No. 51, DHEW Publ. No. (NIH) 79-1622, J.M. Rice, ed. Washington, D.C.: U.S. Government Printing Office.

Vorhees, C.V. 1986. Principles of behavioral teratology. Pp. 23-48 in Handbook of Behavioral Teratology, E.P. Riley and C.V. Vorhees, eds. New York: Plenum.

Vos, J.G., and J.A. Moore. 1974. Suppression of cellular immunity in rats and mice by maternal treatment with 2,3,7,8 tetrachlorodibenzo-p-dioxin. Int. Arch. Allergy 47:777-794.

Walker, W.A., and K.J. Isselbacher. 1974. Uptake and transport of macromolecules by the intestine: Possible role in clinical disorders. Gastroenterology 67:531-550.

Wang, Y.-M., W.W. Sutow, M.M. Romsdahl, and C. Perez. 1979. Age-related pharmacokinetics of high-dose methotrexate in patients with osteosarcoma. Cancer Treat. Rep. 63:405-410.

Warner, A. 1986. Drug use in the neonate: Interrelationship of pharmacokinetics, toxicity and biochemical immaturity. Clin. Chem. 32:721-727.

Watanabe, T., K. Matsuhashi, and S. Takayama. 1990. Placental and blood-brain barrier transfer following prenatal and postnatal exposures to neuroactive drugs: Relationship with partition coefficients and behavioral teratogenesis. Toxicol. Appl. Pharmacol. 105:66-77.

Weil, W.B., Jr. 1955. The evaluation of renal function in infancy and childhood. Am. J. Med. Sci. 229:678-694

Weil, W., M. Spencer, D. Benjamen, and E. Seagull. 1981. The effect of polybrominated biphenyl on infants and young children. J. Pediatr. 98(1):47-51.

Weiss, B. 1983. Behavioral toxicology of heavy metals. Pp. 1-50 in Neurobiology of the Trace Elements: Neurotoxicology and Neuropharmacology, Vol. 2, I.E. Dreosti and R.M. Smith, eds. Clifton, N.J.: Humana. 320 pp.

Weiss, B. 1988. Behavior as an early indicator of pesticide toxicity. Toxicol. Ind. Health 4:351-360.

Weiss, C.F., A.J. Glazko, and J.K. Weston. 1960. Chloramphenicol in the newborn infant: A physiological explanation of its toxicity when given in excessive doses. New Engl. J. Med. 262:787-794.

West, J.R., H.W. Smith, and H. Chasis. 1948. Glomerular filtration rate, effective renal blood flow, and maximal tubular excretory capacity in infancy. J. Pediatr. 32a:10-18.

Wester, R.C., and H.I. Maibach. 1982. Percutaneous absorption: Neonate compared to the adult. Pp. 3-15 in Banbury Report No. 11: Environmental Factors in Human Growth and Development, V.R. Hunt, M.K. Smith, and D. Worth, eds. New York: Cold Spring Harbor Laboratory.

Wester, R.C., P.K. Noonan, M.P. Cole, and H.I. Maibach. 1977. Percutaneous absorption of testosterone in the newborn rhesus monkey: Comparison to the adult. Pediatr. Res. 11:737-739.

Whorton, M.D., and D.L. Obrinsky. 1983. Persistence of symptoms after mild to moderate acute organophosphate poisoning among 19 farm field workers. J. Toxicol. Environ. Health 11:347-354.

Widdowson, E.M., and J.W.T. Dickerson. 1964. Chemical composition of the body. Pp. 1-247 in Mineral Metabolism: An Advanced Treatise, C.L. Comar and F. Bronner, eds. New York: Academic Press.

Windorfer, A., W. Kuenzer, and R. Urbanek. 1974. The influence of age on the activity of acetylsalicylic acid-esterase and protein salicylate binding. Eur. J. Clin. Pharmacol. 7:227-231.

Winneke, G., U. Beginn, T. Ewert, C. Harvestadt, U. Kraemer, C. Krause, H.L. Thron, and H.M. Wagner. 1985. Comparing the effects of perinatal and later childhood lead exposure on neuropsychological outcome. Environ. Res. 38(1):155-167.

Wong, K.-L., and C.D. Klaassen. 1980. Tissue distribution and retention of cadmium in rats during postnatal development: Minimal role of hepatic metallothionein. Toxicol. Appl. Pharmacol. 53:343-353.

Woods, W.G., M. O'Leary, and M.E. Nesbit. 1981. Life-threatening neuropathy and hepatotoxicity in infants during induction therapy for acute lymphoblastic leukemia. J. Pediatr. 98:642-645.

Wu, M.-F., J.R. Ison, J.R. Wecker, and L.W. Lapham. 1985. Cutaneous and auditory function in rats following methyl mercury poisoning. Toxicol. Appl. Pharmacol. 79:377-388.

Wynn, R.M. 1963. Relationship of menadiol tetrasodium diphosphate (synkavite) to bilirubinemia and hemolysis in the adult and newborn rat. Am. J. Obstet. Gynecol. 86:495-503.

Ziegler, E.E., B.B. Edwards, R.L. Jensen, K.R. Mahaffey, and S.J. Forman. 1978. Absorption and retention of lead by infants. Pediatr. Res. 12:29-34.

Zimmerman, J.A., L.D. Trombetta, T.H. Carter, and S.H. Weisbroth. 1982. Pancreatic carcinoma induced by *N*-methyl-*N*'-nitrosourea in aged mice. Gerontology 28:114-120.

Zwiener, R.J., and C.M. Ginsburg. 1988. Organophosphate and carbamate poisoning in infants and children. Pediatrics 81:121-126.

4

Methods for Toxicity Testing

THE PURPOSE OF THIS CHAPTER is to familiarize the reader with the testing that is currently conducted by a manufacturer prior to and during the process of submitting a petition to register a pesticide. Codified toxicologic evaluation of potential pesticides has been a requirement in the United States for approximately 50 years. The testing requirements and guidelines continue to evolve based on new science. This chapter identifies the current testing that is pertinent to the young animal and young human as well as aspects of testing that are needed to fill the data gaps to better ensure the protection of infants and children. The current testing guidelines can be found in *Pesticide Assessment Guidelines* issued by the Environmental Protection Agency (EPA, 1991a,b).

Data, including those derived from toxicity testing, crop residue analyses, environmental fate testing, and ecotoxicology testing, are generated by the manufacturer of a pesticide to meet the mandatory requirements of the Federal Insecticide, Fungicide, and Rodenticide Act (FIFRA) for pesticide registration. Although these data are essential to the EPA's registration process, other data generated by EPA itself, as well as by other government institutions and academia, are considered in the registration decision-making process.

EPA has issued 194 registration standards on 350 chemicals used as active and inert ingredients in pesticide products. These standards are published by EPA and are intended to upgrade and update the data base on a previously registered pesticide or class of pesticide products. They call for additional studies in the areas of toxicity testing, crop residue analyses, environmental fate, and ecotoxicology testing. This testing must be conducted within an EPA-mandated time frame to allow for the contin-

TABLE 4-1   End Points for Various Toxicity Studies

Study	End Points
Developmental toxicity	Fetus: mortality, growth retardation, skeletal variations, gross external malformations, soft tissue/internal organ defects Female parent: general toxicity
Reproductive toxicity	Male parent: general toxicity, effects on fertility, reproductive organ changes Offspring: effects on viability, sex ratio, growth, behavior
Carcinogenicity	Tumor development and general toxicity
Neurotoxicity	Behavior, function, and motor activity deficits; microscopic nervous tissue changes
Mutagenicity	Heritable lesions leading to altered phenotypes

SOURCE: EPA, 1984.

ued registration of a given product. The list of pesticides for which registration standards have been issued is referred to as List A and can be found in Appendix I of the *Federal Register* notice of February 22, 1989. Under the FIFRA Amendments of 1988, the data bases on the remaining registered pesticide products are being upgraded in five phases over a 9-year period.

The first sections of this chapter describe in detail the present toxicity testing procedures for pesticides in relation to their registered use patterns and EPA's proposed changes or additions to these procedures. The conclusions and recommendations of the committee for further changes and additions to the toxicity testing battery to allow for more adequate consideration of the special testing needs for infants and children are presented.

CURRENT METHODS: GENERAL CONSIDERATIONS

Toxicity studies are required to assess potential hazards to humans through the acute, subchronic, and chronic exposure of laboratory animals to pesticides. The more specific types of toxicity that are determined include carcinogenicity; developmental (including teratogenicity in offspring) and reproductive toxicity; mutagenicity; and neurotoxicity (Table 4-1). Detailed information on the metabolism or biotransformation of the pesticide is also obtained. Consideration is given to testing individual metabolites in animals, and in or on pesticide-treated plants to which humans could be exposed through their diet. The extent of metabolite testing required depends on the level of potential toxicity and environmental persistence of the metabolite. With the exception of the acute toxicity tests, most tests are conducted to determine the nature of any toxicity that can be produced by repeatedly dosing animals over an extended period. The results enable toxicologists to estimate the safety of a material for humans (Loomis, 1978).

Weil (1972) published the following set of guidelines, which reflected a consensus among toxicologists. These should be considered before initiating a toxicity test:

1. Use, wherever practical or possible, one or more species that biologically handle the material qualitatively and/or quantitatively as similarly as possible to man. For this, metabolism, absorption, excretion, storage and other physiological effects might be considered.
2. Where practical, use several dose levels on the principle that all types of toxicologic and pharmacologic actions in man and animals are dose-related. The only exception to this should be the use of a single, maximum dosage level if the material is relatively nontoxic; this level should be a sufficiently large multiple of that which is attainable by the maximum applicable hazard exposure route, and should not be physiologically impractical.
3. Effects produced at higher dose levels (within the practical limits discussed in 2) are useful for delineating mechanism of action, but for any material and adverse effect, some dose level exists for man or animal below which this adverse effect will not appear. This biologically insignificant level can and should be set by use of a proper uncertainty factor and competent scientific judgment. . . .
4. Statistical tests for significance are valid only on the experimental units (e.g., either litters or individuals) that have been mathematically randomized among the dosed and concurrent control groups. . . .
5. Effects obtained by one route of administration to test animals are not a priori applicable to effects by another route of administration to man. The routes chosen for administration to test animals should, therefore, be the same as those to which man will be exposed. Thus, for example, food additives for man should be tested by admixture of the material in the diet of animals.

In general, Weil's guidelines are considered by EPA in its toxicity testing requirements and subsequent evaluation of results for pesticides. One exception to Weil's points is found in his guideline 3. EPA does not recognize the existence of a dose level at which a carcinogen will not exert its effect. For carcinogens, EPA generally accepts a risk of 10^{-6}, as extrapolated from bioassays using the nonthreshold modification of the linearized multistage model of Armitage and Doll (1954), as adequate for the protection of humans.

The selection of animal species for toxicity tests depends on life span, behavior, availability, and overall costs. EPA recommends using rats for subchronic, chronic, carcinogenicity, and reproduction studies; mice for carcinogenicity studies; and dogs for subchronic and chronic studies. Rats are routinely used for acute oral and inhalation studies and rabbits for eye and skin irritation studies and acute dermal studies. One exception to this is the use of guinea pigs for dermal sensitization testing. The rat and rabbit are recommended for developmental toxicity (teratogenicity) testing. Justification must be provided for the use of species other than those outlined above.

The number of animals to be tested in each dose group depends on a number of factors, including the purpose of the experiment, the required sensitivity of the study, the reproductive capacity and the fertility of the species, economic aspects, and the availability of animals (IPCS, 1990). Table 4-2 lists the minimum number of animals required by EPA for some toxicity studies. For the most part, these numbers are consistent with those recommended by the International Programme on Chemical Safety (IPCS).

The selection of dose levels for subchronic studies should be based on the results of acute toxicity testing, on range-finding studies, and on pharmacokinetic (metabolism, including rate in various tissues) data. For subchronic studies, four dose groups of animals should be included: a control group; a low-dose group (a dose that produces no compound-related toxicity); a mid-dose group (a dose that elicits some minimal signs of toxicity); and a high-dose group (a dose that results in toxic effects but not in an incidence of fatalities that would prevent a meaningful evaluation; for nonrodents, there should be no fatalities) (EPA, 1984). This same guidance is relevant to chronic toxicity and reproduction studies. For teratology studies, the highest dose tested should elicit some signs of maternal toxicity, but the toxicity should not obscure the results.

The one notable exception to this guidance pertains to carcinogenicity studies. The highest dose levels for these studies should be at a maximum-tolerated dose (MTD), as determined in 90-day toxicity studies in the appropriate test species and from pharmacokinetic information on the material being tested. The Committee on Risk Assessment Methodology of the National Research Council (NRC) recently examined the criteria for the MTD and other doses used in carcinogenicity studies (NRC, 1993). The EPA has issued its own guidance for the selection of this dose level. Some of the factors to consider in selecting an MTD are: 10% decrement in body weight gain in a 90-day study; observation of potential life-threatening lesions during microscopic examination of organs, e.g., liver necrosis; significant inhibition of cholinesterase activity in two biological compartments, such as brain and plasma; and significant signs of anemia or other biologically relevant effects on blood.

TABLE 4-2   Animal Model Requirements in Toxicity Studies

Study Type	Minimum No. of Animals Required Males	Females	Dosing	Age at Start of Study
Acute oral (rat), dermal, or inhalation (rat)	5	5	Single	Young adult
Eye and skin irritation (rabbit)	6[a]		Single	Young adult
Dermal sensitization (guinea pig)	[b]		Repeated	Young adult
21-Day dermal (rat, rabbit, or guinea pig)	5	5	Repeated	Rat, 200-300 g; rabbit, 2.0-3.0 kg; guinea pig, 350-450 g
90-Day oral (rat)	10	10	Repeated	6-8 weeks
90-Day inhalation (rat)	10	10	Repeated	Young adult
90-Day dermal (rat, rabbit, or guinea pig)	10	10	Repeated	Rat, 200-300 g; rabbit, 2.0-3.0 kg; guinea pig, 350-450 g
90-Day or chronic (1 year) oral (dog)	4	4	Repeated	4-6 months
Reproduction (rat)[c]	20	20	Repeated	8 weeks
Teratology				
Rat		20[d]	Repeated	Young adult
Rabbit		12[d]	Repeated	Young adult
Chronic toxicity (1 or 2 year) (rat)	20	20	Repeated	6-8 weeks
Oncogenicity (lifetime) (rat and mouse)	50[e]	50[e]	Repeated	6-8 weeks

[a] Either males or females may be used in this test.

[b] The number of animals used depends on the method used. Several different experimental methods are acceptable.

[c] EPA prefers that one male rat be housed with one female during mating.

[d] Number of pregnant females required.

[e] 50 rats and 50 mice of each sex.

SOURCE: EPA, 1984.

In general, EPA has set a cap on dosing of 1.0 g/kg/day for toxicity tests other than acute studies. This dose level is referred to as the limit dose and corresponds to approximately 20,000 ppm in the diet of rats, 7,000 ppm in the diet of mice, and 40,000 ppm in the diet of dogs.

The duration of exposure for toxicity testing of a pesticide depends on the expected duration of human exposure to the pesticide in practice. The typical length of various toxicity tests and the number of doses administered are shown in Table 4-2. Repeated dosing refers to dosing once per day for the designated number of days. When the material is given to the test animals in their diet, dosing is usually continuous for 7 days a week. If the material is administered by gavage (oral bolus dose), by dermal application, or by inhalation, doses are frequently given 5 days a week, which is acceptable to EPA because of practical considerations (EPA, 1984).

The type of statistical analysis performed on the toxicity data resulting from these studies depends on the type of data under consideration (see, for example, Gad and Weil, 1982, for a review). Interpreting the meaning of statistical significance for any particular parameter depends on the dose level at which it was achieved, the biological significance of the finding, and the normal spontaneous occurrence of this finding in the strain and species being tested.

For regulatory purposes, the no-observed-effect level (NOEL) is defined as a dose level at which no effects attributable to the pesticide under test can be found. A no-observed-adverse-effect level (NOAEL) can also be determined for each study; however, EPA does not routinely use the NOAEL to regulate pesticide usage. To establish a NOAEL, the toxicologist must determine what is and what is not an adverse effect, which can be defined differently by different scientists. For example, effects such as hair loss can be considered adverse by some and not by others. Plasma and red blood cell cholinesterase inhibition can be viewed as either an adverse effect or simply as a marker of exposure to a pesticide.

EPA uses the NOEL to calculate the acceptable daily intake (ADI) of the pesticide under consideration. More recently, the EPA has replaced the ADI with the reference dose, or RfD. Chronic studies, such as reproduction studies and studies lasting 1 year or longer in the rat or dog are used for this purpose. EPA does not routinely use the NOEL determined from teratology (developmental toxicity) studies for calculating ADIs because the observed effects are not considered chronic; however, these NOELs can be used to support the calculated ADI. EPA does routinely use developmental toxicity NOELs for other types of risk assessments, such as calculating the risk from acute, daily dietary or occupational exposure or from exposure of homeowners to a developmental toxicant.

EPA's toxicity testing requirements for food and nonfood use pesticides have been published in 40 CFR Part 158. In general, for a food use chemical with maximum human exposure, the following toxicity tests are required:

- acute oral toxicity
- acute dermal toxicity
- acute inhalation toxicity
- primary eye irritation
- primary dermal irritation
- chronic feeding toxicity

- dermal sensitization
- acute neurotoxicity
- 90-day toxicity
- 21-day dermal toxicity
- 90-day neurotoxicity study
- reproduction study

- carcinogenicity
- developmental toxicity

- mutagenicity tests
- general metabolism study

More than 30% of the toxicity tests for pesticides submitted to EPA in the past have been rejected. Those rejected must be resubmitted until they are in conformance with EPA criteria before registrations can be granted. The criteria for rejection are summarized in Table 4-3. Some of them fall in the category of regulatory policy; others involve scientific concerns. The most commonly cited reason for noncompliance is lack of characterization of the test material. To improve the quality of testing and incorporate new scientific methods in its testing requirements, EPA is currently revising the 40 CFR Part 158 data requirements for food and nonfood use pesticides. The proposed revisions to these requirements can be found in Table 4-4.

ACUTE TOXICITY STUDIES

General Description

Acute toxicity studies provide information on the potential for health hazards that may arise as a result of short-term exposure. Determination of acute oral, dermal, and inhalation toxicity is usually the initial step in evaluating the toxic characteristics of a pesticide. In each of these tests the animal is exposed to the test material only once on 1 day. Together with information derived from primary eye and primary dermal irritation studies (also 1 dose on 1 day), which assess possible hazards resulting from pesticide contact with eyes and skin, these data provide a basis for precautionary labeling and may influence the classification of a pesticide for restricted use. Acute toxicity data also provide information used to determine the need for child-resistant packaging, for protective clothing requirements for applicators, and for calculation of farm worker reentry intervals. A minimum number of animals, usually adults, are used in these studies and only the end points of concern are monitored, i.e., mortality, observable skin or eye effects, dermal sensitization, and observable neurotoxic behavioral changes. One exception is the inclusion of microscopic examination of neural tissues in the newly required acute neurotoxicity study.

EPA's Proposed Changes

Guideline number 81-1 (EPA, 1984), acute oral study in the rat, would be revised to include special visual system testing, which would be required for all organophosphate pesticides and other pesticides known to affect the visual system.

TABLE 4-3 Summary of EPA Rejection Factors

Guideline	Rejection Factor
Acute Oral Toxicity (81-1)	Lack of characterization of the test material Inadequate dose levels to calculate LD_{50}
Acute Dermal Toxicity (81-2)	Lack of characterization of the test material Inadequate percentage of body surface area exposed No quality assurance statement Improper number of animals tested per dose group Only one sex tested Omitted source, age, weight, or strain of test animal
Acute and 90-Day Inhalation (81-3 and 82-4)	Less than 25% of particles were <1 μm; LC_{50} could not be calculated; highest concentration did not produce toxicity Inadequate reporting of exposure methodology Protocol errors Lack of characterization of the test material Compound preparation Chamber concentration not measured
Primary Eye Irritation (81-5)	Lack of characterization of the test material
Primary Dermal Irritation (81-5)	Lack of characterization of the test material No quality assurance statement and/or no Good Laboratory Practice (GLP) statement Improper test material application/preparation Omitted source, age, weight, or strain of test animal Missing individual/summary animal data
Dermal Sensitization (81-6)	Control problems Dosing level problems Lack of characterization of the test material Unacceptable protocol or other protocol problems Individual animal scores or data missing Scoring method or other scoring problem Reporting deficiencies or no quality assurance statement
90-Day Feeding—Rodent (82-1(a))	A NOEL was not established Lack of characterization of the test material or incorrectly reported Lack of clinical chemistry and/or lack of histopathology
90-Day Feeding—Nonrodents (82-1(b))	Reporting deficiencies Lack of characterization of the test material A NOEL was not established An investigational parameter missing

TABLE 4-3 *(Continued)*

Guideline	Rejection Factor
90-Day Feeding—Nonrodents (cont.) (82-1(b))	Information on the pilot study and other problems associated with dose level selection
	An investigational parameter missing
	Information on the pilot study and other problems associated with dose level selection
21-Day Dermal Toxicity (82-2)	Lack of characterization of the test material
	Raw data analyses incomplete or missing
	A systemic NOEL was not established
	Inadequate percentage of body surface area exposed in each dose group
	Insufficient number of dose levels tested
90-Day Dermal Toxicity (82-3)	Lack of characterization of the test material
	A systemic NOEL was not established
	Incomplete/missing raw animal data analyses
	Insufficient number of dose levels tested
	Poorly controlled test environment
Chronic Feeding/ Carcinogenicity—Rats (83-1(a) and 83-2(a))	Missing histopathology information
	Missing information in study reports
	MTD was not achieved
	Missing historical control data
	Lack of characterization of the test material
	Deficiencies in reporting the study data
Carcinogenicity—Mice (83-2(b))	Histopathology information missing
	MTD was not achieved
	Lack of historical control data
	Information missing in study reports
	Lack of characterization of the test material
	Deficiencies in reporting of study data
Developmental Toxicity— Rodents (83-3(a))	Missing historical controls
	Lack of characterization of the test material
	Information missing or requiring clarification of the laboratories' methods
	Information missing or requiring clarification of the laboratories' results
	A NOEL was not established
	Statistical problems
	Did not use conventional assessments for skeletal or visceral examinations
Developmental Toxicity— Nonrodents (83-3(b))	Clarification of laboratory procedures or interpretation of the data
	Individual maternal or fetal data missing
	Missing historical controls
	Lack of characterization of the test material
	Excessive maternal toxicity

TABLE 4-3 *(Continued)*

Guideline	Rejection Factor
Developmental Toxicity— Nonrodents (cont.) (83-3(b))	A NOEL was not established Statistical problems
Reproduction (83-4)	Information missing from laboratory results Lack of characterization of the test material Information missing or requiring clarification of laboratory methods or results Missing historical controls A NOEL was not established due to effects at the lowest dose tested Low fertility and/or inadequate number of animals were used per dose level A NOEL was not established in the absence of reproductive effects
Metabolism (85-1)	Inadequate or missing data on identification of metabolites Improper methodology or dosing regimen Inadequate number of animals were used in the dose groups No individual animal data Improper reporting Inadequate or missing tissue residue analysis data Testing at only one dose level Only one sex of animal used Lack of an intravenous dose group No collection of $^{14}CO_2$
Dermal Penetration (85-2)	Incomplete/missing data evaluation Improper test material preparation/application Raw data missing and incomplete summary tables No signed quality assurance statement Missing purity or concentration of test material

SOURCE: P. Fenner-Crisp, EPA, personal communication, 1992.

The additional acute study proposed in guideline number 81-4 is acute neurotoxicity testing in the rat. This study would be required for all pesticide registrations (food and nonfood) and experimental use permits (EUPs), and it would include assessments of function and activity as well as histopathological (microscopic) examination of selected neural tissue. EPA presently requires that this study be conducted by manufacturers wishing to reregister.

SUBCHRONIC TOXICITY STUDIES

General Description

Subchronic exposures do not elicit effects that have a long latency period (e.g., carcinogenicity). However, they do provide information on health hazards that may result from repeated exposures to a pesticide over a period up to approximately 30% of the lifetime of a rodent. Subchronic tests also provide information necessary to select proper dose levels for chronic studies, especially for carcinogenicity studies for which an MTD must be selected. According to EPA (1984), rats selected for these studies should be started on the test material shortly after weaning, "ideally before the rats are 6 and, in any case, not more than 8 weeks old." For dogs, dosing should begin when they are 4 to 6 months of age and "not later than 9 months of age."

Most subchronic toxicity studies monitor clinical or behavioral (neurological) signs of toxicity, body weight, food consumption, eye effects, certain plasma or serum and urine parameters, organ weights, and gross and microscopic pathology. Clinical and behavioral signs of toxicity are observed and recorded daily. They can consist of activity, gait, excreta, hair coat, and feeding and drinking patterns. Body weight and food consumption data are routinely recorded throughout the study at intervals (usually weekly) determined by the length of the study. Ophthalmoscopic examinations are conducted at the beginning of the study and, typically, just before it terminates. The laboratory parameters typically examined are summarized in Table 4-5.

The results of hematology testing indicate whether, for example, the chemical affects blood cell formation and survival, clotting factors, and platelets. Clinical chemistry and urinalysis results can indicate possible kidney, liver, pancreas, and cardiac function or toxicity as well as any electrolyte imbalance. Urinalysis results can indicate adequacy of kidney, liver, and pancreas function.

After necropsy, the weights of certain organs are also recorded. These organs generally include brain, gonads, liver, and kidneys, which are the four required according to EPA testing guidelines (EPA, 1984). If toxicity is known to occur in another organ from previous testing, the weight of this organ should also be reported. For thyroid toxicity, for example, the weight of the thyroids should be recorded. Changes from untreated control animals are generally an indication of potential toxicity in this organ.

A complete necropsy is performed after sacrifice or death of the test animal. Generally all tissues are examined, and those saved for microscopic examination are aorta, jejunum, peripheral nerve, eyes, bone marrow, kidneys, cecum, liver, esophagus, colon, lung, ovaries, duodenum,

TABLE 4-4 Toxicity Data Requirements Proposed by EPA for Food and Nonfood Uses of Pesticides

Kinds of Data Required	EPA Guideline Number	Comments (see Corresponding Numbers Beginning on Page 00)	Data Requirements		Test Substance Data to Support	
			Food Uses[a]	Nonfood Uses[b]	Manufacturing-Use Product	End-Use Product
Acute Testing						
Acute oral toxicity—rat	81-1	1, 36, 37	[R][c]	[R]	MP and TGAI[d]	TGAI, EP[e], and possibly EP dilution
Acute dermal toxicity—rabbit, rat, or guinea pig	81-2	1, 2, 37	[R]	[R]	MP and TGAI	TGAI, EP, and possibly EP dilution
Acute inhalation toxicity—rat	81-3	3	[R]	[R]	MP and TGAI	EP and TGAI
Primary eye irritation—rabbit	81-4	2	[R]	[R]	MP	EP
Primary dermal irritation—rabbit	81-5	1, 2	[R]	[R]	MP	EP
Dermal sensitization—guinea pig	81-6	4	[R]	[R]	MP	EP
Delayed neurotoxicity (acute)—hen	81-7	5	[CR][f]	[CR]	TGAI	TGAI
Acute neurotoxicity—rat	81-8	6	[R]	[R]	TGAI	TGAI
Subchronic Testing						
90-day oral—two species, rodent and nonrodent	82-1	7, 8, 9, 10, 11, 36	[R]	CR	TGAI	TGAI
21-day dermal—rat, rabbit, or guinea pig	82-2	9, 12	R	None	TGAI	TGAI
90-day dermal—rat, rabbit, or guinea pig	82-3	8, 9, 13, 14	CR	R	TGAI	TGAI
90-day inhalation—rat	82-4	8, 9, 15	CR	CR	TGAI	TGAI
28-day delayed neurotoxicity—hen	82-6	16	CR	CR	TGAI	TGAI
90-day neurotoxicity—rat	82-7	6, 8	[R]	R	TGAI	TGAI

Chronic Testing						
Chronic feeding—two species: rodent and nonrodent	83-1	9, 10, 17, 18, 19, 36	[R]	CR	TGAI	TGAI
Carcinogenicity—two species: rat and mouse preferred	83-2	9, 19, 20, 21	R	CR	TGAI	TGAI
Developmental Toxicity and Reproduction						
Developmental toxicity—two species: rat and rabbit preferred	83-3	22, 23, 24, 25	[R]	R	TGAI	TGAI
Reproduction—rat	83-4	26, 27	[R]	CR	TGAI	TGAI
Postnatal developmental toxicity—rat and/or rabbit	83-6	28, 29	CR	CR	TGAI	TGAI
Mutagenicity Testing						
Salmonella typhimurium reverse mutation assay	84-2	30	[R]	R	TGAI	TGAI
Mammalian cells in culture	84-2	30, 31	[R]	R	TGAI	TGAI
In vivo cytogenetics	84-2	30, 32	[R]	R	TGAI	TGAI
General Metabolism—rat	85-1	33	R	CR	PAI or PAIRA[g]	PAI or PAIRA
Special Testing						
Domestic animal safety	85-2	34	CR	CR	Choice[h]	Choice
Dermal penetration	85-3	35	CR	CR	Choice	Choice
Visual system studies	85-4	36	[CR]	CR	TGAI	TGAI

[a] Food use includes terrestrial food and feed, aquatic food, greenhouse food, and indoor food.

[b] Nonfood use includes terrestrial nonfood, aquatic nonfood outdoor, aquatic nonfood industrial, aquatic nonfood residential, greenhouse nonfood, forestry, residential outdoor, indoor nonfood, indoor medical, and indoor residential.

[c] R = required; brackets [ ] indicate data requirements that apply when an experimental use permit is being sought.

[d] MP = manufacturing-use product; TGAI = technical grade of the active ingredient.

[e] EP = end-use product.

[f] CR = conditionally required.

[g] PAI = pure active ingredient; PAIRA = pure active ingredient, radiolabeled.

[h] Choice of several substances, depending on studies required.

Notes for Table 4-4: Specific Conditions, Qualifications, or Exceptions to the Designated Test Procedures

1. Not required if test material is a gas or highly volatile.

2. Not required if test material is corrosive to skin or has pH <2 or >11.5; such a product will be classified as toxicity category 1 on the basis of potential eye and dermal irritation effects.

3. Required when the product consists of, or under conditions of use will result in, an inhalable material (e.g., gas, volatile substances, or aerosol/particulate).

4. Required unless repeated dermal exposure does not occur under conditions of use.

5. Required for uncharged organophosphorus esters, thioesters, or anhydrides of organophosphoric, organophosphonic, or organophosphoramidic acids or of related phosphorothioic, phosphonothioic, or phosphorothioamidic acids, or other substances that may cause the neurotoxicity sometimes seen in this class.

6. Additional measurements such as cholinesterase determinations for certain pesticides (e.g., organophosphates and carbamates) may also be required. The route of exposure should correspond to a primary route of human exposure.

7. Required if intended use of the pesticide is expected to result in human exposure via the oral route.

8. All 90-day subchronic studies can be designed to simultaneously fulfill the requirements of the 90-day neurotoxicity study.

9. Studies must include additional end points so as to provide an immunotoxicity screen in the rodent. An equivalent independent study may fulfill the requirements for an immunotoxicity screen.

10. In most cases, where the theoretical maximum residue contribution (TMRC) exceeds 50 percent of the reference dose (RfD), a 1-year (or longer) interim report on a chronic (2-year) feeding study is required to support a temporary tolerance. This report is to be in addition to the 90-day feeding studies in rodents and nonrodents.

11. If the pesticide is found to leach into groundwater or may contaminate drinking water, a 90-day drinking water study may be required unless data demonstrate that there are no significant differences in toxicity observed when the test material is administered in feed versus when the test material is administered in drinking water. This study may be requested in addition to any 90-day oral studies that may be required.

12. Required if intended use of the pesticide is expected to result in human exposure via the dermal route and data from a subchronic 90-day dermal toxicity study are not required.

13. For nonfood uses, a 90-day dermal toxicity study is required, since intended use of the pesticide is expected to result in repeated dermal exposure of humans.

14. For food uses, required if: (a) the active ingredient of the product is known or expected to be metabolized differently by the dermal route of exposure than by the oral route, and a metabolite of the active ingredient is the toxic moiety; (b) the active ingredient of the product is classified as toxicity category I or II on the basis of acute dermal toxicity data.

15. Required if the active ingredient is a gas at room temperature or if use of the product results in respirable droplets and use may result in repeated inhalation exposure at a concentration likely to be toxic, regardless of whether the major route of exposure is inhalation.

16. Required for substances when statistically or biologically significant effects were seen in the acute study (Guideline 81-7), or if other available data indicate that the substance can cause this type of delayed neurotoxicity.

17. Required if either of the following criteria is met: (a) use of the pesticide is likely to result in repeated human exposure over a significant portion of the human life span (e.g., products intended for use in and around residences, swimming pools, and enclosed working spaces or their immediate vicinity); (b) the use requires a tolerance for the pesticide or an exemption from the requirement to obtain a tolerance for the pesticide or an exemption from the requirement to obtain a tolerance, or requires issuance of a food additive regulation.

18. Based on acute and subchronic neurotoxicity studies, and/or on other available data, a functional observational battery, an assessment of motor activity, and perfusion neuropathology may be required.

19. Studies designed to simultaneously fulfill the requirements of both the chronic feeding and carcinogenicity studies (i.e., a combined study) may be conducted. Minimum acceptable study durations for chronic feeding and carcinogenicity studies are as follows: chronic rodent feeding study (food use pesticide)—24 months; chronic rodent feeding study (nonfood pesticide)—12 months is usually sufficient; chronic nonrodent (i.e., dog) feeding study—12 months; mouse carcinogenicity study—18 months; and rat carcinogenicity study—24 months.

20. Required active ingredients or any of their metabolites, degradation products, or impurities are structurally related to a recognized carcinogen, cause mutagenic effects as demonstrated by in vitro or in vivo testing, or produce a morphologic effect in any organ (e.g., hyperplasia, metaplasia) in subchronic studies that may lead to neoplastic change. The use requires a tolerance for the pesticide or exemption from the requirement to obtain a tolerance or requires the issuance of a food additive regulation. Use of the pesticide product is likely to result in exposure of humans over a portion of the life span that is significant in terms of either the timing or duration of exposure (e.g., pesticides used in treated fabrics for wearing apparel, diapers, or bedding; insect repellents applied directly to the skin; swimming pool additives; or constant-release indoor aerosol pesticides).

21. Range-finding studies of at least 90 days duration in rats and mice are generally required to determine dose levels adequate to demonstrate an MTD in carcinogenicity studies. A subchronic 90-day oral study conducted in accordance with Guideline 82-1 may also be acceptable for this purpose.

Notes for Table 4-4 *(Continued)*

22. Testing in two species is required for food uses. For products intended for nonfood uses, testing in two species is required if significant exposure of human females of child-bearing age may reasonably be expected. For other nonfood uses, testing in at least one species is required. A study in one species is required to support a temporary tolerance.

23. Testing in a second species is required if significant developmental toxicity is observed after testing in the first species.

24. The test substance or vehicle is usually administered by oral intubation, unless the chemical or physical characteristics of the test substance or pattern of human exposure suggest a more appropriate route of administration.

25. Under certain conditions where a pesticide is determined to be a developmental toxicant (e.g., after oral dosing), additional testing via other routes (e.g., dermal) may be required.

26. Required to support products intended for food and nonfood uses if the use is likely to result in exposure of humans over a portion of the life span that is significant in terms of the frequency, magnitude, or duration of exposure (e.g., pesticides used in treated fabrics for wearing apparel, diapers, or bedding; insect repellents applied directly to the skin; swimming pool additives; or constant-release indoor pesticides used in aerosol form). Also may be required for nonfood uses if adverse effects on organs of the reproductive system are observed in 90-day or other studies, and/or if developmental toxicity is demonstrated by available data (Guideline 83-3).

27. In most cases, where the TMRC exceeds 50% of the RfD, a first-generation (or longer) interim report on a multigeneration reproduction study is required to support a temporary tolerance.

28. Conditionally required to more fully assess any of the manifestations of developmental toxicity. These studies permit assessment of potential functional deficits that cannot be evaluated in the classical developmental toxicity study (Guideline 83-3). Protocols for these studies are usually designed on a case-by-case basis.

29. On the basis of acute and subchronic neurotoxicity studies, and/or on other available data, a developmental neurotoxicity study may be required. For this type of postnatal study, a guideline is available.

30. An initial battery of mutagenicity tests with possible confirmatory testing is minimally required. Also, results from other mutagenicity tests that may have been performed and as complete a reference list as possible shall be submitted. Subsequent testing may or may not be required based on the evidence available to EPA's Office of Pesticide Programs in accordance with the objective and considerations for mutagenicity testing. Current protocols for tests in the initial battery and other mutagenicity tests are given in the EPA's Office of Pesticides and Toxic Substances Health Effects Testing Guidelines (40 CFR Part 798, Subpart F—Genetic Toxicity). Because of the rapid improvements in the field, applicants are encouraged to discuss test selection, protocol design, and results of preliminary testing with the agency.

31. Choice of assays using either mouse lymphoma L5178Y cells, thymidine kinase (tK) gene locus, maximizing assay conditions for small colony expression and detection; Chinese hamster ovary (CHO) or Chinese hamster lung fibroblast (V79) cells, hypoxanthine-guanine phosphoribosyl transferase (*hgprt*) gene locus, accompanied by an appropriate in vitro test for clastogenicity; or CHO cells strain AS52, xanthine-guanine phosphoribosyl transferase (*xprt*) gene locus.

32. Choice of assays; initial consideration usually given to rodent bone marrow, using either metaphase analysis (aberrations) or micronucleus assay.

33. Required for all food uses and when chronic and/or carcinogenicity studies are required. Also may be required if significant adverse effects are observed in toxicology studies (e.g., reproduction and developmental toxicity).

34. May be required, on a case-by-case basis, to support registration of an end-use product if cats, dogs, cattle, pigs, sheep, horses, or other domesticated animals will be exposed to the pesticide product, including, but not limited to, exposure through direct application for pest control and consumption of treated feed.

35. Dermal penetration studies are required for compounds that have serious toxic effects, as identified in oral or inhalation studies, and for which a significant route of human exposure is dermal. Thus, this study is required when any of the following exposure studies are required: passive dosimetry—dermal exposure (Guidelines 133-3, 231 or 233), foliar dislodgeable residue dissipation (Guideline 132-1), soil dislodgeable residue dissipation (Guideline 132-1), and indoor surface residue dissipation, unless the toxicity studies (including Guidelines 82-6, 82-7, 83-1, 83-2, 83-3, 83-4, and 83-6) that triggered the need for these exposure studies were conducted via the dermal route of dosing. Registrants should work closely with the agency in developing an acceptable protocol for performing dermal penetration studies.

36. Special testing (acute, subchronic, and/or chronic) is required for organophosphates, and may be required for other cholinesterase inhibitors and other pesticides that have demonstrated a potential to adversely affect the visual system. Registrants should consult with the agency for development of protocols and methodology prior to initiation of studies.

37. Testing of the end-use product dilution is required if it can be reasonably anticipated that the results of such testing may meet the criteria for restriction to use by certified applicators specified in 40 CFR 152.170(b) or the criteria for initiation of special review specified in 40 CFR 154.7(a)(1).

SOURCE: Code of Federal Regulations, Title 40, Parts 150 to 189,1992.

TABLE 4-5   Laboratory Parameters Measured in Various Data Categories

Data Type	Parameter Measured	
Hematology	Erythrocyte count	Leukocyte count
	Hemoglobin	Differential count
	Hematocrit	Platelet count (or clotting parameter)
Clinical chemistry	Alkaline phosphatase	Total protein
	Alanine aminotransferase	Albumin
	Creatinine kinase	Urea nitrogen
	Lactic dehydrogenase	Inorganic phosphate
	Glucose	Calcium
	Bilirubin	Potassium
	Cholesterol	Sodium
	Creatinine	Chloride
	Aspartate aminotransferase	
Urinalysis	Blood	Total bilirubin
	Protein	Urobilinogen
	Ketone bodies	Sediment
	Appearance	Specific gravity
	Glucose	Volume

SOURCE: EPA, 1984.

lymph nodes, oviduct, brain, stomach, pancreas, skin, mammary gland, rectum, heart, spleen, spinal cord, testes, musculature, thyroid/parathyroid, pituitary, epididymis, salivary glands, ileum, adrenals, thymus, trachea, urinary bladder, accessory sex organs, and gallbladder.

The data described above are not required for all subchronic studies. For the 21-day dermal study, for example, only limited necropsy data are required.

EPA's Proposed Changes

For the 90-day oral study (Guideline 82-1; EPA, 1984) in the rodent and other test species, three changes are proposed:

• the studies would be required for all pesticide uses that could result in oral exposure of humans and would not depend on frequency, magnitude, or duration of exposure;
• this study could be modified to include additional end points for neurotoxicity and immunotoxicity; and
• a separate drinking water study, in addition to the other dietary studies, could be required if the pesticide were found to leach into groundwater or contaminate drinking water.

Either a 21-day dermal (Guideline 82-2; EPA, 1984) or a 90-day dermal study (Guideline 82-3; EPA, 1984) would be required (not conditionally required as in the past) to support all registrations. A 21-day study would be required for all food uses, except when acute dermal toxicity is observed. A 90-day dermal study would ordinarily be required for all nonfood uses. Special tests for neurotoxicity or immunotoxicity could be added to these studies if these toxicity end points are not studied in other tests required for a particular pesticide.

A 90-day inhalation study (Guideline 82-4; EPA, 1984) would be required more frequently whether or not the major route of exposure is inhalation, especially for a nonfood use pesticide that is a gas or whose use generates respirable droplets. The requirement for a 21-day inhalation study for a tobacco use pesticide would be deleted. Special tests for neurotoxicity or immunotoxicity could be added if those end points are not studied in other EPA-required toxicity studies.

The conditionally required 90-day neurotoxicity study in the hen or mammal (Guideline 82-5; EPA, 1984) would be replaced by two new studies:

- a 28-day delayed neurotoxicity study in the hen (Guideline 82-6; EPA, 1984), which would be conducted under the same conditions as the 90-day study; and
- a 90-day neurotoxicity study in the rat (Guideline 82-7; EPA, 1984) (previously required only conditionally) to support all registrations and food/feed use EUPs (this is now required only for organophosphate and carbamate pesticides).

Testing would include assessments of function (functional observation battery), motor activity, and histopathological examination of the nervous system.

CHRONIC TOXICITY STUDIES

General Description

Information derived from chronic studies is used to assess potential hazards resulting from prolonged and repeated exposure to a pesticide over a large portion of the human life span. These studies usually last 12 to 24 months. Of particular importance are long-term carcinogenicity studies, the purpose of which is to observe the test animals for the development of neoplastic lesions after a lifetime of exposure at dose levels up to and including the MTD determined from subchronic testing.

The emphasis of the carcinogenicity study is the detection of tumors in animals. For these studies, both concurrent and historical control data are used to evaluate the relevance of tumors. Historical control data should be derived from studies in the same species and strain and, preferably, in the same laboratory as used in the study under consideration. Carcinogenicity studies should be 24 months long in rats and 18 months long in mice. The age of test animals in carcinogenicity (rat and mouse) studies and other chronic (rat and dog) studies is determined by the same criteria as for subchronic toxicity studies. The parameters to be examined in carcinogenicity studies are also generally the same as those discussed above for subchronic and chronic studies, except that clinical chemistry and urine parameters are not required and only limited hematology data are required.

EPA's Proposed Changes

Modifications to chronic feeding studies in two species (rodent and nonrodent; Guideline 83-1; EPA, 1984) may be required to include additional end points for neurotoxicity or immunotoxicity or special visual system toxicity (for organophosphates) if these were not tested in other studies.

Range-finding studies of at least 90 days duration in rats and mice will generally be required to determine dose levels that are adequate to test the carcinogenicity (Guideline 83-2; EPA, 1984) of a pesticide. Studies conducted to satisfy the requirement for Guideline 82-1 (EPA, 1984) will also be acceptable to satisfy this 90-day study requirement.

DEVELOPMENTAL TOXICITY STUDIES

General Description

Developmental toxicity studies are designed to assess the potential of developmental effects in offspring resulting from the mother's exposure to the test substance during pregnancy. These effects include death of the developing organism, structural abnormalities, altered growth, and functional deficiencies. In addition to the classic teratology (now called developmental toxicity) study, a postnatal study is required by the EPA on a case-by-case basis. It is in this study that functional deficiencies are best studied.

The EPA prefers that the rat and the rabbit be used in these studies; however, hamster and mouse are also acceptable. Doses should be administered over the period of major organogenesis (major visceral and skeletal formation) in the fetus. The maternal animals only are dosed in this study and only for specified periods. When day 0 is the day that evidence of mating was observed, the rat and mouse are dosed on days 6 through 15; the rabbit, days 6 through 18; and the hamster, days 6 through 14.

Dosing is usually administered by gavage (oral bolus dose). The pregnant animal should be observed daily for signs of toxicity. Maternal body weight should be monitored at least every 2 to 3 days during gestation. At sacrifice, the maternal animals should be examined for any abnormalities or pathological changes that may have influenced the pregnancy. The uterus is then removed and examined. The number of corpora lutea and live and dead fetuses should be recorded. The sex of the fetuses should be determined. Each fetus is then weighed and examined externally and malformations recorded by litter along with weight and sex. A certain percentage (depending on the animal species used) of the fetuses are then prepared for visceral examination and the remainder for examination of skeletal anomalies. Although the litter is considered the most relevant unit for statistical analysis, data should also be presented and assessed for each fetus.

Historical control data are also useful for determining the biological importance of visceral or skeletal anomalies that are elevated to a statistically significant level by treatment. Again, only historical control data from studies on the same species and strain of animal should be used for comparison purposes.

EPA's Proposed Changes

At least one developmental toxicity (formerly teratogenicity) study (Guideline 83-3; EPA, 1984) would now be required for all nonfood uses. In the past it was required only if there was expected exposure of women of childbearing age. A second study could be required if concerns are raised from the results of the first study. For food use EUPs accompanied by a temporary tolerance request, a second study could also be required, depending on the results of the first study.

A postnatal developmental toxicity study (Guideline 83-6; EPA, 1984) is proposed as a conditional requirement. This study could be required to more fully assess the manifestations of developmental toxicity, especially potential deficits in function or developmental neurotoxicity.

The parameters that need to be studied in a postnatal study depend on the effects seen in the prenatal study. Guidelines are presently being developed by EPA.

REPRODUCTION STUDIES

General Description

Multigeneration reproduction studies are designed to provide information concerning the general effects of a test substance on overall reproductive capability. Such studies may also provide information about the effects of the test substance on neonatal morbidity and mortality and about the meaning of preliminary data for developmental toxicity. EPA requires that the study include a minimum of two generations and that one litter be produced each generation. Dosing of both parents should begin when they are 8 weeks old and continue for 8 weeks prior to mating. Dosing of parental males should continue at least until mating is completed. Dosing of parental females continues through a 3-week mating period and pregnancy and up to the time of weaning 3 weeks after delivery of the pups. Dosing of pups selected for mating to produce the second generation should begin at weaning and continue as discussed above. The dosing and breeding schedule is clarified in the timeline presented in Table 4-6.

Parental animals should be observed daily for signs of toxicity. This is especially important for females during pregnancy in order to detect signs of difficult or prolonged parturition. Weights of parental animals are recorded weekly. The duration of pregnancy should be determined from the time evidence of mating was first observed. Each litter should be examined for the number of dead and live pups and for gross abnormalities. Live pups should be individually weighed on days 0 (optional), 4, 7 (optional), 14, and 21 after birth. A complete gross necropsy should be performed on all parental animals, all pups found dead prior to day 21 (weaning), and all weanlings not selected as parental animals for a next generation. Pups culled on day 4 do not have to undergo gross necropsy. Histopathology is required for reproductive and target organs (those known from previous studies to be adversely affected by the test material) for all control and high-dose parental animals and should be conducted on weanling animals (except for those selected as parental animals in the next generation) as described for parental animals (EPA, 1988).

EPA's Proposed Changes

The addition of a fertility assessment of parental males is recommended by EPA if fertility or reproductive parameters are found to be affected by the test chemical. The parameters to be examined or reported in this assessment include weight of reproductive organs, spermatid count, total cauda epididymal sperm count, assessment of sperm morphology and motility, examination of epididymal fluid for debris and unexpected cell types, and additional histopathology of the testes. A reproduction study (Guideline 83-4; EPA, 1984) could also be required to support nonfood uses if adverse effects on the reproductive system or developmental toxicity are observed in other studies.

TABLE 4-6 Approximate Dosing and Breeding Schedule for a Rat Two-Generation Reproduction Study

Weeks of Study	P_1	F_1	F_2
0	Dosing of P_1 males and females begins		
8-14	P_1 mating period		
11-17	Dosing of P_1 males may end at week 25; P_1 females are killed		
22-23		F_1 mating; dosing of F_1 males may end at week 40; F_1 males killed	
25-37		Remaining F_1 females are killed	F_2 born. Litter sizes randomly adjusted to 8 each F_2 offspring are killed

SOURCE: EPA, 1984.

MUTAGENICITY STUDIES

General Description

A battery of mutagenicity tests is required to assess the potential of each test chemical to affect genetic material. The test selection criteria focus on the test's ability to detect, with appropriate assay methods, the capacity of the chemical to alter genetic material in cells. When mutagenic potential is demonstrated, these findings are considered in the assessment of potential heritable effects in humans, in the weight-of-the-evidence evaluation for carcinogenicity, and in the decision to require submission of a carcinogenicity study if otherwise conditionally required. Mutagenicity results per se are not used by themselves for risk assessment purposes, even when results suggest possible heritable genetic effects in humans.

EPA's Proposed Changes

EPA has already published changes to the 40 CFR Part 158 data requirements for mutagenicity (EPA, 1984).

As described in *Pesticide Assessment Guidelines: Subdivision F* (EPA, 1984), the original mutagenicity test battery consisted of three assays: one for gene mutations, one for structural chromosome aberrations, and one for other genotoxic effects. Other testing included DNA damage and repair. The revised guidelines would require an initial battery of tests consisting of:

- *Salmonella typhimurium* reverse mutation assay;
- mammalian cells in culture forward gene mutation assay allowing detection of point mutations, large deletions, and chromosome rearrangements; and
- in vivo cytogenetics.

Results derived from these assays could trigger the requirement for further mutagenicity testing. The type of additional required testing would depend on the observed results from the initial battery and other toxicity testing results. For example, testing could involve cytogenetic testing in spermatozoa if other test results suggest that they are targets.

GENERAL METABOLISM STUDIES

General Description

Data from studies on the absorption, distribution, bioaccumulation, excretion, and metabolism of a pesticide may also allow more meaningful evaluation of test results and more appropriate risk assessment (as a result of more meaningful extrapolation from data on animals to humans). Such data may also aid in designing more relevant toxicology studies. Information on metabolites formed in laboratory animals is also used to determine whether further toxicity testing is required on plant metabolites. If a major metabolite forms in the plant but not in the test animal, separate toxicity testing on the plant metabolite could be necessary. The extent of testing required depends on the level of concern raised by the initial battery of toxicity tests (acute and subchronic studies, one teratology study, and a battery of mutagenicity tests).

As presently designed, the metabolism study consists of four separate parts: a single low, intravenous dose of radiolabeled test material (not required if the test material is insoluble in water or normal saline solution); a single low, oral dose of radiolabeled test material; 14 consecutive daily low, oral doses of unlabeled test material followed by a single low dose of radiolabeled material; and a single high, oral dose of radiolabeled test material. Selection of the low dose is based on the NOEL. The high dose should elicit some signs of toxicity but not be so high that it results in mortality. The test species of choice is the rat.

Urine, feces, and expired air are collected for 7 days after administration of the radiolabeled material or until >90% of the radioactivity is recovered. Bone, brain, fat, testes, heart, kidney, liver, lung, blood, muscle, spleen, residual carcass, and tissues showing pathology in this or prior tests should be examined for radioactivity for all animals except those given the intravenous dose. This is done to determine if the test material or radiolabeled metabolite accumulates in any particular organ and to relate this information to the findings in toxicity studies.

In addition, quantities of radiolabeled material in feces, urine, and expired air must be monitored for all dose groups at appropriate intervals up to 7 days after dosing. Furthermore, urinary and fecal metabolites must be identified.

EPA's Proposed Changes

A metabolism study would also be required when significant adverse effects are observed in toxicology studies, including reproduction and developmental studies (Guideline 85-1; EPA, 1984). EPA is currently rewriting the guidelines for conducting metabolism studies and is including a tiered approach for study design and conduct.

NEUROTOXICITY STUDIES

General Description

Neurotoxicity studies are required to evaluate the potential of each pesticide to adversely affect the structure or function of the nervous system. The objectives of these studies are to detect and characterize the following:

- effects on the incidence and severity of clinical signs, the alteration of motor activity, and histopathology in the nervous system following acute, subchronic, and chronic exposures;
- the potential of cholinesterase inhibiting pesticides and related substances to cause a specific organophosphate-pesticide-type induced delayed neurotoxicity;
- other neurotoxic effects based on screening studies on certain chemical classes; and
- effects on organisms exposed prior to birth or weaning.

Results from these studies may be used for qualitative and quantitative risk assessment. The guidelines for these studies were published in March 1991 as addendum 10 to the EPA guidelines (EPA, 1991a).

EPA's Proposed Changes

The changes in the requirements for neurotoxicity testing were described above under "Acute Toxicity" and "Subchronic Toxicity."

SPECIAL TESTING

EPA intends to develop better definitions of the conditions under which domestic animal safety (Guideline 85-2; EPA, 1984) testing and visual system studies (Guideline 85-4; EPA, 1984) would be required for all organophosphates and other pesticides shown to affect the visual system. These studies could be of acute, subchronic, or chronic duration, whichever is deemed appropriate for the pesticide under study. Since guidelines have not been formulated for these studies, they will be designed in conjunction with EPA scientists.

CONCLUSIONS AND RECOMMENDATIONS

Conclusions

Current and past studies conducted by registrants are designed primarily to assess pesticide toxicity in sexually mature animals. The protocols for these studies have evolved over several decades and have included some testing paradigms that allow extrapolation to infant and adolescent animals. These studies have produced some valuable information on toxicity and exposure. After reviewing EPA's current and proposed toxicity testing guidelines, however, the committee concluded that current studies do not directly address the following areas:

- toxicity of pesticides in neonates and adolescent animals;
- metabolism of pesticides in neonates and adolescent animals; and
- exposure during early developmental stages (after the second trimester through adolescence) and sequelae in later life.

Recommendations

- Studies should be redesigned and expanded in scope to elucidate the differences in the metabolism and disposition of pesticides in the infant, adolescent, and young adult.

Current metabolism studies are designed to provide information about sex-related differences, metabolic pathways and excretion, bioaccumulation in tissues, and tissue distribution in adult rats. EPA uses the data to determine whether toxicity testing needs to be conducted on individual plant or animal metabolites in addition to the parent compound.

- The metabolism of pesticides in newborn animals needs to be more thoroughly investigated.

Greater knowledge in this area would make it possible to develop computer programs for physiological pharmacokinetic modeling to forecast how information about metabolism in infant animals could be extrapolated to infant humans. The committee realizes that this is a very difficult area of investigation and application. Nevertheless, it urges that such investigations be pursued, since the resulting information could provide more realistic systemic exposure scenarios for risk assessment.

- A study should be conducted to compare the toxicity of several representative classes of pesticides in both adult and immature animals.

Results of such a broad-range study designed to specifically address the infant and young adult animal should indicate whether comparative studies of this nature should routinely be required by EPA. This study should be designed to examine several critical end points in the developing animal, including neural (functional and behavioral), immune, and endocrine systems to cite a few examples. Because the battery of acute toxicity tests now required by EPA is generally performed in adult animals, very little information is available on acute toxicity in immature animals. Such data are important in determining dietary risk to infants and children for acutely toxic pesticides such as organophosphates and carbamates. The committee recognizes that some of these data can be obtained from multigeneration studies if specific observation requirements are added to the current studies.

- Test animals should be exposed to the chemical of concern early in their lives so the risks of exposure of infants and children to the compound can be more adequately assessed.

The committee recognizes the difficulty in dosing animals during lactation and is aware that testing requirements would have to be modified to accomplish these studies. EPA Guideline 83-5 for a chronic toxicity/carcinogenicity study states that exposure of rats to pesticides should begin at approximately 6 to 8 weeks of age, essentially when they are adolescents (EPA, 1984). Because the effects of the pesticide in the rat are not determined early in its lifetime, chronic toxicity/carcinogenicity studies in adolescent animals may not be representative of the responses of younger animals. Current reproduction studies (Guideline 83-4; EPA, 1984) partially address this period in the life of a rat, but the effects of early exposure are not addressed past 21 days of age for second-generation pups or past the death of the second-generation parents (first-generation pups used for mating to produce the second generation). The protocol does not indicate whether exposure early in life has any impact on the adults or whether continuous exposure from birth to young adulthood influences the severity of the toxicity over a lifetime. FDA has used the multigeneration studies to include the F_2A or F_3A generation of laboratory animals for direct and indirect food additives (Becci et al., 1982).

• To obtain lifelong data on rodents for a given pesticide, the committee recommends that the testing guideline for a rat chronic toxicity/carcinogenicity study be modified to include in utero exposure during the last trimester, exposure through the mother's milk, and after weaning, oral exposure through diet.

This would mean that weanlings from the F_1A or F_2A generation would be selected from each dose group and tested throughout their lifetimes (see Table 4-6). In addition to this group, another smaller group of rats from the F_1 generation would be killed at 6 months and 1 year and necropsied to examine the same parameters normally measured at the end of a lifetime feeding study.

The NTP tested three chemicals using a similar protocol and their standard protocol for an carcinogenicity study. One of the chemicals was ethylenethiourea, which is a breakdown product and metabolite of the ethylenebisdithiocarbamate fungicides and a thyroid toxicant (decreases T3 and T4). In utero exposure did not affect the occurrence of liver tumors in male and female mice, but did result in a sex-dependent increase in the number of malignant thyroid tumors in mice and rats (NTP, 1992).

• Measurement of the serum thyroid hormones T3 and T4 and serum TSH should be routinely added to the EPA chronic/carcinogenicity study protocol or to the subchronic toxicity protocol for the rat so that adverse effects on thyroid function can be determined earlier.

When examining the parameters currently measured in the EPA chronic/carcinogenicity study, the committee found that endocrine function was adequately covered for all but the thyroid. Although the thyroid is saved in these studies for microscopic examination and its weight is recorded, the committee believes that changes in the functioning capabilities of this organ can occur regardless of whether there are organ weight or histopathologic changes.

• If abnormalities are found during histopathologic examination of the spleen, lymph nodes, thymus, and bone marrow, more detailed and specific studies should be conducted on a case-by-case basis relevant to the types of effects initially seen in immune system tests.

EPA has developed protocols for immunotoxicity testing for some pesticides that affect the immune system, and the agency is considering developing a generic testing protocol. The committee believes that because the human immune system is one of the most robust of systems in terms of resistance to pesticides or other chemical toxicity, initial evaluation using current histopathologic examination of spleen, lymph nodes, thymus, and bone marrow should be sufficient unless abnormalities are noted.

• A modified reproductive/developmental toxicity study in the rat is suggested for registration of all food-use pesticides.

One set of dams in this study would be dosed continuously with the test material from day 6 of gestation through birth of the pups and until weaning of their offspring at 21 days of age. A developmental assessment would be performed on the pups as described in EPA's recently published developmental neurotoxicity testing guidelines (EPA, 1991a). In addition, a set of pups from each dam would undergo gross and histopathologic examination at day 60 post partum. The second set of dams would be dosed from day 6 of gestation to term; however, these animals would not be allowed to deliver but, rather, would be subjected to cesarean section as in a routine teratology study. The fetuses would be subjected to skeletal and visceral examination, as described for a teratology study (Guideline 83-3) designed to examine the prenatal development of pups. This study design allows a determination of the reversibility of postnatal significance of findings seen in fetuses at the time of cesarean section. EPA has indicated in its proposed changes to Part 158 that a similar study be required; however, the committee recommends that this study be made a requirement for registration of all food-use pesticides.

• Because neurotoxicity is such an important consideration for the newborn, EPA should continue to revise its published guidelines on developmental and functional neurotoxicity testing as new information emerges from the actual conduct of preregistration studies and from ongoing research in rodent neurotoxicity.

The committee supports EPA's proposed requirement for acute and subchronic neurotoxicity testing for pesticides and encourages the agency to make this a general requirement for all food-use pesticides—not just for organophosphate and carbamate pesticides. New approaches to neurotoxicity testing are described in the report *Environmental Neurotoxicology* (NRC, 1992).

• EPA should develop a general guideline for visual system toxicity testing that can be modified and applied on a case-by-case basis.

The eye is exquisitely sensitive to changes in glucose metabolism, blood flow, and neuronal function, and several pesticides have been shown to be visual system toxicants (hexachlorophene, naphthalene, 2,4-DNP, and some organophosphates). In the past, scientists have examined the effects of chemicals that may irritate the eye by accidental contact. More recently, however, researchers have been examining the effects of chemicals on specific sections of the visual system, such as the optic nerve, iris, retina, and lens. The guideline proposed by the committee should be applied to species in which this type of testing appears to be appropriate, e.g., EPA has recently considered protocols for visual system testing in dogs.

Recent studies indicate that visual system damage may be associated with dietary exposure to some cholinesterase inhibiting compounds. Thus the committee supports EPA's proposed testing (the sensory evoked potential test) of such pesticides for visual system toxicity. However, it does not believe that a single protocol would suffice to cover all classes of compounds because different classes would affect different parts of the visual system.

REFERENCES

Armitage, P., and R. Doll. 1954. The age distribution of cancer and multi-stage theory of carcinogenesis. Br. J. Cancer 8:1-12.

Becci, P.J., K.A. Voss, F.G. Hess, M.A. Gallo, R.A. Parent, K.R. Stevens, and J.M. Taylor. 1982. Long-term carcinogenicity and toxicity study of zearalenone in the rat. J. Appl. Toxicol. 2(5):247-254.

EPA (U.S. Environmental Protection Agency). 1984. Pesticide Assessment Guidelines, Subdivision F: Hazard Evaluation—Human and Domestic Animals. Revised Ed. November 1984. PB-86-108958. Washington, D.C.: U.S. Environmental Protection Agency.

EPA (U.S. Environmental Protection Agency). 1988. FIFRA Accelerated Reregistration Phase 3 Technical Guidance. US EPA 540/09-078. Washington, D.C.: U.S. Environmental Protection Agency.

EPA (U.S. Environmental Protection Agency). 1991a. Pesticide Assessment Guidelines, Subdivision F: Hazard Evaluation—Human and Domestic Animals, Carcinogenicity of Ethylene Thiourea [CAS No. 96-45-7] in F/344 Rats and B6C3F1 mice. Addendum 10—Neurotoxicity. Washington, D.C.: U.S. Environmental Protection Agency.

EPA (U.S. Environmental Protection Agency). 1991b. Pesticide Assessment Guidelines, Subdivision F: Hazard Evaluation—Human and Domestic Animals Series 84. Addendum 9—Mutagenicity. PB-91-158394. 540/09-91-122. Washington, D.C.: U.S. Environmental Protection Agency.

Gad, S.C., and C.S. Weil. 1982. Statistics for toxicologists. Pp. 273-320 in Principles and Methods of Toxicology, A.W. Hayes, ed. New York: Raven Press.

IPCS (International Programme on Chemical Safety). 1990. In Environmental Health Criteria 104: Principles for the Toxicological Assessment of Pesticide Residues in Food. Geneva, Switzerland: World Health Organization.

Loomis, T.A. 1978. Essentials of Toxicology. Philadelphia, Pa.: Lea & Febiger.

NRC (National Research Council). 1992. Environmental Neurotoxicology. Washington, D.C.: National Academy Press.

NRC (National Research Council). 1993. Issues in Risk Assessment. Washington, D.C.: National Academy Press.

NTP (National Toxicology Program). 1992. NTP Technical Report on the Perinatal Toxicology and Carcinogenesis Studies of Ethylene Thiourea (CAS No. 96-45-7) in F3441N Rats and B6C3F$_1$ Mice (Feed Studies). NTP TR 388. Research Triangle Park, N.C.: National Toxicology Program.

Weil, C.S. 1972. Guidelines for experiments to predict the degree of safety of a material for man. Toxicol. Appl. Pharmacol. 21:194-199.

5

Food and Water Consumption

D IETARY EXPOSURE OF CHILDREN to pesticides can be estimated by combining data on levels of pesticide residues in foods with data on food consumption patterns for infants and children. Risks to health can be assessed by combining estimates of dietary exposure with information on the toxic potential of pesticides. Various data bases are available for use in these calculations. In this chapter, the committee reviews the dietary surveys conducted to assess patterns of food consumption by the U.S. population, including infants and children. It examines the methods used in these surveys and evaluates their relative strengths and weaknesses. The committee then describes the age-related differences in food consumption patterns demonstrated by survey data and discusses food consumption estimates. The committee also notes various factors and limitations that must be considered in determining the basis for estimating exposure to pesticide residues in food and assessing the risk to infants and children.

FOOD CONSUMPTION SURVEYS

Small-scale studies on food intake and nutrition were first conducted toward the end of the nineteenth century, when processing techniques were leading to rapid changes in the food supply. The variety of foods available to consumers again increased when in-home refrigerators and freezers became generally available, more sophisticated preservation techniques were introduced, and manufactured foods found their way into the retail market. By 1960 approximately 60% of the food items on supermarket shelves had come into existence during the preceding 15 years, that is, since the end of World War II (Hampe and Wittenberg, 1964).

It therefore became of great interest to researchers to determine what the people of this country were actually eating. In 1909 the U.S. Department of Agriculture (USDA) began to identify changes in the foods available to the civilian public by determining the disappearance of foods into the wholesale and retail markets. This is still done annually by subtracting data on exports, year-end inventories, nonfood uses, and military procurement from data on total production, imports, and beginning-of-the-year inventories. Overestimates result from this method, however, because losses that occur during processing, marketing, and home use are not taken into account. Thus, the resultant information is sometimes called *availability* or *use* of foods or nutrients (Stamler, 1979).

The USDA estimates national per capita use of foods or food groups by dividing the total available food by the total U.S. population. These data provide information on overall trends in available foods, but they do not indicate how use varies among population subgroups or individuals.

Since 1935 the USDA's Human Nutrition Information Service (HNIS) has conducted a series of Nationwide Food Consumption Surveys (NFCS). The first four (1935, 1942, 1948 [urban only], and 1955) surveyed household food use over a 7-day period. No record was made of waste or difference in use among household members. Surveys conducted in 1965-1966, 1977-1978, and 1987-1988 included information on the kinds and amounts of foods eaten by individuals in the household in addition to household food use. For the reasons given later in this chapter, the 1977-1978 survey served as the major source of consumption data used by the committee in the present study.

The USDA has conducted a planned series of surveys since 1985 solely concerned with individual food intake (USDA, 1985, 1986a,b, 1987a,b,c, 1988). The results of the most recent surveys have not yet been published. In these surveys, called Continuing Surveys of Food Intakes of Individuals (CSFII), data are collected on three separate samples: women 19 to 50 years old and their children 1 to 5 years old (the core group); a sample of low-income women and their children; and in 1985 only, men 19 to 50 years old. In the 1989, 1990, and 1991 surveys, data were collected on all individuals.

The National Center for Health Statistics (NCHS), a division of the Department of Health and Human Services, has conducted the National Health and Nutrition Examination Surveys (NHANES) since 1971. The purpose of these surveys is to monitor the overall nutritional status of the population of the United States through comprehensive health and medical histories, dietary interviews, physical examinations, and laboratory measurements. The committee opted not to use the results from these surveys, however, because the number of observations within the age and demographic categories of interest were inadequate for the purposes of the present study.

The only national study of average intakes of pesticides, toxic substances, radionuclides, and industrial chemicals is the Total Diet Study conducted by the Food and Drug Administration (FDA). Four times a year, foods considered to be representative of the average U.S. diet are purchased from grocery stores across the United States and individually analyzed in FDA laboratories for the constituents mentioned above. Until April 1982 the food items used in the Total Diet Study were based on data from the 1965-1966 NFCS. Since then, they have been based on data from the 1977-1978 NFCS and the second NHANES, which was conducted during 1976-1980. (The Total Diet Study is discussed further in Chapter 6, which focuses on pesticide residues in food.)

Infants, defined as less than 1 year of age, represent a separate and extremely critical population group with respect to purposes of this study. The number and classification of foods consumed by infants are almost exclusively processed and manufactured by a limited number of companies. Dietary intake studies have been conducted by Gerber Products Company and Ross Laboratories to evaluate intake for these populations, and those dietary studies were used by the committee.

SURVEY METHODOLOGY

Three basic types of methods are used to gather data in food consumption surveys (Burk and Pao, 1976; Dwyer, 1988): retrospective, prospective, and a combination of retrospective and prospective. *Retrospective* methods include the *24-hour (or 1-day) food recall* and the *food frequency questionnaire*. *Prospective* methods require the use of food records or diaries—the *weighed food record* and the *estimated food record* (sometimes called the *household measure food record*). In combination surveys, investigators use both *prospective* and *retrospective* methods, e.g., the recall and weighed food record. (For a comprehensive overview, see Dwyer, 1988.)

Retrospective Methods

The 24-Hour (or 1-Day) Recall Method

In surveys conducted with this method, interviewers ask subjects to recall the quantities of particular foods and beverages consumed on the previous day or during the preceding 24 hours. More precise estimates of portions are obtained when the respondents are provided with measurement guides such as food models, abstract shapes, and tableware (DHHS, 1983); measuring cups, spoons, and rulers (USDA, 1987); and pictures (Frank et al., 1977; Posner et al., 1982).

Madden et al. (1976) found no significant differences between mean intake data obtained from the 24-hour recall and actual observed intake over that period. They noted, however, that the data were not representative of an individual intake averaged over different days. Recalls of food consumption over a period longer than 24 hours produce more representative data on general intake patterns, but are less precise than the 24-hour recall because they depend to a greater extent on memory (Block et al., 1986).

In general, the validity of the recall method and the extent of bias depend on a variety of factors related to collection methods (Backstrom and Hursh-Cesar, 1981). In face-to-face interviews it is easier to elicit a response. Such interviews are more efficient when there are several family members or a long questionnaire; however, they are expensive primarily because of travel costs. In contrast, telephone interviews are relatively inexpensive, quick, easy to monitor, unlimited geographically, and can ensure a greater degree of anonymity (Wilson and Rothschild, 1987). Telephones may be used in combination with, or as a follow up to, other interview techniques. They do have several disadvantages, however: subjects can easily terminate an interview prematurely, long or complicated questionnaires are difficult to administer, and many low-income families lack telephones. In all cases, selection and training of the interviewers are critical to the success of the survey.

The single, short (15- to 30-minute) guided interview produces higher response rates than all other methods. Moreover, accuracy is enhanced in several ways: the recall period is short and precise, measurement aids may be used to increase the accuracy of consumption estimates, and the quantities reported can be easily converted to nutrient equivalents.

Mail interviews are inexpensive, and they may be used in combination with other interview methods (Posner et al., 1982). Furthermore, interviewer bias is avoided. Negative aspects include low response rates, exclusion of illiterate persons, and lack of control over who responds and how they respond.

Food Frequency Questionnaires

In its simplest form, the food frequency questionnaire consists of a checklist of foods or food groups and a set of categories indicating daily, weekly, or monthly frequency of food consumption during a specified period—weeks, months, or a year. The checklists may contain as few as 20 items or more than 100. The questionnaire can be administered in person, over the telephone, or by mail, with the attendant advantages and disadvantages described above (Willett et al., 1985). Subjects may use a visual aid to estimate portion sizes (Chu et al., 1984). When the quantities have been estimated, the nutrient content of the foods consumed may in turn be estimated.

The food frequency questionnaire is a quick, inexpensive, and simple method for obtaining information on food intake from large numbers of subjects (Sampson, 1985), and its administration does not require highly trained personnel. Epidemiologists have found the questionnaire useful in studying the relationship between diet and disease risk. Attempts have been made to validate the accuracy of consumption frequency estimates by comparing them to food records (Chu et al., 1984; Willett et al., 1985; Freudenheim et al., 1987).

The limitations of this method outweigh its strengths. Its accuracy may be compromised by the long recall period, estimates of past intake may be biased by current intake, and respondent burden is heavy if the checklist is long and complex. Moreover, the questionnaire may not be appropriate for people who consume unusual diets or for children. In the future, a combination of methods may be used to overcome these limitations.

Prospective Methods

Food Records or Diaries

These usually self-administered reports of current food intake can cover periods ranging from 1 day to as long as 1 year (Basiotis et al., 1987). Subjects or their surrogates, e.g., parents, record the portions of all foods and beverages ingested immediately after each eating occasion.

The foods are either weighed (the *weighed food record*) (Acheson et al., 1980; Anderson and Blendis, 1981; Marr and Heady, 1986) or measured with a cup, ruler, or other aid (the *estimated food* or *household measure record*) (McGee et al., 1982; Acosta et al., 1983; Elahi et al., 1983). Foods eaten away from home must be estimated. The latter method is used most often, but the weighed food record is widely regarded as the most accurate procedure and is often used to validate other methods. Both methods require trained personnel to demonstrate proper weighing, or measuring, and recording techniques. Written instructions are also provided (Sempos et al., 1984).

Reliance on memory is minimal, the recall period is precise, interviewer bias is avoided, omissions of foods and beverages consumed tend to be minimal, and the accuracy of portion estimates is enhanced by the weighing and measuring techniques. However, respondents must be literate and willing to accept the heavy burden of participation—factors that can bias the sample. In long-term surveys, boredom or fatigue may lead to a decline in accurate reporting.

Combined Retrospective and Prospective Methods

A combination of recall and record methods is sometimes used to obtain multiple-day intake data for individuals in a large survey or study

(Schnakenberg et al., 1981; USDA, 1983). Because the limitations of one method offset the limitations of the other, a greater accuracy of mean intake estimates can be expected. Examples of combinations include 1-day recall and 2-day record (Patterson, 1971; USDA, 1983), 3-day recall and 4-day record (Futrell et al., 1971), and 1-day recall and 14- to 17-day record (Schnakenberg et al., 1981). Diet histories may also involve a combination of methods (Dwyer, 1988).

Methods Used in USDA Surveys

Although the USDA has sought similar information in its various dietary intake surveys, it has changed the methods used in an attempt to obtain better measures of average intake. The agency has advanced from a 1-day recall only in spring (1965-1966 Household Food Consumption Survey) to a 3-day combination (1-day recall and 2-day record) during all four seasons in the 1977-1978 and 1987-1988 NFCS. The 3 days of intake data for more than 30,000 people, obtained by using this combination method, produced a better measure of an individual's average intake than did the 1-day measure used in 1965.

In the 1985 and 1986 CSFII, 6 nonconsecutive days of intake data were collected by trained interviewers who administered the 1-day recall method every 2 months over the course of the year. The first interview was conducted in person; the remaining five were accomplished over the telephone. Mothers provided the recall information for their children. This system proved costly, and the drop-out rates over the course of the year were high (Table 5-1). The substantial decrease in participation (approximately 50%) can be attributed to either the 145 respondents (10% of the sample) who moved to another geographical region or to such socioeconomic characteristics as being younger, having a low income, being in poor health, being on a special diet, having one or more children, being a suburban dweller, or working.

As a result of the drawbacks in the 1985 and 1986 CSFII, the combination of 1-day recall and 2-day record used in the 1977-1978 NFCS was reinstituted in the 1987-1988 NFCS and the 1989 and subsequent CSFII to obtain 3 consecutive days of dietary intake data. In both the NFCS and later CSFII, dietary information was collected on all individuals—not just the sex-age groups surveyed in the 1985 and 1986 CSFII.

SURVEY DESIGN

To determine the validity of a survey sample, it is necessary to consider a variety of factors such as the survey design, sample weighting, and comparison of resulting data to standards. The committee began by exam-

TABLE 5-1   Unweighted Counts of Individuals for the 1985 and 1986 CSFII

Year	Income Category	Subjects	Number of Subjects in Survey[a]	
			First Day	All 6 Days
1985	All incomes	Children	489	161
		Women	1,459	692
		Men	658	NA[b]
1985	Low incomes	Children	1,190(714)[c]	221
		Women	2,081(1,322)[c]	547
1986	All incomes	Children	509	219
		Women	1,451	751
1986	Low incomes	Children	762	307
		Women	1,320	595

[a] The numbers in these columns are not additive since they represent different study groups.

[b] Men were sampled for only 1 day.

[c] More responses than expected were received for day 1, and they included a large number of low-income households. To reduce the number of low-income households to the targeted number of 1,200 for interviewing on days 2-5, systematic subsamples were drawn for both women and children. The numbers in parentheses refer to those subsamples.

SOURCE: USDA, 1988.

ining the design of the CSFII and the NFCS. The 1985 and 1986 CSFII and the 1977-1978 and 1987-1988 NFCS were designed to provide a multistage, stratified, probability sample that was representative of the 48 conterminous states.

A *multistage* sample is drawn by selecting random groups in stages. At each stage, groups of individuals are selected from increasingly smaller segments of the population. The term *stratified* indicates that the population is divided into mutually exclusive subsets, or strata, before the sample is drawn. Taken together, these subsets represent the total population that is being examined. However, the sampling plan is applied separately within each stratum. Because the strata are defined by geographic location, the sampling within each stratum is called an *area* sample.

In a *probability* sample, every group has a known probability of selection. Thus, every element in the population must be enumerated before the sample is drawn to facilitate determination of the likelihood of selecting any group of individuals into the sample. These probabilities may or may not be equal for all groups. The use of uniform criteria for each group helps minimize the extent of enumeration required to determine selection probabilities. Each group sampled, therefore, has a known probability of selection. Although this method is fairly complex, it provides data that

TABLE 5-2 Distribution of Strata within the Conterminous United States as Defined by the Bureau of Census Geographic Divisions

Region	Census Division	Number of Strata Central City	Suburban	Nonmetropolitan
Northeast	New England	1	1	1
	Middle Atlantic	3	5	1
Midwest	East North Central	3	6	2
	West North Central	1	1	2
South	South Atlantic	2	5	3
	East South Central	1	1	2
	West South Central	2	3	2
West	Mountain	1	1	1
	Pacific	3	5	1
Total		17	28	15

SOURCE: Adapted from the U.S. Department of Agriculture, 1985.

are statistically projectable, with known sampling error, to the entire conterminous United States. Nonprobability samples, where groups are not enumerated, are generally easier to obtain but do not provide data that are statistically projectable to the general population.

The 48 states were grouped into the nine census geographic divisions, which in turn were divided into three classifications: central city, suburban, and nonmetropolitan (Table 5-2). From these 27 superstrata, 60 strata (17 central city, 28 suburban, and 15 metropolitan) were obtained. Both the CSFII and the NFCS surveys consisted of four stages. During the first stage, the probability proportion to size (PPS) technique was used to select two primary sampling units (PSUs) from each of the 60 strata in both the CSFII and NFCS. These 120 PSUs included counties, cities, or parts of cities, and were relatively homogenous with regard to demographic characteristics.

In designing food surveys, care must be taken to ensure that the sample of consumers is representative of the general population. Even a scientifically designed probability sample may not be representative due to nonresponse and other practical problems. This could lead to bias in the estimates of mean intake, especially if there is a systematic component to the nonresponse. An accepted method for adjusting the estimates of intake would be to weight the data in such a way that they more closely reflect the general population. Fuller (1991) discusses using regression estimation for adjusting intake when appropriate auxiliary information exists. Fuller et al. (1991) adjusted intake for sociodemographic factors for the 1987-1988 NFCS and found a significant difference between the resulting mean intake as compared with mean intake calculated without such weighting.

Nusser et al. (1991) considered characterizing usual daily intake distributions as opposed to mean daily intake. Usual intake is useful in providing information about nutritional deficiencies occurring over a long period. They show that use of the distribution of mean intake as an estimate of the distribution of usual intake can lead to erroneous inferences regarding nutritional status. In particular, the variance of mean intakes contains intraindividual variability and is thus greater than the variance of usual intakes. Other parameters of the two distributions may differ as well.

SAMPLE WEIGHTS

Although the CSFII and NFCS samples were designed to be self-weighting, adjustments to the samples were required because not all eligible households participated, not all eligible individuals in eligible households were interviewed, and not all interviews yielded complete dietary information. Weighting factors were developed for each individual participating in the survey and were applied to data from completed intake records to adjust for these sources of nonresponse. Other weighting considerations included economic homogeneity, geographic heterogeneity, and age.

SAMPLE SIZE

Sample size is an important determinant of sample variation and statistical precision. Decisions regarding sample sizes therefore depend on the level of precision desired for the data needed to estimate a population parameter of interest.

The size of the sample required to achieve an established goal of precision for an estimate can be determined through calculations that depend on the coefficient of variation (CV) and an assumed probability distribution for the data. The coefficient is the standard error of an estimate expressed as a fraction of the sample mean. As the CV of an estimate increases, its accuracy decreases. In general, NFCS and CSFII intake estimates with CVs exceeding 50% are not published. Estimates with CVs of 15% to 50% are published by USDA with caveats regarding their accuracy.

COMPARISONS OF INTAKE DATA WITH STANDARDS

Intake data are often compared with a widely accepted value such as the Recommended Dietary Allowances (RDAs). The RDAs are intended to meet "the known nutrient needs of practically all healthy persons" in the United States (NRC, 1989); however, since nutrient needs vary among people, margins of safety are built into the RDAs for many nutrients. For this reason, analysts have selected a fixed cutoff point, such as two-thirds or three-fourths of the RDA, as a point of comparison. One committee convened by the Food and Nutrition Board (FNB) of the National Research Council suggested that distributions of requirements or tolerances would be more appropriate for this purpose (NRC, 1986). This FNB committee recommended that a probability approach be used to estimate the prevalence of inadequate intake, i.e., the probability that a specific intake is inadequate to meet an individual's requirement (NRC, 1986).

VALIDATION OF FOOD CONSUMPTION DATA

Food consumption data must be validated before they are used to calculate risk, in part because the integrity of formulated and combined foods is compromised when those foods are broken down into their constituents. The method of calculation used by the EPA to estimate risk converts consumption data directly into food components, thereby eliminating the need to follow all the separate validation steps. (See "Quantification of Consumption Data," below, for a further discussion of the EPA method.)

Substantial uncertainty is inherent in food consumption data because of a variety of factors. Principal among these are recording errors and biases. It is therefore difficult to extrapolate results to the general U.S. population.

Of the different protocols that have been used in dietary surveys, none is uniformly better than all others. Even the most extensive dietary sampling schemes may be subject to biases or practical limitations. Because of this, a variety of survey methods have evolved to address different objectives, each with its own strengths and weaknesses.

Two primary objectives of food consumption surveys are to assess the mean intake of a group of individuals or the mean intake of a particular individual. Because food consumption varies markedly both within individuals and between individuals (Beaton et al., 1979; Beaton et al., 1983; Todd et al., 1983), methods that are appropriate for one objective may not be appropriate for the other.

The validity of the results relative to the study objectives is of central importance (Block, 1982). Validation of data on food consumption is a difficult task. Ideally, all types and quantities of food consumed by the survey respondents would be recorded in a complete and accurate fashion; this record of actual consumption could then be used as a reference value against which to compare estimates of consumption based on different survey protocols. Kim et al. (1984) studied a method whereby individuals are asked to set aside a duplicate portion of the food they ate for future analysis. The investigators found that the process of setting aside the extra food actually led to smaller quantities consumed than in the usual pattern of intake. To avoid biases associated with self-reported food consumption data, an inconspicuous observer could maintain a diary of foods consumed—a technique used by Madden et al. (1976), Gersovitz et al. (1978), and Krantzler et al. (1982). However, these studies were carried out in controlled environments such as dormitories, dining halls, or other congregate meal sites, thus making it difficult to extrapolate conclusions to the general population.

Because of the difficulties in measuring actual intake, validation to a large extent tended to focus on the comparison of results obtained from different survey methods (Block, 1982). For example, relatively new survey methods such as questionnaires have been compared with established methods such as multiple-day records (Jain et al., 1980; Axelson and Csernus, 1983; Willett et al., 1985; Byers et al., 1987; Krall and Dwyer, 1987; Pietinen et al., 1988). Methods are also sometimes compared to determine which one will produce the most reliable estimates of intake of specific nutrients such as vitamin A (Young et al., 1952; Russell-Briefel et al., 1985; Sorenson et al., 1985).

THE STRENGTHS AND WEAKNESSES OF THE FOOD CONSUMPTION DATA BASES IN ESTIMATING PESTICIDE EXPOSURE OF CHILDREN

The 1977-1978 NFCS data have the following major limitations.

- Over the years since that survey was conducted, average overall dietary patterns may have changed in response to advanced food technology, advertising, taste, and health consciousness, among other variables.
- The sample of nursing infants was small ($n = 106$).
- The 3-day survey period reflects too brief a consumption period, even though it was conducted over all four seasons.
- Water consumption was not considered.

Despite these substantive limitations, the NFCS has provided the only comprehensive data currently available for comparisons of food consumption by all age classes in our population, which is the primary reason the committee chose the 1977-1978 NFCS as the basis for this report. Because the EPA also relies on the same food intake data, the committee's findings can easily be compared with the risks estimated by that agency. The committee compensated for the limited sample size for nursing infants by reviewing survey data reported by Purvis and Bartholmey (1988).

Appraisal of these results validated the similarity of results, even though methodology and sample selection were different.

The 1987-1988 NFCS data were not used by the committee because of the low response rate (34%) and resulting procedural problems. Furthermore, the committee found that the sampling units were so small that in some cases they were not representative of certain age or geographical sectors. At a later date, the U.S. Government Accounting Office (GAO, 1991) conducted an independent evaluation of the 1987-1988 NFCS data and came to similar conclusions. Thus, the data obtained in this survey were not adequate for use by the committee in fulfilling its mandate.

The CSFII provides a more recent view of consumption patterns; however, these surveys experienced high drop-out rates, as described above, and are limited to children between the ages of 1 and 5 years (USDA, 1985, 1986a). They provide no data on infants less than 1 year old or on children between the ages of 6 and 18 years.

To gain a clearer view of the consumption patterns of infants, the committee examined the results of an Infant Nutrition Survey conducted by the Gerber Products Company in 1986. This was one in a series of surveys conducted by Gerber since 1969 to monitor infant feeding practices, nutrient intake, and nutrition contribution and changes (Purvis, 1973; Johnson et al., 1981; Purvis and Bartholmey, 1988).

With the exception of minor refinements, the same method for collection and compilation of data has been used in each of the Gerber surveys to facilitate comparisons over time (Purvis, 1973; Johnson et al., 1981). Initial contact was made through mail questionnaire. The consumption data were collected in a 4-day diary maintained by a parent who had been given instructions for recording intake information and obtaining additional assistance. After the diaries had been completed and returned, interviewers used the telephone to clarify information when needed. The data were analyzed to determine usual nutrient intake, portion size, food preferences, and age at which supplemental foods were introduced.

The 1986 Gerber sample initially consisted of 1,000 infants between the ages of 2 and 12 months. Balance of age and geographic distribution was achieved to the extent possible through random sample generation (Table 5-3). There were 637 satisfactorily completed diaries—a return rate of 64% (G. Purvis, Gerber Products, personal commun., November 14, 1990).

The results of the Gerber survey were found to reflect the same consumption patterns for infants and children as did the USDA surveys (Figure 5-1). Furthermore, a comparison of data from the CSFII and the NFCS showed that some of the foods that dominate the diets of children have changed little in the decade between the 1977-1978 NFCS and the 1985 and 1986 CSFII. With this additional confidence in the larger NFCS

TABLE 5-3 Sample Selected in the 1986 Gerber Infant Nutrition Survey

Region	Age, Months											Totals
	2	3	4	5	6	7	8	9	10	11	12	
Northeast	16	17	17	16	17	17	17	16	17	17	17	184
North Central	22	22	22	22	22	22	22	22	22	22	23	243
South	32	31	31	32	31	32	31	32	31	32	31	346
West	21	21	20	21	21	20	21	21	21	20	20	227
Totals	91	91	90	91	91	91	91	91	91	91	91	1,000

SOURCE: Based on data from Gerber Products Company, personal communication, 1992.

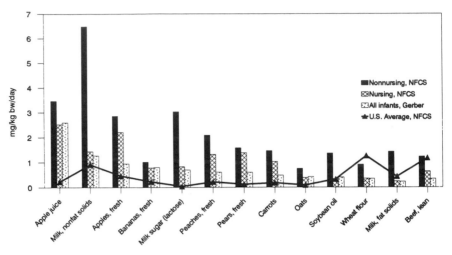

FIGURE 5-1 A comparison of infant intake data (on raw agricultural commodities) from Gerber Products, 1988, unpublished, and from USDA, 1983.

data base, the committee decided to use the results of the 1977-1978 NFCS as the basis for its consumption estimates. The 1985 and 1986 CSFIIs were used to account more fully for intraindividual variation and differences among ages for children less than 5 years of age.

WATER INTAKE

The National Cancer Institute (NCI) conducted an extensive analysis of the 1977-1978 NFCS data to develop more accurate estimates of water intake (Ershow and Cantor, 1989). Three types of water were considered:

- water intrinsic to food,
- tapwater added to food during preparation in the home, and
- tapwater consumed by itself.

Since total food moisture was not part of the NFCS data base, the NCI investigators calculated the intake of water from this source by applying the following formula:

water content (g/100 g food) =

$$100 - \frac{\text{total carbohydrate (g) + total protein (g) + total fat (g) + total ash (g) + total ethanol (g)}}{100 \text{ g food}}$$

Data on tapwater intake were collected as part of the NFCS survey. When the amount of water added to foods was not clear in the USDA Food Code Description Files, the NCI investigators used standard dilutions (e.g., for canned soups, frozen juices) or consulted cookbooks. The investigators acknowledged that their estimates of total water intake by nursing infants were low. In the absence of data on human milk consumption, they derived their estimates from other water sources. Tapwater intake estimates may also be low because ready-to-feed infant formula was assumed when the survey results did not specify the type of formula consumed. In fact, some of the infant formula not clearly identified may have been powder or concentrate to which tapwater had been added.

Dietary Sources of Water

The dietary sources of total water and tapwater reported by NCI are shown in Figure 5-2 (Ershow and Cantor, 1989). Relatively few beverage and food items contributed to total water and tapwater intake for most age groups. For infants less than 1 year old, formula provided 32.7% of total water intake, milk and milk drinks contributed 24.7%, and drinking water, 16.1%. For the 1- to 10-year age group, infant formula was no longer a factor, drinking water increased to 30.3%, and milk and milk products remained at levels similar to those for infant intake (25%). Of the tapwater intake by infants less than 1 year old, 69% was provided by drinking water and 11.9% by formula. For the 1- to 10-year age group, 64.8% was provided by drinking water, and 13.6% by fruit juices, tomato juice, and noncarbonated drinks—up from 4.5% for the infants.

Water Intake Estimates

As shown in Table 5-4, mean total water intake by infants during the first 6 months of life is 1,014 ± 294 g/day. Intake increases to 1,258 ± 322 g/day before their first birthday. These estimates correspond to 189 ± 73.5 and 141.7 ± 43.0 g/kg body weight (bw)/day, respectively. Mean tapwater intakes for these age groups are 272 ± 247 g/day (52.4 ± 53.2 g/kg bw/day) and 328 ± 265 g/day (36.2 ± 29.2 g/kw bw/day).

There is a steady increase in both total water and tapwater intake into adulthood and a gradual decrease after the age of 65 years. When viewed on a grams-per-kilogram-of-body-weight basis, however, the highest intakes are found for infants during the first 6 months of life (Figures 5-3 and 5-4). Daily total water intake decreases from 189 ± 73.5 g/kg bw during the first 6 months to 41.9 ± 15.6 g/kw bw for the 11- to 14-year age group (Table 5-4).

Males generally had higher mean intakes than females ($p \leq 0.05$) for all age groups over 1 year of age after adjusting for race, region, season, body weight, urban residence, and age. As shown in Figures 5-5 and 5-6, these differences were relatively small for young children, and consump-

Total Water

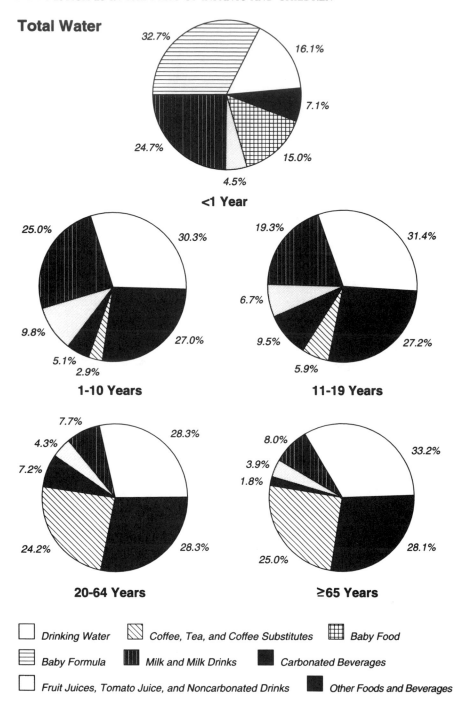

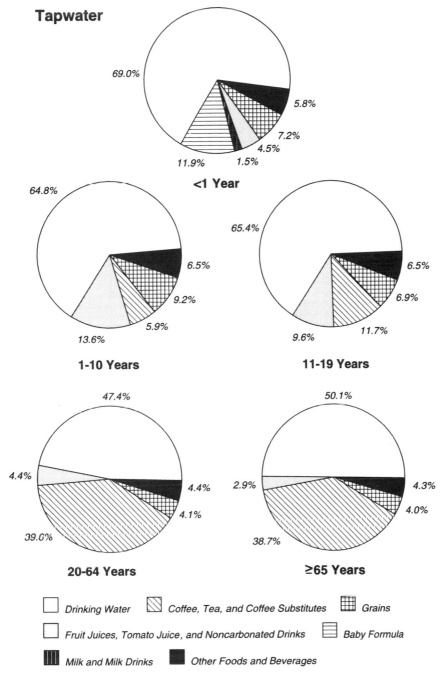

FIGURE 5-2   Dietary sources of total water and tapwater by age, expressed as percentages. SOURCE: Ershow and Cantor, 1989, pp. 28-29.

TABLE 5-4 Mean Intake of Total Water and Tap Water by All Ages, Both Sexes, All Regions, and All Seasons

Age group, year	Total Water		Tap Water	
	g/day ± SD	g/kg bw/day ± SD	g/day ± SD	g/kg bw/day ± SD
<0.5	1,014 ± 294	189.0 ± 73.5	272 ± 247	52.4 ± 53.2
0.5-0.9	1,258 ± 322	141.7 ± 43.0	328 ± 265	36.2 ± 29.2
1-3	1,356 ± 450	99.7 ± 35.7	646 ± 390	46.8 ± 28.1
4-6	1,520 ± 482	77.7 ± 27.6	742 ± 406	37.9 ± 21.8
7-10	1,711 ± 508	58.7 ± 20.9	787 ± 417	26.9 ± 15.3
11-14	1,918 ± 651	41.9 ± 15.6	925 ± 521	20.2 ± 11.6
15-19	2,049 ± 768	33.8 ± 12.3	999 ± 593	16.4 ± 9.6
20-44	2,171 ± 839	32.1 ± 12.5	1,255 ± 709	18.6 ± 10.7
45-64	2,359 ± 826	33.5 ± 12.5	1,546 ± 723	22.0 ± 10.8
65-74	2,249 ± 739	32.9 ± 11.5	1,500 ± 660	21.9 ± 9.9
>75	2,103 ± 697	32.9 ± 11.4	1,381 ± 600	21.6 ± 9.5
All Ages	2,072 ± 803	41.8 ± 27.4	1,193 ± 702	22.6 ± 15.9

SOURCE: Based on data from Ershow and Cantor, 1989, pp. 42, 51, 65, and 74.

TOTAL WATER

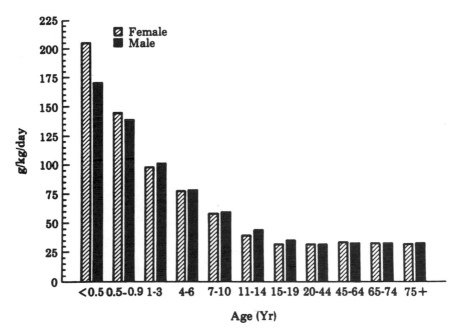

FIGURE 5-3   Mean daily intake of total water per unit of body weight by age group and sex. SOURCE: Ershow and Cantor, 1989, p. 26.

tion figures for female infants were actually higher than those for males of the same age.

Variations between regions and smaller seasonal differences were observed for all age groups (Figures 5-7 and 5-8). For infants, total water and tapwater consumption were lowest in the northeast, tapwater was highest in the west, and total water was highest in the midwest. Intake of both tapwater and total water was highest in the summer in all regions for all age groups.

Water consumption data reported by Gerber showed a wide variation among the various age groups. In general, however, the data supported those reported by NCI and described above (Gerber Products, personal commun., 1992).

QUANTIFICATION OF CONSUMPTION DATA

The 1977-1978 NFCS results were reported as daily consumption of individual foods. Gerber reported consumption data for foods as formulated and processed for feeding. Tolerance levels established by the EPA for pesticide residues in foods, however, are based on crop-based compo-

TAPWATER

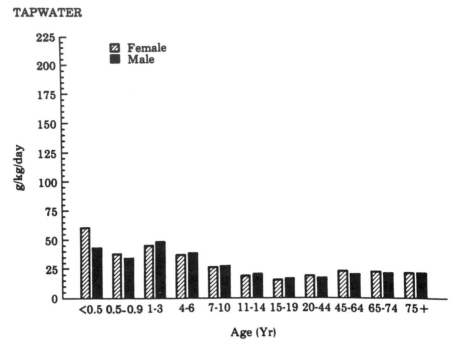

FIGURE 5-4   Mean daily intake of tapwater per unit of body weight by age group and sex. SOURCE: Ershow and Cantor, 1989, p. 26.

nents of the foods reported in the 1977-1978 NFCS. Therefore, to assess pesticide exposures and to compare those exposures with established reference doses, it is necessary to break down the foods consumed into raw agricultural commodities (RACs)—the components used by the regulatory agencies. For example, pizza is broken down into wheat flour, water, yeast, tomato paste, tomato sauce, cheese, and other ingredients and expressed in grams in order to match the form in which pesticide tolerances are reported in the Code of Federal Regulations (1986a,b). (RACs are discussed in more detail in Chapter 6.)

EPA developed and uses the Dietary Residue Evaluation System (DRES) to estimate dietary exposures of humans to pesticides through the diet. This system is based on pesticide residue concentrations found in RACs, which are sampled from harvested agricultural crops and analyzed at the farm gate. The residue estimates are then multiplied by the food consumption estimates to assess the extent of human exposure and to allow the development of new tolerance concentrations. DRES includes recipe files for each food tested. Furthermore, DRES may be used to estimate consumption of various foods in 22 different population groups based on such characteristics as age, nursing or nonnursing status, and ethnic background from survey information available in the NFCS. The DRES method, formerly called the Tolerance Assessment System (TAS), and the use of RACs as derived by EPA from the NFCS to estimate the exposure of humans to pesticides have been described in detail in other publications (e.g., Research Triangle Institute, 1983; Saunders and Petersen, 1987).

Several problems are associated with this method of calculating food intake, however. The processing effects cannot be accurately considered because there is no compensation for the fractionation of food components (e.g., the stripping of soybean oil). Individual components of specialized foods cannot be identified when only the total of the components is presented. For example, the term *milk solids* applies to fresh milk, milk in formulated foods, and milk in infant formula. This practice fails to compensate for ingredient selection and processing differences for specialized infant foods.

With these caveats in mind, the committee used these data to develop tables showing the predominant foods in the average U.S. diet and in the diets of various subgroups: nursing and nonnursing infants (<1 year old),

TOTAL WATER

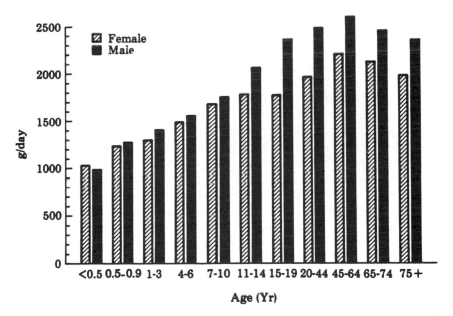

FIGURE 5-5 Mean daily intake of total water by age group and sex. SOURCE: Ershow and Cantor, 1989, p. 24.

TAPWATER

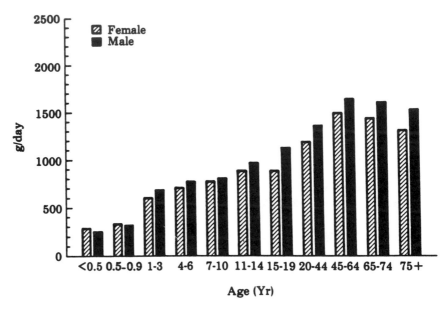

FIGURE 5-6   Mean daily intake of tapwater given in grams by age group and sex. SOURCE: Ershow and Cantor, 1989, p. 24.

children from 1 to 6 and 7 to 12 years old, teenagers from 13 to 19 years old, and adults over the age of 19. The result is a clear picture of the most commonly consumed foods in the diets of infants and children and how these differ from those consumed most frequently by adults, as discussed in the next section.

Because EPA presented intake data for milk as RACs (that is, as the dry constituents such as fat and nonfat milk solids and lactose), comparisons of intake to other foods consumed could not be reasonably made without making an adjustment for the water content of the milk. To accomplish this, the committee estimated the solid content of milk at 15% and its water content at 85% and adjusted the data for the dry constituents accordingly. In this way, percentages of the total diet for each of the age groups could be calculated to present a more representative approximation of their dietary patterns.

The 85% was derived from the DRES recipe files, which show that milk and infant formula are approximately 85% water, depending on the product. Other estimates are quite close. For example, Petersen and Associates (1992) reported the water content of milk-based baby formula as 87%, and of light cream, 74.1%. Clearly, the water component of milk products consumed by infants and children is an important consideration in developing dietary comparisons.

AGE-RELATED DIFFERENCES IN DIETARY PATTERNS

Table 5-5 lists the 17 foods (expressed as RACs) that comprise more than 1% of the average U.S. diet, as reported in the 1977-1978 NFCS. Individual age categories were compared to the U.S. average and multiples determined. Only two age categories had multiples in excess of 2: nonnursing infants and 1- to 6-year-old children. Table 5-6 shows percentages for the foods comprising more than 1% of the average diets of various age groups. Table 5-7 presents the multiples of the intake of those foods compared with the U.S. average (e.g., 2.00 means twice the U.S. average; 5.25 means 5-and-a-quarter times the U.S. average). The information in these tables is also presented as RACs.

The numbers in these tables were derived from EPA intake data based on milligrams per kilogram of body weight and are presented as RACs consistent with regulatory practice. Water intake was not considered, except for the water component of milk products, as noted above.

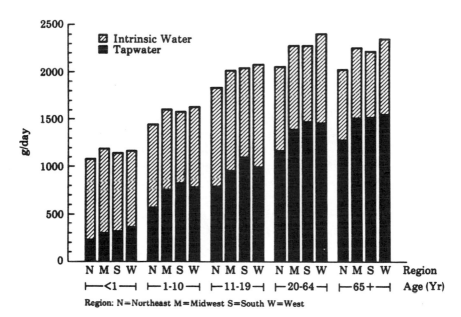

FIGURE 5-7 Mean daily total intake of water by source, age group, and region. SOURCE: Ershow and Cantor, 1989, p. 27.

SEASON

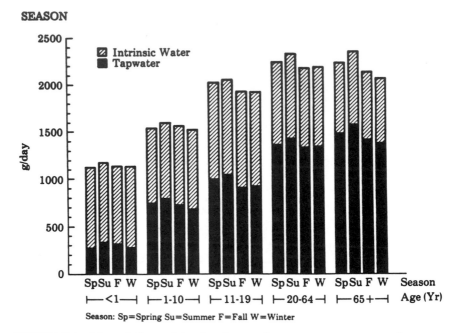

FIGURE 5-8   Mean daily total intake of water by source, age group, and season. SOURCE: Ershow and Cantor, 1988, p. 27.

In the process of reviewing the data and developing the comparisons, the committee noted variances in dietary patterns across age categories, within age categories, within individuals, and over time. All these factors must be considered in estimating exposure to pesticide residues and assessing risk for infants and children. Clearly, the marked differences between children's diets and the diets of adults and their relationship to the U.S. average have implications for assessing patterns of dietary exposure to pesticides.

Milk in its entirety constitutes an extremely large portion of the U.S. diet—not only for infants and children, but also for adults. This can be attributed to the relatively large quantities of milk contained in many of the foods consumed, e.g., formulated foods and baked goods, and its use as a diluent, e.g., for cereal.

Human milk is a dominant source of nutrition for nursing infants less than 3 months old and a substantial source for those less than 11 months. It is consumed exclusively by more than 60% of nursing infants until they reach 3 to 6 months of age. Data collected by Ross Laboratories in 1992 indicate that 53.3% of all mothers in the United States breastfed their

TABLE 5-5 The 17 Foods Comprising More than 1% of the Average U.S. Diet in 1977-1978 and the Age Class Consuming the Highest Multiple of That Average

Age Class of Children Consuming Highest Multiples of U.S. Average Intakes	Food Expressed as Raw Agricultural Commodity	% of Average U.S. Diet, 95th percentile, upper bound	Multiple of Average U.S. Consumption
Nonnursing Infants	Milk, nonfat solids	24.9	7.3
	Milk, fat solids	11.8	3.6
	Apples, fresh	1.9	6.9
	Soybean oil	1.3	4.6
	Apple juice	1.0	16.7
1- to 6-Year Olds	Wheat, flour	5.1	2.3
	Beef, lean	4.8	1.8
	Orange juice	4.6	3.1
	Potato, white, pulp	3.3	2.2
	Cane sugar	3.0	2.5
	Eggs, white	2.3	2.3
	Tomatoes, whole	2.0	1.7
	Pork, lean	1.6	1.9
	Chicken, with skin	1.6	2.2
	Beef, fat	1.5	1.8
	Potato, white, whole	1.5	1.8
	Beet sugar	1.4	2.5

[a] The EPA intake figures for fat and nonfat milk solids were divided by 0.15 to derive a percentage that would more accurately reflect the contribution to the total diet made by milk with an 85% water content. See text for further discussion.

SOURCE: Based on data from the 1977-1978 Nationwide Food Consumption Survey.

TABLE 5-6 Foods Comprising More than 1% of the Average Diet of Age Groups Indicated

Foods Expressed as Raw Agricultural Commodity	% of Average Diet of Age Group Indicated, 95th Percentile, Upper Limit					
	Infants		Age, years			
	Nursing (n = 109)	Nonnursing (n = 457)	1-6 (n = 3,633)	7-12 (n = 4,309)	13-19 (n = 5,130)	≥20 (n = 16,801)
Milk, nonfat solids	28.4	44.2	30.4	28.5	28.6	18.2
Apple juice	8.5	4.4	1.9	—	—	—
Apples, fresh	7.0	3.2	2.2	2.1	1.6	1.7
Orange juice	4.5	3.2	5.7	4.6	4.2	4.4
Pears, fresh	4.3	1.9	—	—	—	—
Milk, fat solids	4.9	10.4	13.4	13.3	13.2	10.5
Peaches, fresh	4.1	2.3	—	—	—	—
Carrots	3.3	1.6	—	—	—	—
Beef, lean	3.0	1.5	3.5	4.3	5.4	6.1
Milk sugar (lactose)	2.8	3.4	—	—	—	—
Bananas, fresh	2.7	1.1	1.3	—	—	—
Rice, milled	1.8	1.5	—	—	—	—
Peas, succulent, garden	1.3	—	—	—	—	—
Beans, succulent, green	1.3	—	—	—	—	—
Oats	1.2	—	—	—	—	—

Soybean oil	1.1	1.5	1.0	1.2	1.4	1.5
Coconut oil	—	1.4	—	—	—	5.7
Wheat flour	—	1.0	4.6	5.8	6.1	3.1
Cane sugar	—	—	3.1	3.4	3.7	3.7
Potato, white, pulp	—	—	2.9	3.4	3.9	3.0
Eggs, whole	—	—	2.1	1.8	2.2	2.0
Chicken, with skin	—	—	1.4	1.4	1.6	2.8
Tomatoes, whole	—	—	1.4	1.7	2.1	1.4
Beet sugar	—	—	1.4	1.5	1.7	2.1
Pork, lean	—	—	1.2	1.3	1.8	2.0
Beef, fat	—	—	1.1	1.4	1.7	2.0
Potato, white, whole	—	—	1.0	1.2	1.4	1.0
Corn, sweet	—	—	—	1.1	1.1	1.5
Lettuce, head	—	—	—	—	—	1.1
Fish, fin, saltwater	—	—	—	—	—	1.1
Pork, fat	—	—	—	—	—	

NOTE: EPA intake data for fat and nonfat milk solids were divided by 0.15 to derive a percentage that would more accurately reflect the contribution to the total diet made by milk with an 85% water content. No percentages are given for foods comprising less than 1% of the diet.

SOURCE: Based on data from the 1977-1978 Nationwide Food Consumption Survey.

TABLE 5-7 Foods Comprising More Than 1% of the Average Diets of Different Age Groups and Multiple of U.S. Average Consumption

| | Multiple of U.S. Average Consumption | | | | | |
| | Infants | | Age, years | | | |
Foods Expressed as Raw Agricultural Commodity	Nursing (n = 109)	Nonnursing (n = 457)	1-6 (n = 3,633)	7-12 (n = 4,309)	13-19 (n = 5,130)	≥20 (n = 16,801)
Milk, nonfat solids	2.0	7.3	3.0	1.8	0.96	0.47
Apple juice	14.9	16.7	4.8	—	—	—
Apples, fresh	6.3	6.9	2.8	1.7	0.76	0.58
Orange juice	1.7	62.9	3.1	1.6	0.84	0.63
Pears, fresh	14.4	15.0	—	—	—	—
Milk, fat solids	7.3	3.6	2.8	1.8	1.00	0.57
Peaches, fresh	7.7	10.6	—	—	—	—
Carrots	7.9	9.1	—	—	—	—
Beef, lean	1.1	1.3	1.8	1.4	1.01	0.81
Milk sugar (lactose)	27.9	79.1	1.8	—	—	—
Bananas, fresh	4.8	5.0	3.3	—	—	—
Rice, milled	4.7	8.7	—	—	—	—
Peas, succulent, garden	3.1	—	—	—	—	—
Beans, succulent, green	2.6	—	—	—	—	—
Oats	5.9	—	—	—	—	—

Soybean oil	1.5	4.6	1.9	1.2	0.96	0.75
Coconut oil	—	49.8	—	—	—	—
Wheat flour	—	0.8	2.3	1.7	1.06	0.70
Cane sugar	—	—	2.5	1.7	1.10	0.64
Potato, white, pulp	—	—	2.1	1.6	1.09	0.73
Eggs, whole	—	—	2.3	1.2	0.85	0.82
Chicken, with skin	—	—	2.2	1.4	0.93	0.79
Tomatoes, whole	—	—	1.7	1.3	0.93	0.88
Beet sugar	—	—	2.5	1.7	1.10	0.64
Pork, lean	—	—	1.8	1.3	0.98	0.84
Beef, fat	—	—	1.8	1.4	1.02	0.82
Potato, white, whole	—	—	1.8	1.2	0.88	0.89
Corn, sweet	—	—	—	1.8	0.99	0.67
Lettuce, head	—	—	—	—	—	1.07
Fish, fin, saltwater	—	—	—	—	—	0.89
Pork, fat	—	—	—	—	—	0.80

NOTE: No multiples are provided for foods comprising less than 1% of the diet of the age groups indicated. The multiples for fat and nonfat milk solids take into consideration their water content (see NOTE under Table 5-6).

SOURCE: Based on data from the 1977-1978 Nationwide Food Consumption Survey.

TABLE 5-8 Percentage of Mothers Breastfeeding Newborn Infants in the Hospital and at 5 or 6 Months of Age in the United States in 1992, by Ethnic Background and Selected Demographic Variables

Category	Total		White		Black		Hispanic[a]	
	Newborns	5-6 mo Infants	Newborns	5-6 mo Infants	Newborns	5-6 mo Infants	Newborns	5-6 mo Infants
All mothers	53.3	19.7	59.2	22.6	25.8	7.3	51.8	16.1
Parity								
Primiparous	54.5	17.0	59.6	19.1	26.9	6.8	55.1	15.0
Multiparous	52.3	22.0	58.8	25.9	24.9	7.6	49.1	17.0
Marital status								
Married	61.2	24.5	63.0	25.4	40.0	13.3	58.0	19.8
Unmarried	31.4	7.5	41.6	10.0	19.3	4.6	41.4	10.0
Maternal age								
<20 yr	32.2	6.5	37.9	7.6	16.7	3.2	39.3	8.7
20-24 yr	45.7	12.4	50.8	13.9	22.8	5.5	50.6	13.1
25-29 yr	57.8	21.4	62.1	23.5	29.9	9.1	56.1	18.8
30-34 yr	65.2	29.3	69.3	32.0	37.0	12.6	59.0	22.7
>35 yr	66.7	35.1	71.6	39.1	37.8	14.3	57.5	24.3
Maternal education								
No college	42.6	13.2	48.2	15.2	20.3	5.3	45.9	13.4
College[b]	70.2	29.8	74.0	32.5	41.8	12.9	67.0	23.3

Family income								
<$10,000	33.3	8.5	39.9	10.2	18.4	4.4	40.8	10.5
$10,000-$14,999	45.7	13.9	49.9	15.3	25.8	7.4	51.3	15.8
$15,000-$24,999	53.3	19.3	56.8	21.4	31.8	8.7	54.4	17.8
>$25,000	67.1	27.4	68.4	28.4	46.3	15.1	66.7	22.9
Maternal employment								
Full time	54.0	11.6	57.7	12.4	32.3	7.0	55.9	10.9
Part time	60.1	23.3	64.2	25.6	29.7	7.9	58.9	18.6
Not employed	51.2	22.6	58.3	26.9	22.3	7.3	48.9	17.9
U.S. census region								
New England	55.9	21.3	57.2	22.3	33.8	9.1	57.0	13.3
Middle Atlantic	48.3	18.2	52.9	21.3	33.3	10.2	42.3	10.5
East North Central	48.3	18.0	53.8	20.1	23.9	6.9	44.7	13.0
West North Central	56.4	20.2	58.9	21.3	25.2	7.1	55.0	18.8
South Atlantic	46.4	15.7	55.8	19.7	22.8	6.2	52.9	13.4
East South Central	37.5	11.6	46.0	14.4	15.3	3.8	39.9	14.1
West South Central	48.4	15.2	56.9	18.6	19.3	4.6	42.9	12.2
Mountain	70.0	29.8	73.8	31.9	38.5	12.2	57.7	20.7
Pacific	69.5	27.8	76.2	32.7	46.1	13.9	59.2	20.4

[a] Hispanic is not exclusive of white or black.
[b] College includes all women who reported completing at least 1 year of college.

SOURCE: From Fritz Krieger, Ross Laboratories, personal communication, 1992.

TABLE 5-9   Percentages of Breastfeeding Respondents in the 1988 National Maternal and Infant Health Survey

	Respondents Still Breastfeeding, %	
Infant Age, months	Low-Income Women[a]	U.S. Population
0	39.0	52.4
1	25.8	37.1
2	19.8	29.1
3	14.7	22.9
4	12.0	18.7
5	10.1	16.2
6	7.2	11.8
7	6.1	9.8
8	4.7	7.7
9	3.8	5.8
10	2.9	4.3
11	2.4	3.2
12	1.0	1.5

[a] In this table, low income is defined as total income less than or equal to 185% of the federal poverty line.

SOURCE: Unpublished data from the 1988 National Maternal and Infant Health Survey. Provided by J. Tognetti, Office of Analysis and Evaluation, Food and Nutrition Service, U.S. Department of Agriculture.

newborn children but only 19.7% of them continued the practice when their infants were 5 to 6 months of age (Table 5-8). The highest percentages were noted for white mothers (59.2% for newborns; 22.6% at 5 to 6 months) and Hispanic mothers (51.8% for newborns; 16.1% at 5 to 6 months), in contrast to black mothers (25.8% for newborns; 7.3% at 5 to 6 months). There were also wide differences that correlated with socioeconomic status across all ethnic groups. For all mothers in families with annual incomes of $25,000 or more, breastfeeding rates were 67.1% for newborns and 27.4% for 5- to 6-month-old infants. At incomes less than $10,000, the rates were 33.3% and 8.5%, respectively. Data from the 1988 National Maternal and Infant Health Survey show similar trends: 52.4% of new mothers in the United States were breastfeeding their newborns and that percentage dropped to 11.8% at 6 months and to 1.5% at 1 year (Table 5-9). They also show that breastfeeding rates among low-income women are lower than the national average.

A number of studies demonstrate that the volume of milk intake among healthy, exclusively breastfed infants also ranges widely (Figure 5-9).

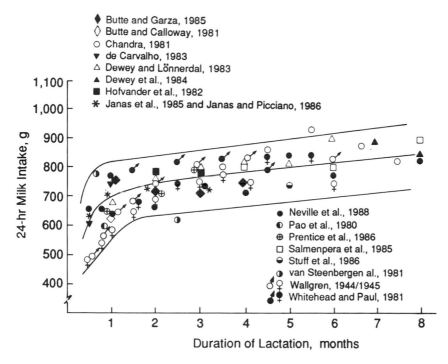

FIGURE 5-9 Milk intakes during established lactation. The lines represent the smoothed mean ± standard deviation. The points represent average intakes obtained in 16 studies that included test weighing, validated exclusive breastfeeding, three or more subjects, and monthly reports on milk transfer. SOURCE: Neville et al., 1988, with permission.

After the first 4 to 5 months, the variance is even greater. For infants who were breast-fed for at least 12 months and given solid foods beginning at 4 to 7 months, milk intake averaged 769 g/day (range, 335 to 1,144 g/day) at 6 months ($n = 56$), 637 g/day (range, 205 to 1,185 g/day) at 9 months ($n = 46$), and 445 g/day (range, 27 to 1,154 g/day) at 12 months ($n = 40$) (Dewey et al., 1990; K. Dewey, personal commun., 1992).

Milk intake is most often determined by weighing the infant before and after feeding. This method leads to underestimations of intake ranging from approximately 1 to 5% (Brown et al., 1982; Woolridge et al., 1985) because of water loss through evaporation from the infant between weighings. Newer techniques based on stable isotopes have been developed to measure breast milk intake (Coward et al., 1982; Butte et al., 1988), but few data have been generated by this method to date (e.g.,

Orr-Ewing, 1986; Butte et al., 1992). For a more detailed review of these data, see *Nutrition During Lactation* (NRC, 1991).

Because the volume of human milk consumption varies widely, the estimates of supplementary food consumption shown in Table 5-6 are conservative, i.e., they are higher than the amounts likely to be consumed by these infants. The 1988 National Maternal and Infant Health Survey contained a number of questions that promise to shed more light on infant food consumption during the first 6 months of life. Unfortunately, the analysis could not be made available to the committee in time for inclusion in this report.

As demonstrated by the preceding discussion and accompanying tables, use of available intake data to assess dietary exposures of nursing infants is complex. Human milk is the major food and source of essential nutrients consumed by these infants during their first year of life, but a vast range of variables must be considered: the age at which supplementary foods are introduced, the selection of foods given to them, and the volume of human milk consumed. Factors greatly affecting the feeding patterns of this group include economic status, ethnic background, region, and the age, marital status, educational level, parity, and employment of the mother.

Infant formula is the sole source of food for nonnursing infants for the first 3 months of life. Milk or milk-based food remains the predominant source of energy and nutrients for all infants throughout their first year of life. Averaged over the first 12 months, nonfat and fat milk solids provide 44.2% and 10.4%, respectively, of their diet. These figures were derived from DRES conversion of formula into its component parts: fat and nonfat milk solids, milk sugar, coconut oil, and soybean oil. Coconut oil represents only 1.4% of the average diet of nonnursing infants, but the consumption is almost 50 times greater than that of the national average (Table 5-6).

The diet of infants is gradually supplemented with specially prepared, predominantly processed foods produced by a small number of manufacturers. When presented as RACs, fruit and fruit juice constitute the highest proportion of the early supplemental foods, accounting for an estimated 16.1% of the diet, averaged over the first year of life. These are led by apple juice (4.4% as RAC) and fresh apples (3.2% as RAC).

Caloric consumption by infants per unit of body weight is higher than that for adults—approximately 2.5 times higher for the very young infant (NRC, 1989). Therefore, comparison of the consumption data for infants and adults on the basis of grams per kilogram of body weight results in an elevated value for infants.

The diets of infants and children are less diverse than those of the general population. The 1977-1978 NFCS reports intakes of

375 foods (reported as RACs) for the entire sample of more than 30,000 people. In contrast, 148 RACs were reported for the 457 nonnursing infants sampled, indicating a relative lack of diversity in the diets of this sub-group. It is therefore important to monitor both the percentage of total diet and the multiple of the national average consumption for each food and for each age group to identify areas relative to dietary exposure to pesticides.

Several factors must be considered when evaluating the consumption data on infants. For example, caution must be exercised not to overestimate introduced foods to avoid estimates of intake that greatly exceed the consumption capacity of the infant. Furthermore, changes resulting from various processing techniques, especially fractionation (e.g., for soybean oil), must not be overlooked. Other influences include many of those noted earlier for nursing infants, e.g., region, socioeconomic status, and other variables.

Milk also predominates in the diet of 1- to 6-year-old children, as shown by the values for nonfat and fat milk solids (30.4% and 13.4%, respectively). Orange juice, fresh apples, apple juice, and bananas together constitute 11.1% of their diet. The data also show that the diets of this group have become more diverse to include wheat, beef, sugar, eggs, and chicken, and more varieties of vegetables are consumed. The number of foods eaten above the U.S. average, and the multiples of their excess, have declined. This is most likely attributable not only to the rapid increase in dietary diversity after the age of 1 year, but also to a diminishing effect of the body weight conversion factor as average childhood weights approach average adult weights.

The diets of children from 7 to 12 years old have attained a greater level of diversity and show changes in proportion. Wheat flour, beef, and potatoes have reached higher percentages of the diet, whereas the intake levels of some foods (e.g., milk constituents and orange juice) have declined slightly. Many of the foods that constitute the greatest percentages of the diet are the same as those for 1- to 6-year-old children, although there are differences in the rankings of the percentages of the total diet of those age groups. As shown by the multiples of the U.S. average consumption, the intakes by 7- to 12-year-old children are approaching those of the overall population.

Among teenagers from 13 to 19 years old, wheat flour, beef, potatoes, and eggs continue their ascendance in dietary predominance over fruits and vegetables with the exceptions of orange juice, apples, and tomatoes. Orange juice ranks fifth among the foods that constitute the highest per-centages of the diet consumed by this age group. The multiples of the U.S. average consumption have declined, and for 10 food items, are below the national average.

There is a dramatic drop in nonfat milk solids for adults 20 years old and older (from 28.6% of the total diet of teenagers to 18.2% for adults). All the foods comprising more than 1% of the diet of teenagers appear again for adults along with lettuce, fin fish, and fat pork. All food items listed except lettuce are consumed in amounts less than the national average, showing the greater diversity of the adult diet.

ISSUES RELATED TO THE EVALUATION OF FOOD MONITORING DATA

Differences in Consumption Among Age Groups

As demonstrated by the NFCS data (Tables 5-5, 5-6, and 5-7), infant and early childhood consumption patterns differ greatly from those of adults and the population as a whole. The most dominant factor that emerges from interpretive evaluation of intake data remains as milk in some form. The proportion of the balance of the diet represented by fruits, by liquids, especially fruit juices, and by vegetables is much higher for infants than for older children or adults. For several fruit juices, this relationship is maintained even when commodity intake is not adjusted by body weight. Thus, use of the U.S. average intake as a basis for estimating pesticide exposures of infants and children may result in underestimates of pesticide residues in their diet.

As noted earlier in this chapter, water intake is considerably higher for infants than for other age categories. The sources of this high intake include foods such as concentrated juices, cereals, and infant formula that are mixed with water prior to consumption. Therefore, water must be a major consideration in estimating of the risk from dietary exposure to pesticides for infants and children.

The Gerber data show several changes in the diet over the first year of life that have potentially important implications for the exposure of infants to pesticides (Johnson et al., 1981). There is a gradual substitution of cow's milk and apple juice for human milk and infant formula. Soybean oil and coconut oil consumption decreases, reflecting the declining intake of infant formula. Intakes of fruits and vegetables increase steadily, and the diet in general begins to diversify.

Because of the young ages of the study population, the 90th percentile of the Gerber sample often shows consumption of such foods as peaches, pears, carrots, and oats for only a few individuals. As the percentage of eaters increases as the children grow older, the 90th percentile commonly exceeds the mean of eaters only and would therefore provide a higher degree of protection if used as the basis of risk projections and tolerance setting.

Differences in Consumption within Age Groups

Variations in consumption patterns within age groups may be considerable. Consider, for example, the great differences among the diets of vegetarians, certain ethnic and religious groups, individuals with medically restricted diets, and populations in different regions of the country. It is quite probable that subgroups within the various age categories consume both average daily and daily high levels of food considerably above the NFC3 values, but an analysis of these variations is not possible with current survey data.

For chronic risk assessment, EPA has traditionally relied on mean consumption estimates for large groups such as the entire U.S. population or children between the ages of 1 and 6 years or 7 and 12 years. The 90th percentile has been avoided, because no individual consumes the 90th percentile of all foods regulated by the FDA and USDA. Some individuals may, however, consistently consume a few foods at the 90th percentile for their age group. Obviously, these clusters would vary among individuals and eventually resolve to the mean for the age group.

To determine the significance of this observation, the committee identified the 25 foods most consumed by 170 1-year-old children in the 1985-1986 CSFII. Results from DRES calculations based on CSFII results illustrate that mean intake data clearly do not reflect the actual intakes of some foods by some subjects and therefore that their use can lead to underestimates of pesticide exposure. For example, 88 of the children (52%) consumed 6 to 10 foods over the mean, and only 5 (2.9%) consumed no foods over the mean. The many foods consumed at levels lower than the mean constituted a minor portion of the diet.

These considerations regarding consumption are used as the basis for the committee's discussions in the remaining chapters of this report.

CONCLUSIONS AND RECOMMENDATIONS

The committee reached its conclusions and recommendations for this chapter after an extensive review of information available on food and water consumption. All major sources of data on food consumption by infants and children compiled by government agencies and the food industry were considered and evaluated. Information on water intake was derived from an extensive evaluation of food survey information conducted for the National Cancer Institute. The committee weighed the strengths and the limitations of each source of information in order to identify the most acceptable data to be used in conjunction with pesticide residue information to determine exposure, as discussed in Chapters 6 and 7.

Information on foods as consumed was broken down into constituents of foods by the same process currently used by government to establish regulatory policy. This conversion process—the Dietary Residue Evaluation System (DRES)—expresses foods in terms of individual raw agricultural commodities (RACs).

Potential dietary exposures to pesticides are related to food and water intake. To identify the differences in exposures of infants and children compared with adults, intake data were grouped into various age categories. Consumption of water, human milk, and processed foods were considered separately in order to assess their respective contributions to dietary pesticide exposure for infants and young children.

Conclusions

• Food consumption patterns for infants and children differ markedly from those of adults.

–Children consume more calories relative to body weight than do adults.
–Dietary diversity increases with age: infants and young children consume fewer distinct foods than do adults.
–On a body weight basis, infants and young children consume notably more of certain foods than do adults.

• Water, as drinking water and as a component of food, is not adequately considered in most consumption surveys.

• Examination of intake data by age categories clearly illustrates the differences in consumption patterns that must be considered when estimating exposure of infants and children to pesticides; however, current information on food and water intake by age category is insufficient to produce credible exposure estimates.

• Processed foods are predominant in the diets of younger age groups.

Recommendations

Knowledge of food consumption is an important consideration in assessing the risk to infants and children from dietary exposures to pesticides. Therefore, more focused, direct, comprehensive, and contemporary dietary information is required for infants and children. Because of the myriad and rapid changes in diet that occur during the developmental stages of life, intake data must be precisely divided into age subdivisions and the sample must be large enough to produce meaningful results.

• A simple, uniform method needs to be developed for conversion of a product as consumed to its components in terms of raw agricultural commodities.

• Those foods most frequently consumed by young children and infants need to be identified and quantified more specifically. Reporting should be specific and discrete foods clearly identified.

• Water intake and food intake should both receive full consideration in estimating dietary exposure and assessing risk, especially for infants and young children.

• Because of the changing nature of children's diets during growth, food consumption surveys should include adequate sample sizes of children aged 0 to 12 months, 13 to 24 months, 25 to 36 months, 37 to 48 months, 49 to 60 months, 5 to 10 years, and 11 to 18 years.

• Intake data and survey methodology need to be standardized to make them more useful in a variety of applications, including estimating exposure and assessing risk.

• Food consumption surveys should be coordinated among the involved organizations and performed on a continuing basis in order to examine trends of food and water consumption, especially by infants and children.

REFERENCES

Acheson, K.J., I.T. Campbell, O.G. Edholm, D.S. Miller, and M.J. Stock. 1980. The measurement of food and energy intake in man: An evaluation of some techniques. Am. J. Clin. Nutr. 33:1147-1154.

Acosta, P.B., C. Trahms, N.S. Wellman, and M. Williamson. 1983. Phenylalanine intakes of 1- to 6-year-old children with phenylketonuria undergoing therapy. Am. J. Clin. Nutr. 38:694-700.

Anderson, G.H., and L.M. Blendis. 1981. Plasma neutral amino acid ratios in normal man and in patients with hepatic encephalopathy: Correlations with self-selected protein and energy consumption. Am. J. Clin. Nutr. 34:377-385.

Axelson, J.M., and M M. Csernus. 1983. Reliability and validity of a food frequency checklist. J. Am. Diet. Assoc. 83:152-155.

Backstrom, C.H., and G.D. Hursh-Cesar. 1981. Survey Research, 2nd ed. New York: Macmillan.

Basiotis, P.P., S.O. Welsh, F.J. Cronin, J.L. Kelsay, and W. Mertz. 1987. Number of days of food intake records required to estimate individual and group nutrient intakes with defined confidence. J. Nutr. 117:1638-1641.

Beaton, G.H., J. Milner, P. Corey, V. McGuire, M. Cousins, E. Stewart, M. de Ramos, D. Hewitt, P.V. Grambsch, N. Kassim, and J.A. Little. 1979. Sources of variance in 24-hour dietary recall data: Implications for nutrition study design and interpretation. Am. J. Clin. Nutr. 32:2546-2549.

Beaton, G.H., J. Milner, V. McGuire, T.E. Feather, and J.A. Little. 1983. Source of variance in 24-hour dietary recall data: Implications for nutrition study design and interpretation. Carbohydrate sources, vitamins, and minerals. Am. J. Clin. Nutr. 37:986-995.

Block, G. 1982. A review of validations of dietary assessment methods. Am. J. Epidemiol. 115:492-505.

Block, G., A.M. Hartman, C.M. Dresser, M.D. Carroll, J. Gannon, and L. Gardner. 1986. A data-based approach to diet questionnaire design and testing. Am. J. Epidemiol. 124:453-469.

Brown, K.H., R.E. Black, A.D. Robertson, N.A. Akhtar, G. Ahmed, and S. Becker. 1982. Clinical and field studies of human lactation: Methodological considerations. Am. J. Clin. Nutr. 35:745-756.

Burk, M.C., and E.M. Pao. 1976. Methodology for Large-Scale Surveys of Household and Individual Diets. HERR No. 40. Washington, D.C.: U.S. Department of Agriculture.

Butte, N.F., W.W. Wong, B.W. Patterson, C. Garza, and P.D. Klein. 1988. Human-milk intake measured by administration of deuterium oxide to the mother: A comparison with the test-weighing technique. Am. J. Clin. Nutr. 47:815-821.

Butte, N.F., S. Villalpando, W.W. Wong, S. Flores-Huerta, M.J. Hernandez-Beltran, E.O. Smith, and C. Garza. 1992. Human milk intake and growth faltering of rural Mesoamerindian infants. Am. J. Clin. Nutr. 55:1109-1116.

Byers, T., J. Marshall, E. Antony, R. Fiedler, and M. Zielezny. 1987. The reliability of dietary history from the distant past. Am. J. Epidemiol. 125:999-1011.

Chu, S.Y., L.N. Kolonel, J.H. Hankin, and J. Lee. 1984. A comparison of frequency and quantitative dietary methods for epidemiologic studies of diet and disease. Am. J. Epidemiol. 119:323-334.

Code of Federal Regulations (CFR). 1986a. Tolerances and Exemptions from Tolerances for Pesticide Chemicals in or on Raw Agricultural Commodities. 40 CFR 1, Part 180.

Code of Federal Regulations (CFR). 1986b. Tolerances for Pesticides in Food Administered by EPA. 21 CFR 1, Part 193.

Coward, W.A., T.J. Cole, H. Guber, S.B. Roberts, and I. Fleet. 1982. Water turnover and measurement of milk intake. Pfleugers. Arch. 393:344-347.

Dewey, K.G., M.J. Heinig, L.A. Nommsen, and B. Lönnerdal. 1990. Low energy intakes and growth velocities of breast-fed infants: Are there functional consequences? Pp. 35-44 in Activity, Energy Expenditure, and Energy Requirements of Infants and Children, B. Schürch and N. Scrimshaw, eds. Lausanne, Switzerland: International Dietary Energy Consulting Group-Nestlé Foundation.

DHHS (Department of Health and Human Services). 1983. Dietary Intake Source Data: United States 1976-80. DHHS Pub. No. (PHS) 83-1681. National Center for Health Statistics. Washington, D.C.: U.S. Government Printing Office. 483 pp.

Dwyer, J.T. 1988. Assessment of dietary intake. Pp. 887-905 in Modern Nutrition in Health and Disease, 7th ed., M.E. Shils and V.R. Young, eds. Philadelphia: Lea and Febiger.

Elahi, V.K., D. Elahi, R. Andres, J.D. Tobin, M.G. Butler, and A.H. Norris. 1983. A longitudinal study of nutritional intake in men. J. Gerontol. 38:162-180.

Ershow, A.B., and K.P. Cantor. 1989. Total Water and Tapwater Intake in the United States: Population-Based Estimates of Quantities and Sources. Bethesda, Md.: Life Sciences Research Office, Federation of American Societies for Experimental Biology.

Frank, G.C., A.W. Voors, P.E. Schilling, and G.S. Berenson. 1977. Dietary studies of rural school children in a cardiovascular survey. J. Am. Diet. Assoc. 71:31-35.

Freudenheim, J.L., N.E. Johnson, and R.L. Wardrop. 1987. Misclassification of nutrient intake of individuals and groups using one-, two-, three-, and seven-day food records. Am. J. Epidemiol. 126:703-713.

Fuller, W.A. 1991. I. Regression estimation for sample survey data. In Regression Estimation for the 1987-1988 Nationwide Food Consumption Survey. Washington, D.C.: U.S. Department of Agriculture.

Fuller, W.A., M.M. Loughin, and H. Baker. 1991. II. Application to the Nationwide Food Consumption Survey. In Regression Estimation for the 1987-1988 Nationwide Food Consumption Survey. Washington, D.C.: U.S. Department of Agriculture.

Futrell, M.F., L.T. Kilgore, and F. Windham. 1971. Nutritional status of Negro preschool children in Mississippi: Evaluation of HOP index. J. Am. Diet. Assoc. 59:218-223.

GAO (General Accounting Office). 1991. Nutrition Monitoring: Mismanagement of Nutrition Survey Has Resulted in Questionable Data. (GAO/RCED-91-117) Washington, D.C.: U.S. General Accounting Office.

Gersovitz, M., J.P. Madden, and H. Smiciklas-Wright. 1978. Validity of the 24-hour dietary recall and seven-day record for group comparisons. J. Am. Diet. Assoc. 73:48-55.

Hampe, E.C., Jr., and M. Wittenberg. 1964. The Lifeline of America: Development of the Food Industry. New York: McGraw-Hill.

Jain, M., G.R. Howe, K.G. Johnson, and A.B. Miller. 1980. Evaluation of a dietary history questionnaire for epidemiologic studies. Am. J. Epidemiol. 111:212-219.

Johnson, G.H., G.A. Purvis, and R.D. Wallace. 1981. What nutrients do our infants really get? Nutr. Today 16:4-10, 23-27.

Kim, W.W., W. Mertz, J.T. Judd, M.W. Marshall, J.L. Kelsay, and E.S. Prather. 1984. Effect of making duplicate food collections on nutrient intakes calculated from diet records. Am. J. Clin. Nutr. 40:1333-1337.

Krall, E.A., and J.T. Dwyer. 1987. Validity of a food frequency questionnaire and a food diary in a short-term recall situation. J. Am. Diet. Assoc. 87:1374-1377.

Krantzler, N.J., B.J. Mullen, H.G. Schutz, L.E. Grivetti, C.A. Holden, and H.L. Meiselman. 1982. Validity of telephoned diet recalls and records for assessment of individual food intake. Am. J. Clin. Nutr. 36:1234-1242.

Madden, J.P., S.J. Goodman, and H.A. Guthrie. 1976. Validity of the 24 hr. recall: Analysis of data obtained from elderly subjects. J. Am. Diet. Assoc. 68:143-147.

Marr, J.W., and J.A. Heady. 1986. Within- and between-person variation in dietary surveys: Number of days needed to classify individuals. Hum. Nutr. Appl. Nutr. 40:347-364.

McGee, D., G. Rhoads, J. Hankin, K. Yano, and J. Tillotson. 1982. Within-person variability of nutrient intake in a group of Hawaiian men of Japanese ancestry. Am. J. Clin. Nutr. 36:657-663.

Neville, M.C., R. Keller, J. Seacat, V. Lutes, M. Neifert, C. Casey, J. Allen, and P. Archer. 1988. Studies in human lactation: Milk volumes in lactating women during the onset of lactation and full lactation. Am. J. Clin. Nutr. 48:1375-1386.

NRC (National Research Council). 1986. Nutrient Adequacy: Assessment Using Food Consumption Surveys. Washington, D.C.: National Academy Press.

NRC (National Research Council). 1989. Recommended Dietary Allowances: 10th ed. Washington, D.C.: National Academy Press.

NRC (National Research Council). 1991. Nutrition During Lactation. Washington, D.C.: National Academy Press.

Nusser, S.M., A.L. Carriquiry, and W.A. Fuller. 1991. A semiparametric transformation approach to estimating usual daily intake distributions. Unpublished.

Orr-Ewing, A.K., P.F. Heywood, and W.A. Coward. 1986. Longitudinal measurements of breast milk output by a $2H_2O$ tracer technique in rural Papua New Guinean women. Hum. Nutr. Clin. Nutr. 40:451-467.

Patterson, L. 1971. Dietary intake and physical development of Phoenix area children. J. Am. Diet. Assoc. 59:106-110.

Petersen and Associates, Inc. 1992. Background: Derivation of Water Consumption Estimates Used by the Tolerance Assessment System. Bethesda, Md.: Petersen and Associates.

Pietinen, P., A.M. Hartman, E. Haapa, L. Rasanen, J. Haapakoski, J. Palmgren, D. Albanes, J. Virtamo, and J.K. Huttunen. 1988. Reproducibility and validity of dietary assessment instruments. II. A qualitative food frequency questionnaire. Am. J. Epidemiol. 128:667-676.

Posner, B.M., C.L. Borman, J.L. Morgan, W.S. Borden, and J.C. Ohls. 1982. The validity of a telephone-administered 24-hour dietary recall methodology. Am. J. Clin. Nutr. 36:546-553.

Purvis, G.A. 1973. What nutrients do our infants really get? Nutr. Today 8:28-34.

Purvis, G.A., and S.J. Bartholmey. 1988. Infant feeding practices: Commercially prepared baby foods. Pp. 399-417 in Nutrition During Infancy, R.C. Tsang and B.L. Nichols, eds. Philadelphia: Hanley & Belfus.

Research Triangle Institute. 1983. Interim Report Number 1: The Construction of a Raw Agricultural Commodity Consumption Data Base. RTI Project No. 252U-2123-7. Research Triangle Park, N.C.: Research Triangle Institute.

Russell-Briefel, R., A.W. Caggiula, and L.H. Kuller. 1985. A comparison of three dietary methods for estimating vitamin A intake. Am. J. Epidemiol. 122:628-636.

Sampson, L. 1985. Food frequency questionnaire as a research instrument. Clin. Nutr. 4:171-178.

Saunders, D.S., and B.J. Petersen. 1987. An Introduction to the Tolerance Assessment System. Hazard Evaluation Division, Office of Pesticide Programs. Washington, D.C.: U.S. Environmental Protection Agency.

Schnakenberg, D.D., T.M. Hill, M.J. Kretsch, and B.S. Morris. 1981. Dietary-interview technique to assess food consumption patterns of individual military personnel. Pp. 180-197 in Assessing Changing Food Consumption Patterns, Committee on Food Consumption Patterns, National Research Council. Washington, D.C.: National Academy Press.

Sempos, C.T., N.E. Johnson, E.L. Smith, and C. Gilligan. 1984. A two-year dietary survey of middle-aged women: Repeated dietary records as a measure of usual intake. J. Am. Diet. Assoc. 84:1008-1013.

Sorenson, A.W., B.M. Calkins, M.A. Connolly, and E. Diamond. 1985. Comparison of nutrient intake determined by four dietary intake instruments. J. Nutr. Educ. 17:92-99.

Stamler, J. 1979. Population studies. Pp. 25-88 in Nutrition, Lipids, and Coronary Heart Disease, R.I. Levy, B.M. Rifkind, B.H. Dennis, and N. Ernst, eds. New York: Raven Press.

Todd, K.S., M. Hudes, and D.H. Calloway. 1983. Food intake measurement: Problems and approaches. Am. J. Clin. Nutr. 37:139-146.

USDA (U.S. Department of Agriculture). 1983. Nationwide Food Consumption Survey. Food Intakes: Individuals in 48 States, Year 1977-78. Report No. I-1. Hyattsville, Md.: Consumer Nutrition Division, Human Nutrition Information Service.

USDA (U.S. Department of Agriculture). 1985. Nationwide Food Consumption Survey. Continuing Survey of Food Intakes of Individuals: Women 19-50 Years and Their Children 1-5 Years, 1 Day, 1985. Report No. 85-1. Hyattsville, Md.: Nutrition Monitoring Division, Human Nutrition Information Service.

USDA (U.S. Department of Agriculture). 1986a. Nationwide Food Consumption Survey. Continuing Survey of Food Intakes of Individuals: Low-Income Women 19-50 Years and Their Children 1-5 Years, 1 Day, 1985. Report No. 85-2. Hyattsville, Md.: Nutrition Monitoring Division, Human Nutrition Information Service.

USDA (U.S. Department of Agriculture). 1986b. Nationwide Food Consumption Survey. Continuing Survey of Food Intakes of Individuals: Men 19-50 Years, 1 Day, 1985. Report No. 85-3. Hyattsville, Md.: Nutrition Monitoring Division, Human Nutrition Information Service.

USDA (U.S. Department of Agriculture). 1987a. Nationwide Food Consumption Survey. Continuing Survey of Food Intakes of Individuals: Low-Income Women 19-50 Years and Their Children 1-5 Years, 1 Day, 1986. Report No. 86-2. Hyattsville, Md.: Nutrition Monitoring Division, Human Nutrition Information Service.

USDA (U.S. Department of Agriculture). 1987b. Nationwide Food Consumption Survey. Continuing Survey of Food Intakes of Individuals: Women 19-50 Years and Their Children 1-5 Years, 1 Day, 1986. Report No. 86-1. Hyattsville, Md.: Nutrition Monitoring Division, Human Nutrition Information Service.

USDA (U.S. Department of Agriculture). 1987c. Nationwide Food Consumption Survey. Continuing Survey of Food Intakes of Individuals: Women 19-50 Years and Their Children 1-5 Years, 4 Days, 1985. Report No. 85-4. Hyattsville, Md.: Nutrition Monitoring Division, Human Nutrition Information Service.

USDA (U.S. Department of Agriculture). 1988. Nationwide Food Consumption Survey. Continuing Survey of Food Intakes of Individuals: Low-Income Women 19-50 Years and Their Children 1-5 Years, 4 Days, 1985. Report No. 85-5. Hyattsville, Md.: Nutrition Monitoring Division, Human Nutrition Information Service.

Willett, W.C., L. Sampson, M.J. Stampfer, B. Rosner, C. Bain, J. Witschi, C.H. Hennekens, and F.E. Speizer. 1985. Reproducibility and validity of a semiquantified food frequency questionnaire. Am. J. Epidemiol. 122:51-65.

Wilson, L.B., and B.B. Rothschild. 1987. Results of an exploratory study of longitudinal measures of individual food intake. Pp. 9-19 in Research on Survey Methodology: Proceedings of a Symposium. USDA Admin. Rep. No. 382. Washington, D.C.: U.S. Department of Agriculture.

Woolridge, M.W., N. Butte, K.G. Dewey, A.M. Ferris, C. Garza, and R.P. Keller. 1985. Methods for the measurement of milk volume intake of the breastfed infant. Pp. 5-21 in Human Lactation: Milk Components and Methodologies, R.G. Jensen and M.C. Neville, eds. New York: Plenum.

Young, C.M., G.C. Hagan, R.E. Tucker, and W.D. Foster. 1952. A comparison of a dietary study methods II. Dietary history vs. seven-day record vs. 24-hr. recall. J. Am. Diet. Assoc. 28:218-221.

6

Pesticide Residues

D ATA ON DIETARY LEVELS of pesticide residues combined with food consumption estimates provide the basis for exposure estimates used by the Environmental Protection Agency (EPA) to assess the risks of pesticide exposure in the diet. Thus, sampling and residue testing methods to estimate levels of pesticide residues in the food supply are extremely important components of the risk assessment process.

The committee examined pesticide usage, residue sampling and testing methods, and the data on pesticide residues to

- understand the relative quality of data sets available to EPA as a foundation for recommending practical improvements in data collection and testing;
- identify the foods in the diets of infants and children with residues of pesticides that cause the greatest public-health concern;
- assess the need for residue sampling methods and residue testing procedures that can provide the data needed to ensure the protection of infants and children;
- recommend residue monitoring methods that could be incorporated into an exposure assessment methodology that would ensure the protection of infants and children;
- identify steps to improve risk assessment and establish priorities for those steps; and
- determine which, if any, data are of sufficient quality to support risk assessment models designed to protect infants and children.

SOURCES OF DATA ON USAGE

Despite the importance pesticides have attained in agricultural production, data on the amount and distribution of their use are remarkably scanty. There is no single, comprehensive data source, derived from actual sampling, on pesticide usage for all crops and all chemicals.

The U.S. Department of Agriculture's (USDA) Economic Research Service (ERS) conducted national surveys of pesticide use in 1964, 1966, 1971, 1976, and 1982; smaller areas and fewer crops have been included in successive surveys. The 1964, 1966, and 1971 surveys included field crops, fruits, vegetables, and livestock. In 1976 fruits and vegetables were excluded from the survey, and in 1982, only major field crops (e.g., corn, soybeans, cotton, wheat, barley, oats, peanuts, tobacco, alfalfa, and hay) were sampled (Osteen and Szmedra, 1989). The foci of later reports on pesticide usage are even narrower: vegetable, melon, and strawberry crops in Arizona, Florida, Michigan, and Texas (USDA, 1991); fruits and nuts in 12 states (USDA, 1992a); and eight field crops (corn, cotton, peanuts, potatoes, rice, sorghum, soybeans, and wheat) in different numbers of states, ranging from 47 states for corn down to 2 states for rice and 1 for durham wheat (Osteen and Szmedra, 1989).

Resources for the Future maintains a county-based file of annual pesticide usage estimates by county and by crop for the 184 widely used pesticides that appear on EPA's list for the National Ground Water Survey and the California Priority Pollutant List (Gianessi, 1986). The usage information was derived from the limited ERS surveys and from the annual California survey (State of California, 1981), which included only restricted-use chemicals until 1991, when the state's reporting system was extended to all pesticides, including unrestricted chemicals. Resources for the Future has also estimated the amounts of pesticides applied to lawns and in nurseries.

The data in Table 6-1 illustrate the variation in the kind and amount of pesticides used on crops in various geographic regions. The corn belt, for example, accounted for 39% of all pesticides used on major crops in 1982. Most of this volume was represented by herbicides; fungicides constituted only 2% of total usage. In contrast, the southeast accounted for only 8% of total pesticide applications but for 66% of fungicides used. There are similar differences in use patterns between other regions.

The implications for residue and exposure estimation are more clearly illustrated in Table 6-2, which focuses on one crop (fall potatoes) and one class of pesticides (fungicides) and their application in the northeast, midwest, and western regions of the United States. In 1991, 96% and 90% of croplands planted with potatoes in the northeast and midwest, respectively, were treated with fungicides, while only 52% of croplands in the west were treated. Fungicides were also applied more times during the growing season in the northeast and midwest. As a result, the northeast, which accounts for 11% of the hectares planted with potatoes, accounts for 30% of all fungicide hectare treat-

TABLE 6-1 Regional Distribution of Pesticide Use on Major Crops in Selected Regions in 1982

	Amounts of Active Ingredient Used (1,000 lbs.) and (Percent) of Total, by Geographic Region									Total Usage	
Pesticide	Northeast	Lake	Corn Belt	No. Plains	Appa-lachia	Southeast	Delta	So. Plains	Mountains	lbs. (1,000)	%
Herbicides	14,727 (3)	62,778 (14)	197,894 (43)	53,107 (12)	34,142 (7)	22,884 (5)	41,168 (9)	17,554 (4)	11,315 (2)	455,569	99[a]
Insecticides	1,915 (3)	3,800 (5)	17,307 (24)	7,784 (11)	5,833 (8)	13,460 (19)	11,567 (16)	7,149 (10)	2,418 (3)	71,233	99[a]
Fungicides	<10 (<1)	80 (1)	147 (2)	38 (1)	849 (13)	4,331 (66)	923 (14)	213 (3)	12 (2)	6,593	102[a]
Other	<10 (<1)	<10 (<1)	72 (<1)	130 (1)	11,540 (47)	2,533 (10)	4,863 (20)	2,422 (12)	2,247 (9)	29,307	102[a]
Total	16,642 (3)	66,658 (12)	215,420 (39)	61,059 (11)	52,364 (9)	43,208 (8)	58,521 (10)	27,838 (5)	15,992 (3)	557,702	100

NOTE: Major crops included corn, soybeans, cotton, wheat, barley, oats, peanuts, tobacco, alfalfa, and hay.

[a] Totals do not add up to 100 due to rounding.

SOURCE: Based on data from Osteen and Szmedra, 1989.

TABLE 6-2   Total Fungicide Use on Fall Potatoes in the United States, 1991

Region[a]	Production		Hectares Treated			
	Planted (1,000 ha)	% of Total Planted Hectares	No. (1,000 ha)	% in Region	Average No. of Applications	% of Total Hectare Treatments[b]
Northeast	51	11	49	96	5	30
Midwest	133	29	119	90	4	42
West	269	59	141	52	2	28
Total	453	99	309		4	100

NOTE: Numbers do not add up to 100 due to rounding.

[a] Northeast: Maine, New York, Pennsylvania; Midwest: Michigan, Minnesota, North Dakota, Wisconsin; West: Colorado, Idaho, Oregon, Washington.

[b] Hectare treatment: number of hectares treated times number of applications per year.

SOURCE: Derived from USDA, 1992c.

ments. In contrast, the west, which accounts for 59% of all hectares planted with potatoes, accounts for only 28% of total fungicide hectare treatments. This variation of pesticides used on the same crop grown in different regions means that the amount and kind of residues will depend not only on the crop, but also on where it is grown.

THE OCCURRENCE AND FATE OF PESTICIDE RESIDUES

Pesticide residues originate when a crop or food animal (commodity) is treated with a chemical or exposed unintentionally by drift, in irrigation water, in feed, or by other routes. The size of the residue depends on the exposure level (treatment rate), its dissipation rate, environmental factors, and its physical and chemical properties. For example, an insecticide sprayed on apples may volatilize into the atmosphere. This is influenced by the insecticide's volatility or vapor pressure and the temperature and wind movement in the orchard. Removal by rainfall or overhead irrigation is governed by the insecticide's water solubility and the amount of rain or irrigation water. The chemical may also degrade (as influenced by the molecular makeup of the insecticide and by such factors as sunlight, moisture, and temperature) or it may dissipate by growth dilution (e.g., as the fruit becomes larger, the residue concentration will decrease even in the absence of physical or chemical dissipation). In farm animals and some plants, metabolism and excretion are the primary mechanisms. The degradation products become the major constituents of the remaining residue. In a few cases, chemical residue concentrations may actually increase over time after exposure ceases. This would result from weight loss by the commodity, e.g., loss resulting from the conversion of grapes to raisins after treatment with a relatively stable, nonvolatile chemical.

The overall dissipation rate is a composite of the rate constants of the individual processes (e.g., volatilization and degradation). Typically, overall residue concentrations (parent plus degradation products) decrease over time after exposure ends. Because most individual dissipation processes follow first-order kinetics, overall dissipation will have the characteristics of first-order kinetics. In first-order decline, the logarithm of concentration is linearly related to time, and a plot of concentration remaining versus time is asymptotic with respect to the time coordinate. Thus, residue concentrations will approach zero over time but in theory will never cease to exist entirely (Zweig, 1970). Stated simply, a commodity treated with or exposed to a pesticide theoretically can never totally be rid of all traces of residue. In time, however, the residue will cease to be detectable because of the limitations of current measuring instrumentation and the continuing asymptotic decline processes. This limit of detection (LOD) will therefore vary according to the sensitivity of the analytical method used. (LODs are described below under "Detection Limits.")

Conventionally, residues in raw commodities are monitored until they have declined to a concentration approximately 1/10th that of the legal maximum—that is, the tolerance or action level. Very little public monitoring is intended to identify the residues that the consumer may ingest, which may range from the legal maximum to 1/10th, 1/100th, 1/1,000th, or smaller fractions of that level on foods prepared for consumption. One can expect that consumers are exposed to small residues if their food was treated with or exposed to pesticides during production, processing, or preparation; however, we do not always know the quantity of those residues either because they are lower than the LOD or because there are no monitoring data available. For these reasons, it is difficult to estimate actual dietary exposure to pesticides and any associated risk with a high degree of certainty.

PESTICIDE REGISTRATION AND THE DEVELOPMENT OF ANALYTICAL METHODS

Early in the development of a pesticide, the manufacturer must identify the analytical methods used to ascertain the concentrations of chemicals in formulations (formulation methods) and the fate of the material on target crops, in laboratory animals and livestock, and in environmental media (soil, water, air) that might be exposed to the chemical (residue methods). Most companies that develop and register chemicals employ staffs to develop these analytical methods, whereas others hire or fund commercial or university laboratories for this purpose.

Development of analytical methods is a lengthy and technically difficult process because the methods must account for the parent chemical or control agent as well as toxicologically significant formulation impurities, metabolites, and environmental conversion products. The impurities and products may not be known early in the development phase and thus must be included later but before registration is sought from EPA. Before a pesticide can be registered under the Federal Insecticide, Fungicide, and Rodenticide Act (FIFRA), a tolerance level must be established for each food use or an exemption granted, e.g., for a pesticide that is essentially nontoxic. To obtain a formal tolerance level, pesticide manufacturers must submit their analytical methods to EPA, which then verifies that the pesticide can be detected at a certain tolerance level for each proposed food use. It is not unusual for a food tolerance level to include the parent chemical and several breakdown products. In such cases, versatile residue detection methods must be available to detect the various tolerance levels in every food or feed product for which registration is being sought. Typically, the primary method will have several variations extending it to soil, water, air, and nontarget organisms such as fish and wildlife.

The manufacturer applies these methods to determine the rate of dissipation or decline of the pesticide on target crops in field trials. The results are submitted to EPA with the registration data for use in establishing a tolerance level for the raw agricultural commodity and determining the interval required between the last application of the pesticide and harvest to achieve residues below that tolerance.

Field trials are conducted in several geographical regions of the United States that typify areas in which the crop is produced, so that different climatic conditions and soil types are represented. The test plots are treated with pesticides in concentrations high enough to eradicate a large percentage of the target pest(s). If the trials are not complete, if the data are too variable, if conversion products are not adequately included, or if the analytical methods themselves are considered imprecise, inaccurate, not sufficiently sensitive, or otherwise deficient, registration may be denied. EPA may base its judgment on the data submitted by the manufacturer, or it may inspect the company's raw data in accordance with the FIFRA provision for data audits. Field trial data are further evaluated in Chapter 7.

Methods must be provided by the manufacturer when requested by any federal or state regulatory agency and may be included in the *Pesticide Analytical Manual, Volume II (PAM II): Methods for Individual Residues,* which was first published by the Food and Drug Administration (FDA) in 1968 but has been updated in a series of revisions since then. These methods do not need to fit within the available multiresidue methods (MRMs) used by the FDA to screen food or feed products entering commerce (see section on "Methods for Sampling and Analysis," below, for a further discussion of MRMs). More recently, EPA has asked pesticide manufacturers to determine whether new compounds are detectable by existing MRMs. If they are not, however, the registration process is not impeded.

Interregional Project Number 4

Use of pesticides on some crops (e.g., strawberries, hops, artichokes, cranberries) may be too limited to provide the economic incentive needed for chemical companies to develop the analytical methods and residue data required for registration. In such cases, this work is performed by Interregional Project Number 4 (IR-4), which operates within State Agricultural Experiment Stations (SAES) with funding from USDA's Cooperative State Research Service (CSRS) and Agricultural Research Service (ARS). The nation's four IR-4 leader laboratories are located at Cornell University, the University of Florida, the University of California at Davis, and Michigan State University. Several participating laboratories are situated at other land-grant institutions and within ARS.

The IR-4 laboratories use the methods provided by manufacturers to EPA for pesticide residues on the major crops listed on the chemical's label. If the method fails on the minor crop, they modify the company method to make it fit the minor crop situation. Occasionally, they develop new methods for minor crops of interest.

SAES or ARS field scientists establish the plots, sample the commodity at harvest, and provide samples to IR-4 laboratories, which then conduct the analyses. All data are submitted to EPA. If the petition is approved, the minor crop is added to the pesticide label. IR-4 actions annually account for approximately half of the petitions processed by EPA.

Universities and the ARS

Several U.S. universities and the ARS conduct research on pesticides to study their field behavior, formation of breakdown products, persistence during food processing and storage, and analytical behavior. Many advances in food residue chemistry (e.g., detection of previously unrecognized toxic metabolites) and new approaches to residue analysis (e.g., the immunoassay) result from this basic research. In addition, this academic environment provides the training ground for pesticide scientists who eventually enter the industrial, government, and commercial sectors.

METHODS FOR SAMPLING AND ANALYSIS

Sampling

Sampling should be conducted

- by persons trained in the practice of sampling;
- randomly, so that all individuals in the population sampled have an equal chance of selection in the final analysis;
- with replication, so that analytical results can be treated statistically;
- in such a manner as to maintain sample integrity by adequate containment, preservation, and prevention of contamination; and
- with care and attention to record keeping, including visual observations, sample preservation, and safeguards against cross-contamination.

Usually omitted from reports are the manner of collecting samples (where, by whom, and how) and information on compositing, subsampling, preservation of samples and subsamples, and other important matters. Lykken (1963) generalizes, however, that all residue monitoring programs operate somewhat as follows:

- Several commodity units (e.g., bunches of grapes, oranges, heads of cabbage) are taken from the field or lot to be sampled.
- These commodity units are composited to form the gross sample.
- The gross sample is reduced in size to produce the composite sample.
- The composite units are then peeled, husked, or further reduced in size by cutting or chopping in accordance with the Code of Federal Regulations, which identifies the portion(s) of the commodity to which the tolerance applies.
- The individual parts of the commodity may then be quartered to reduce bulk and perhaps subdivided to smaller aliquots. These samples are generally frozen or preserved in some other way, transported to the laboratory, and preserved further until analyzed. If a freezer stability test is to be conducted (a recent Good Laboratory Practice [GLP] requirement; 40 CFR Part 160), control samples may be spiked at this point and then handled the same as the treated sample. This is usually done when field plots are sampled to determine residues for registration requirements, but less frequently for monitoring and enforcement of tolerance levels for registered chemicals. Sampling and sample handling for field trials are described by the National Agricultural Chemicals Association (NACA, 1988).

• At the time of analysis, individual subsamples may be more extensively chopped or blended or reduced further in size prior to extraction with a solvent and analysis. (See *PAM I* or *II* for more detailed description.)

The absence of uniform training has likely led to haphazard sampling or bias resulting in samples that are neither random nor representative. To rectify this situation, FDA and most state agencies are taking steps to improve their training and written sampling guidelines. Furthermore, true replication, with three or more field composites, appears not to have been common practice, evidenced by the fact that averages *and* standard deviations are absent from virtually all residue monitoring reports. Sample handling has improved since implementation of GLP protocols; but again, without accompanying quality assurance records, older data must be questioned for reliability. EPA and FDA are now training and certifying field inspectors to ensure proper sampling by all personnel engaged in work to meet FIFRA requirements. In addition, the American Chemical Society's Committee on Environmental Improvement has prepared a comprehensive volume dealing with the basics of environmental sampling (Keith, 1988).

Analysis

Methods for analyzing pesticides are expensive, time consuming, and difficult, and they require a skilled analyst. Furthermore, methods are tailored to specific purposes (e.g., monitoring, enforcement, or registration). As a result, considerable variability is associated with the methodology. In many cases, descriptions of differences among the specific methods used do not accompany the residue data, thus diminishing public confidence in the data. Furthermore, the committee found no analytical program directed toward water specifically as an ingredient of foods or as a component added to foods. This results in an important gap in the residue data, since water represents such a large part of the diets of infants and children.

There are two general types of analytical methods for determining residues in foods: single residue methods and multiresidue methods. These are described in the following sections.

Single Residue Methods

Single residue methods (SRMs) are used for the quantitative determination of a single pesticide (and its toxicologically important conversion products, e.g., through metabolism or degradation) in all foods for which tolerance levels have been established. This is generally the type of method submitted by the manufacturer to EPA and eventually published in *PAM II* after registration is secured. It may also be the method used (sometimes in modified form) for IR-4 petitions.

Multiresidue Methods

Multiresidue methods (MRMs) are capable of detecting and quantifying more than one pesticide in more than one food. These methods are commonly used by government agencies for surveillance and monitoring to determine which pesticides (and how much) are present in a given food sample. FDA's MRMs are published in *PAM I*; the MRMs of state agencies, foreign governments, private industry, and academia are published in the open literature or in special reports. Some MRMs are rapid; others are more comprehensive and therefore more time consuming. In general, MRMs may be used for screening and quantitation. In *screening*, MRMs are used to determine rapidly if any pesticide is present near or above the tolerance level. This approach usually precedes a more detailed analysis. Cholinesterase enzyme inhibition tests screen for organophosphorus and carbamate insecticides; insect bioassays screen for any insecticide residue. Immunoassays may be used in the future for targeted chemicals or classes of chemicals. In *quantitation*, MRMs are used to detect and measure multiple pesticide residues and their metabolites that might be present in a given sample. These MRMs are usually based on gas or liquid chromatography or both. FDA and other agencies often use simplified versions of MRMs in their surveillance program to determine if violations exist in given samples before proceeding to full quantitation with a more elaborate version. Because all MRMs can accommodate only a limited number of chemicals, agencies use SRMs for targeted pesticides that are not included in the MRM. They also use SRMs in special circumstances such as when public health is endangered by a single pesticide or when a single pesticide comes under special review and, thus, special scrutiny is required for its presence in foods.

Most laboratories improvise when using an MRM, and the improvisations are often not subject to peer review or published. Requesting the latest method from an agency is usually the only way to obtain up-to-date information on the method being used, the number of pesticides it can accommodate, and its LOD. MRMs used by regulatory laboratories are frequently modified in response to changing availability of solvents and analytical instrumentation within the laboratory and the need to expand the MRM's coverage or lower its detection limit.

Criteria for Selecting a Method

Single Residue Methods or Multiresidue Methods?

SRMs are selected when the sample is known or believed to contain the residue of a chemical not included in the MRM. MRMs are used when the residue history of the sample is unknown and the presence and quantity of pesticide residues must be determined. MRMs will provide information on a much broader range of pesticides than an SRM for the same investment of time, energy, and resources.

Breadth of Applicability

MRMs most commonly used by the FDA can determine roughly 50% of the approximately 300 pesticides with EPA tolerances and other chemicals for which no tolerances have been established. Some of the MRMs can also detect many metabolites, impurities, and alteration products of pesticides with and without tolerances (FDA, 1991). Typically *not* included are polar chemicals of high water solubility (e.g., paraquat, glyphosate), very volatile chemicals (e.g., fumigants), and compounds that are unstable to Florisil chromatography (e.g., some carbamates). An aliquot of the sample (or its extract) must be analyzed separately so that these chemicals can be included in the analytical report.

Detection Limits

All analytical methods have a limit below which the chemical could not be detected even if present. This limit of detection (LOD) is the lowest concentration that can be determined to be statistically different from a blank. Elsewhere in this report, the committee refers to the limit of quantification (LOQ), which differs from the LOD in that it refers to the concentration above which quantitative results may be obtained with a specified degree of confidence.

The LOD is influenced by extraneous, background material that is always present in the sample and the sensitivity of the instrumentation used for detection and quantification. Moreover, the LOD may vary according to application. LODs are determined by analyzing background (untreated) samples of the food products of interest and spiked samples, which contain known amounts of the chemicals. LODs for a given method will vary with the type of sample, the chemical, and the extent of sample cleanup provided.

LODs can be as low as twice the background reading. That is, a signal that is twice the background could be measured and result in a calculated residue value. In practice, however, most laboratories set an LOQ that is several times higher than the theoretical LOD. Keith (1983) provides general guidelines for establishing the LOQ, but in practice, the criteria for setting the LOQ varies among laboratories.

To be of regulatory use, detection limits must be below established tolerance levels. The California Department of Food and Agriculture sets LODs at approximately one-tenth the tolerance level; FDA generally sets them at 0.1 to 0.01 ppm, depending on the chemical; and Florida's Department of Agriculture and Consumer Services sets them at or just below tolerance in order to screen large numbers of samples for clear violations. Unfortunately, LODs are not always specified in residue reports so that samples with no detectable residue levels cannot be assigned an upper limit of finite residue content. Furthermore, the reports do not clearly describe the extent to which residues below tolerance but above LOD are quantitated and confirmed, and they may not include a complete list of the pesticides that were *not* found but could have been detected had they been present. A report of pesticides not found (i.e., below the LOD) is usually not included in descriptions of the results of the overall programs but is done when there is special regulatory interest in specific pesticide residues (McMahon and Burke, 1987). Reporting only positive findings leads to a bias in the residue results.

Accuracy and Precision

Accuracy refers to agreement between a measured value and the true value. In residue methods, accuracy is often defined as the percentage recovery. Acceptable residue methods will give 80% to 120% recovery, indicating that if 1 ppm of a chemical were present, the analytical method would yield results between 0.8 and 1.2 ppm. Precision refers to reproducibility and the variability existing in a set of replicate measurements. Precision errors caused by variable reproducibility in residue methods tend to run high, with relative standard deviations (expressed as a percent of the mean) of 25% or more. The total error (accuracy plus precision) ideally should not exceed 100%. This must be assessed by the analyst running replicate spiked samples through the method.

Speed and Cost

Regulatory agencies require fast-response methods that can produce results in an 8-hour workday or less so that produce does not spoil when awaiting the results of an analysis. These faster methods are less expensive because they require less of an analyst's time. Needless to say, however, the quickest, least expensive analytical method may not be the best one in terms of other criteria. As a result, many of the methods used are compromises of speed for quality.

Instrumentation

Most regulatory agencies rely on element-selective gas chromatography determination. The more rigorous methods based on mass spectrometry are not practical, especially for screening, given the cost of the equipment. The expense is becoming less of a barrier, however, as analytical laboratories acquire more sophisticated instruments and as the cost of technician time overtakes capital costs of instruments as the primary budgetary consideration.

Validation

Methods should be validated before they are used routinely for regulatory purposes. The most rigorous level of validation is a collaborative study of the method by several different laboratories. The Association of Official Analytical Chemists conducts such studies and publishes the validated method as "official" in the *Official Methods of Analysis*. Because this is a time-consuming process, most methods are validated less rigorously—perhaps by one cross-check either by investigators in the same laboratory or by one outside laboratory. For example, the MRM used by the California Department of Food and Agriculture was developed in-house and had not been subjected to outside collaborative validation when put into service.

MONITORING

The following discussion of monitoring activities for pesticide residues is based primarily on information that existed for 1988 and earlier. The committee realizes that changes in the design and scope of monitoring programs have occurred after 1988 but, unfortunately, information on more recent developments was not generally available for inclusion in the committee's discussion.

Federal Activities

Four federal agencies have primary jurisdiction over pesticide residues in food—the EPA, the FDA, and the USDA's Food Safety and Inspection Service (FSIS) and Agricultural Marketing Service (AMS). Their efforts are supported by USDA's ARS and CSRS and the U.S. Fish and Wildlife

Service (FWS). CSRS has its own programs through land-grant universities. All these agencies have some analytical capability associated with pesticide monitoring in foods (FDA, FSIS) or pesticide research.

Following is a list of the principal responsibilities of each agency:

The Environmental Protection Agency
• registers pesticides under FIFRA;
• sets pesticide tolerance concentrations for individual commodities, including meat and poultry, and for processed foods (the tolerance concentration for individual commodities is established by EPA for raw produce at the farm gate; as produce is processed into finished foods, pesticide concentrations may either decrease, increase, or remain the same);
• serves as lead agency for enforcement;
• reviews manufacturers' registration data, including analytical methods; and
• conducts research on the environmental fate of residues.

The Food and Drug Administration
• enforces compliance with residue tolerance concentrations in food, except meat and poultry, and feed;
• monitors residues in domestic and imported food; and
• develops analytical methods for monitoring.

The Food Safety and Inspection Service (USDA)
• enforces residue tolerance concentrations in meat and poultry;
• monitors residues in meat and poultry;
• develops analytical methods for monitoring; and
• gathers information on the incidence and concentrations of pesticide residues in the food supply.

The Agricultural Marketing Service (USDA)
• enforces compliance with residue tolerance concentrations and monitors residues in raw egg products.

The Agricultural Research Service (USDA)
• conducts research on pesticides, including efficacy and residue fate, and
• conducts ARS segment of IR-4 and National Agricultural Pesticide Impact Assessment Program (NAPIAP) operations.

The Cooperative State Research Service (USDA)
• oversees and funds NAPIAP and IR-4, both of which operate at land-grant universities.

The Fish and Wildlife Service
• monitors pesticides in fish and wildlife.

The accomplishments and shortcomings of all programs in sampling and analyzing pesticide residues in foods were reviewed by the Office of Technology Assessment (OTA, 1988). Much of the following description is based on that report.

EPA does not monitor pesticide residues in food. The agency's residue chemistry section within the Office of Pesticide Program's Registration Division reviews registration data compiled by pesticide manufacturers, and its laboratories in Beltsville, Maryland, and Bay St. Louis, Mississippi, may test the methods with spiked samples. Acceptable methods are submitted for publication in *PAM II*. EPA has for some time been studying the feasibility of combining the wide array of data bases in existence to maximize their utility for scientific and regulatory purposes. In 1989, for example, the agency contracted with Dynamac Corporation to compile and summarize residue data obtained from Agriculture Canada, FDA's state monitoring program, and the National Food Processors Association (NFPA). Together these sources provided data on 286 pesticides in an estimated 49,857 samples. In a report prepared for the EPA, Dynamac (1989) noted the difficulties encountered in attempts to compare these data bases, especially the differences in information reported and sampling methods. It made three fundamental recommendations intended to improve the utility of the data with minimum cost increases: that standard residue sampling protocols be used by state and federal agencies to facilitate comparison, that certain minimal information be provided with each sample (for example, identification of sample, purpose of sampling, and analytical method used), and that a standard data coding system and data base format be used.

EPA recognizes the need for uniform record keeping, sampling, and analytical methods in order to determine exposure and assess risk, especially for infants and children. It is important to this effort that EPA information be made compatible with FDA data. The utility of the data for estimating exposure and risk varies with the intended purpose of the monitoring programs. At present, EPA is evaluating the feasibility of drawing on the many and varied food intake and dietary exposure data bases to improve assessments of total human exposure. This information-gathering activity is one component of a larger effort to design a national human exposure survey that, among other things, will measure the route, magnitude, duration, and frequency of human exposure to environmental chemicals.

In May 1988, FDA's MRMs included 316 pesticides for which tolerance levels had been set, 74 pesticides with temporary and pending tolerances, 56 pesticides with no EPA tolerance levels (those previously canceled or those used only in foreign countries), and 297 metabolites, impurities,

FDA Pesticide in Food Monitoring Program in 1987

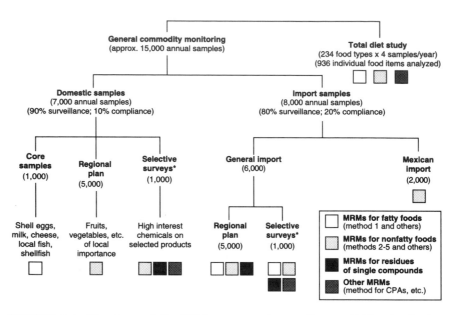

FIGURE 6-1 Structure of the FDA program to monitor pesticides in foods. SOURCE: McMahon and Burke, 1987. Reprinted from the *Journal of the AOAC*, Volume 70, Number 6, pages 1072-1081, 1987. Copyright 1987 by AOAC International.

inert ingredients, and other pesticide-associated chemicals (OTA, 1988). Of these, 35% to 40% were covered by five primary FDA MRMs: one liquid chromatographic method used primarily for N-methyl carbamate pesticides and four gas chromatographic methods—one for organophosphorus pesticides and metabolites, one for both polar and nonpolar pesticides involving a variety of selective detectors, and two for nonpolar (primarily organochlorine and organophosphorus) pesticides in fatty and nonfatty foods.

In its monitoring program in 1987, FDA analyzed approximately 15,000 commodity samples in its 16 laboratories (Figure 6-1). Most samples were collected at random; the remainder were taken from targeted food sources after a violation or suspected violation. Approximately 7,000 of the samples are domestic and 8,000 are imported.

Since the early 1960s, FDA has gathered its information on pesticide residues through its Total Diet Study (TDS), also called the Market Basket Study, in which the dietary intakes of pesticide residues (as well as some

industrial chemicals, toxic substances, and essential minerals) are estimated for eight age and sex groups from infants to senior citizens. To
accomplish this, FDA personnel purchase foods from local supermarkets
or grocery stores four or five times per year in three cities in each of four
different geographic regions of the United States. The cities are changed
each year. Each market basket contains 234 food items intended to be
representative of the diet of the U.S. population. The foods are prepared
for consumption, e.g., by peeling bananas or making beef and vegetable
stew, and then are analyzed for pesticide residues. The results, combined
with food consumption data, provide a model of dietary intakes. Because
of the limitation of the food intake data and residue monitoring methods,
coverage is not complete. Human exposure to all pesticides cannot be
estimated because some pesticides cannot be detected by the analytical
methods used. It is even more difficult to derive estimates of human
exposure for population subgroups.

FDA's program has been criticized as being too slow in terms of analyses, in need of better sampling and enforcement of imported foods (GAO,
1992), and too limited in the numbers and types of pesticides detected
(GAO, 1986a,b,c). Despite these criticisms and the age of the data, this
program proved to be an important source of information for the committee's purposes. If the TDS were improved, risk from exposure to dietary
residues could be better assessed. Increased funding for TDS would almost
certainly be required to improve it. Later in this chapter, the committee
addresses how the sampling program might be restructured to provide
the data necessary to estimate the exposures of infants and children.

FSIS in its National Residue Program annually analyzes approximately
50,000 samples for about 100 residues of pesticides, animal drugs, and
environmental contaminants in meat, poultry, and raw egg products.
About one-third of the samples are analyzed for pesticides. Most samples
are collected at random by FSIS inspectors located at slaughterhouses.
The results are reported in the *Journal of the Association of Official Analytical
Chemists* and in a series of annual reports entitled "Residues in Foods"
issued by FSIS.

State Activities

Thirty-eight states monitor pesticide residues in food but vary widely
in the number of samples they process and the purposes of their programs.
Figure 6-2 shows the range of sampling activity for 10 states in 1987.
California has the largest and oldest program, which is designed to monitor the major raw commodities produced in, or imported into, the state.
Its purpose is to enforce tolerance levels for residues on both domestic and
imported commodities. This program is administered by the California

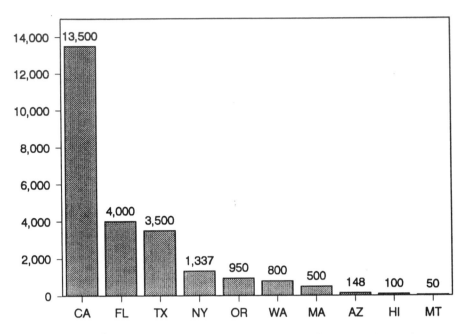

FIGURE 6-2 The number of food samples analyzed for pesticide residues in monitoring programs conducted by 10 states in 1987. SOURCE: OTA, 1988.

Department of Food and Agriculture, which in 1987 analyzed 13,500 samples of fresh fruits, nuts, and vegetables (State of California, 1987). Although routine postharvest monitoring is the largest component of the California program, efforts also include considerable preharvest monitoring (for early detection and deterrence), focused monitoring (to determine levels of specific chemicals of primary health concern), and foods to be processed (samples destined for processing and taken up to the point of actual processing).

Florida's Department of Agriculture and Consumer Services began monitoring raw agricultural commodities for pesticide residues in 1960. That program is targeted to potential problem areas in contrast to the random sampling conducted in California.

Many states have much more extensive pesticide analysis programs than is apparent in Figure 6-2, since they incorporate nonfood pesticide monitoring such as programs for farm worker health and safety and for groundwater contamination. To coordinate pesticide residue data from states and the FDA, the Mississippi State Chemical Laboratory, with FDA funding, has developed two U.S. data collection and dissemination programs: FeedCon provides information on contaminants in animal feeds;

FoodContam provides similar information for human foods. Participation in these programs is voluntary. At present, 21 states participate in Food-Contam. There is an effort to enlist all states and agencies as data sources (Minyard et al., 1989).

Food Processing Industry Activities

The food processing industry has a special interest in pesticide residues in produce. If processed and packaged foods are found to contain illegal residues or residues with the potential to cause adverse health effects, entire lots may need to be recalled from distribution centers and even from grocer's shelves. Tolerance levels must be established for any residue that concentrates during food processing (e.g., milling, cooking, and dehydrating). The behavior of residues during processing is evaluated based on processing studies required of the pesticide manufacturer before registration. Nevertheless, continual surveillance is needed to detect any unanticipated behavior, such as the formation of a previously unrecognized metabolite (Elkins, 1989).

The National Food Processors Association (NFPA) represents nearly 600 companies, including most of the major food processing companies in the United States. Approximately 450 of these companies are involved in processing common foodstuffs. Since 1960 the food processing industry has used the NFPA Protective Screen Program—detailed recommendations for preventing illegal and unnecessary residues, published annually in the *Almanac of the Canning, Freezing and Preserving Industry* (see, e.g., Judge, 1992). The recommendations involve washing, blanching, and processing steps, and emphasize proper use of chemicals at the farm source. NFPA operates National Food Labs in Dublin, California, in Washington, D.C., and in Seattle, Washington, to which food processors may submit fresh and processed commodities. Many food manufacturers and trade associations maintain analytical programs and data bank resources to accomplish the same purpose wherever possible and generally use the same multiresidue procedures as FDA. For example, NFPA uses the Luke method (*PAM* 212-3) to analyze for chlorinated hydrocarbons, organophosphates, carbamates, and substituted ureas.

In 1988 EPA enlisted the help of NFPA and other groups to determine the availability and quality of data on pesticide residues in foods. Data were collected from 16 companies, including 3 major baby food companies, on more than 86,000 samples. Results indicated that 81% of these had no detectable residues (E. Elkins, NFPA, personal commun., 1992). More than half of the samples (47,737) were classified as fruit. Vegetables (14,037), tomatoes (8,424), and meat (8,215) were the next highest categories. Potatoes (964) were given a separate category. The balance of the samples (6,947) were combined under miscellaneous; these were mostly flour samples in which the pesticide of interest was ethylene dibromide (EDB). In the fruit category, 8,573 of the samples were fresh or juice concentrates that were analyzed for Alar from 1985 to the present. Of these, 8,056 were negative for Alar.

From 1988 to the present, NFPA has been building a new pesticide residue data base that currently contains almost 74,000 samples, 97.5% of which had residues below the limit of quantification (LOQ) of the method used (E. Elkins, NFPA, personal commun., 1992). In the current data base, only 20% of the samples are raw foods. The balance is processed. Data from NFPA members usually pertain to crops on which the pesticide is known to have been used, but other data are random.

The data have been obtained from the food industry, from NFPA, and from other sources. NFPA plans to add data from the FDA/USDA APHIS monitoring and expects to eventually have more than 100,000 samples.

The pesticide residues reported so far include aldicarb on bananas, potatoes, and oranges; benomyl on apples and tomatoes; guthion on tomatoes and peaches; methamidophos in tomatoes; malathion on wheat; and diazenon on apples and tomatoes. The concentrations reported were at or near the LOQ.

Private, for-profit certification programs serve as third parties to verify the retailer's claims regarding residues on their produce. Some of the programs also certify growers who comply with a full disclosure statement of chemical usage and allow sample testing of their produce at random. Some large farming operations and commodity groups are beginning to do their own monitoring to ensure acceptability of their produce.

Private Laboratories

Commercial analytical laboratories have flourished in recent years, especially in analyses for pesticides and other toxicants in foods, water, soil, and waste sites. These laboratories must undergo a rigorous EPA certification procedure if they are to analyze groundwater/drinking water contamination by the pollutants of priority interest to the agency. There has been no similar certification for food residue analyses by either the FDA or any other agency. However, some states (such as California) are now instituting laboratory certification programs that involve inspection, performance standards, and adherence to GLPs. As a result of these certification programs, the data generated by these laboratories are becoming more consistent than in the past and promise greater reliability and more standardized operating procedures in the future.

QUALITY CONTROLS

All analytical laboratories (government, academic, and private) performing work of a regulatory nature must comply with GLPs as set forth in the *Federal Register*. In practice, this requires written protocols for

- sample preservation;
- sample processing and extraction;
- sample preparation, including cleanup, concentration, and derivatization;
- sample determination, including instrument tune-up and operating parameters, and confirmatory measures; and
- data handling, including maintenance and interpretation of records.

Quality assurance/quality control (QA/QC) procedures specify that the method be validated by spiking-recovery experiments designed to assess each portion of the overall method from sample preservation through tabulating and interpreting numerical results. A QA officer must oversee the analyst and analysis. Records must be kept so that an outsider can reconstruct the analysis, including calculation of final results.

Because the GLPs and the quality assurance and control procedures are relatively recent developments, the quality and completeness of recent data may differ considerably from those produced from analyses done 3 or more years ago (Garner and Barge, 1988). In the past, for example, a zero was entered into the NFPA data base when no residue was detected, and the zero was averaged with the positive values. Today, the term *None Detected* replaces the zero and the method's detection limit is stated. Furthermore, NFPA data were sometimes obtained by nonstandard methods (e.g., methods not in *PAM*) or without stated recoveries, detection limits, or quality assurance measures. At present, however, NFPA is requiring that each contributor of data complete a residue report form for each reported pesticide-food combination. Among the information required are the analytical method used, detection limits, quantitation limits, and recovery information. In addition, there has been no coordination with federal and state monitoring efforts but, as noted above, NFPA plans to include FDA and USDA in the future. Thus, NFPA's findings have added considerably to the data base on exposure to residues in food because the samples were "as-served," finished products, and the ongoing effort promises to be even more useful because of the new requirements.

MRMs used by individual agencies or laboratories may be modified, usually are not peer reviewed, and usually are not externally validated. Although this does not necessarily make them less reliable or of lesser quality than standard methods, the lack of external quality assurance makes it difficult to assess their reliability. Improvements can be seen as a growing number of analytical laboratories adopt GLP. Compliance data may have the best quality because they require more stringent methodological control.

In general, the lack of quality control procedures limits assessments of data quality. Confidence intervals cannot be assigned to monitoring results because of the general lack of sample replication, and degrees of uncertainty cannot be clearly defined.

LIMITATIONS OF THE DATA

Sources of Error

All residue determinations are subject to error caused by limitations in the training or skill of the analyst, in laboratory glassware and other equipment, in the reference standards used as the basis for quantitation, and in the instruments (chromatographs and spectrometers) used in the final determination. Error is also partially attributable to difficulty in identifying a relatively low response above a high (and variable) background noise from nonpesticide materials in the sample. As noted previously, total errors of 50% or more are not uncommon, even for standard procedures in the hands of experienced analysts. For example, a method that gives 80 ± 20% recovery (20% relative error and 20% relative standard deviation) yields a total error of 60%. In most cases, this is quite acceptable for laboratory validation of a new method. Such errors can be dealt with by running a larger number of replicate samples—a costly solution that is not often used.

Another source of error or uncertainty is the large, difficult-to-gauge variability due to sampling. This variability may be small in well-designed field experiments in which a chemical is applied by a calibrated sprayer to a uniform stand of grapes, or it may be unknown (but almost certainly large) when samples are taken from large shipping bins of grapes from several fields treated with different chemicals and applied by a variety of spray equipment. It is difficult to estimate the sampling error in random sampling of a field or lot, especially because of differences in the training of personnel and sampling protocols. A sampling error of 100% is probably conservative. When added to an analytical error of 50 to 100%, the overall uncertainty in a given analysis of a single commodity may be 200% or more. This is in addition to the inherent variability of residues due to uneven field applications of pesticides to the commodity of interest. Thus, a residue analysis that yields a 1-ppm level should be assigned a range of 0.3 to 3 ppm in the absence of any accompanying data relating to the actual error limits (Keith, 1983; Hance, 1989). This error tends to swamp

TABLE 6-3   Consequences of Averaging Fictitious
Residue Data Sets in Which Nondetects[a] Are Used at
LOQ, 0, and 0.5 of LOQ

	Case 1. Low Residues			Case 2. High Residues		
	LOQ	0	0.5 of LOQ	LOQ	0	0.5 of LOQ
	1.5	1.5	1.5	5.5	5.5	5.5
	0.5	0.5	0.5	0.5	0.5	0.5
	0.1	0.1	0.1	0.1	0.1	0.1
	<0.1	0	0.05	<0.1	0	0.05
	<0.1	0	0.05	<0.1	0	0.05
	<0.1	0	0.05	<0.1	0	0.05
	<0.1	0	0.05	<0.1	0	0.05
	<0.1	0	0.05	<0.1	0	0.05
Average	0.32	0.26	0.29	0.83	0.76	0.79

[a] Samples for which analyses show no residue detected at or above
the LOQ.

the generally small variation in the data sets in which LOQ, 0, or 0.5 of
LOQ are used to compute averages. This can be seen in Table 6-3, which
shows the consequences of averaging fictitious residue data when nondetects are assumed to be at the LOQ, zero, or one-half the LOQ.

Other Limitations

The limitations of the residue data derive also from the lack of consis-
tency among methods used for sampling, analyzing, and processing residue data. The lack of commonality among the analytical methods used
by agencies to monitor food for residues, in the number of chemicals
included in the studies, especially metabolites and degradation products,
or in the limits of detection of the methods impedes comparability of data
and limits the utility of the data for exposure assessment. Thus the basic
validity of the sample, including the extent to which it represents the
population sampled, is frequently difficult to assess. This contributes to
the uncertainty in the final residue report.

Federal, state, and industry groups differ considerably in their processing and reporting of data on residue levels. Residues that exceed
established EPA tolerance levels are subject to confirmation and are included in the residue reports. Residues below the tolerance levels are
usually reported but are not always confirmed. When a given method
includes chemicals that are not present above the LOD, these absences
are not always specified in the final residue report.

In calculating averages, some laboratories use only positive data (above the LOD), some include the LOD for nondetectables, and others enter a zero for nondetectables. The rigor with which positives and nondetectables are recorded, and positives are confirmed, varies from laboratory to laboratory. Thus it is difficult to judge the quality associated with individual residue results or averages. These are some of the reasons why it is not always possible to assess a given data set or to include its data in calculating average dietary exposures of the U.S. population.

Data are collected for different reasons and from clearly different populations. *Random sampling* of consumption data reflects overall estimates of dietary and pesticide intake. *Surveillance* or *compliance samples* are directed to problem areas suspected of violating tolerance levels and therefore involve intense appraisal of products to which the compounds have been applied.

The biases that exist in terms of the number (frequency) and types of samples collected are apparent. Negative observations are more frequent in a truly random sampling program. On the other hand, surveillance, targeted, or compliance sampling may provide primarily positive values. Residues are expected in these types of sampling because they are designed primarily to intercept violations of FIFRA and related federal and state codes. Identification, separation, and calculation of the discrete observation categories are essential, because data are drawn from clearly different populations. Yet many residue report compilations do not discriminate between the categories, or reasons, for sampling.

The results of many residue data sets are clearly skewed due to inadequate sampling plans or a heavy emphasis on pesticides or products that have high potential for violation or health risk. Sampling may be biased to seek positive results when application of the pesticides is known. Sampling intensity tends to decrease when the potential for a measurable residue does not exist, for example, when the pesticide is not used. This impedes assessment of residue exposure of children or infants, whose primary food items may not be sampled frequently enough to provide a broad data base of residue information.

A lack of detectable residues may be an important signal that the compound is not used because of a lack of approval, local practice, or lack of a need to control pests. Several factors may lead to uneven, or no, use of a pesticide, even when it has been granted regulatory clearance. Prominent among these are climatic conditions, agricultural practices, and economic exclusion. Because pesticide costs account for a relatively large portion of the total cost of growing crops, unneeded pesticides or excessive quantities of pesticides are rarely used and expensive products are replaced by less costly compounds when available. Furthermore, pesticide usage may be governed by recommendations issuing from a state university extension program or a licensed pest control advisor and by the preference of the grower.

Agricultural practices can also be influenced by regulatory matters, consumer attitude (e.g., toward Alar), and manufacturer decision (e.g., about benomyl). Approvals may have been terminated for a specific use of a compound. Examples include ethylenebisdithiocarbamates (EBDCs) and 1,3-dichloropropene (1,3-D). 1,3-D was suspended by only one state (California); its use was continued in others. Integrated pest management programs also favor some chemicals over others, or result in the increasing substitution of nonchemical control measures for chemical compounds.

When applied according to recommendation and practice, some pesticides are used to maintain soil conditions and are not absorbed by the plants. For example, when applied to soil according to specifications, aldicarb maintains soil activity and is not absorbed systemically by all produce. Some pesticides dissipate so rapidly that residues are usually not detectable at harvest.

Knowledge of the relationship between residue level and the amount actually consumed is incomplete. This stems from a generally inadequate understanding of residue behavior during the storage, processing, and preparation of foods as influenced by biological constituents, pH, temperature, moisture levels, and heat treatment. Data on residue behavior provided by registrants are not adequate. Extrapolation of the limited data to population subgroups may be subject to major sources of error.

PESTICIDES IN WATER

Because more water is consumed per kilogram of body weight than any other item in the diet (see Chapter 5), it is an important medium to consider in assessing total dietary exposure. For the pesticides examined in the *Nonoccupational Pesticide Exposure Study* (Immirman and Schaum, 1990), "exposure [to pesticides] from drinking water appeared to be minimal." (See Chapter 7, section on "Nondietary Exposure to Pesticides.") Unfortunately, however, the contribution of pesticide residues in drinking water is difficult to assess because of the variety of water sources (surface water and groundwater; public and private water systems) and because of the geographic differences and seasonal variations in pesticide use and consumption patterns. Moreover, no single survey of pesticides in food commodities has included both surface and groundwater sources of drinking water. As a consequence, it is not yet possible to estimate with any degree of certainty all the variations that must be considered in assessing dietary exposure to pesticide residues in water used in the processing and preparation of foods. The data that have been produced are discussed in the following paragraphs.

Groundwater

Approximately 53% of the U.S. population (more than 97% in rural areas) draws its drinking water from groundwater sources (USGS, 1988), which supply 40.1% of the water in public systems (USGS, 1990). This source of water has been the subject of several studies.

Hallberg (1989) reported that residues of 39 pesticides and their degradation products have been detected in the groundwater of 34 states and Canadian provinces. The sources of these data ranged from controlled field studies to ongoing programs to monitor public water systems. The pesticides most frequently reported were mobile or volatile compounds used in soil treatments, such as aldicarb and its products, which were detected in 24 states from California to Maine. EDB was found in 12 states, 1,2-dichloropropane (1,2-D) in 7 states, and dibromochloropropane (DBCP) in 5 states. Also prominent were herbicides widely used in the humid regions of the corn belt. These included alachlor, atrazine and its products, cyanazine, dicamba, dinoseb, metolachlor, metribuzin, simazine, trifluralin, and 2,4-D. The most frequently detected pesticides were the triazine herbicides atrazine, cyanazine, and simazine.

Most of the herbicides found in groundwater in these studies are still widely applied (USDA, 1992b); however, many of the fumigants and nematicides are no longer in use. According to a study conducted for the EPA by Williams et al. (1988), registrations of 16 pesticides have been canceled or their use severely restricted, including the fumigants and nematicides 1,2-D, DBCP, and EDB.

In a compilation of data from groundwater monitoring studies conducted by pesticide registrants, universities, and government agencies, Williams and colleagues confirmed detections of 46 pesticides in the groundwater of 26 states resulting from normal agricultural use and 32 pesticides in 12 states attributed to point sources or pesticide misuses. Most frequently reported were atrazine (normal use, 13 states; point source, 7 states) and alachlor (normal use, 12 states; point source, 7 states). The median and maximum concentrations reported as a result of normal use were 0.90 ppb and 113 ppb for alachlor and 0.50 ppb and 40 ppb for atrazine.

In mid-1987, the Monsanto Company, manufacturer of alachlor, initiated the National Alachlor Well Water Survey (Holden and Graham, 1990). Its purpose was to estimate the proportion of private rural domestic wells that contained detectable residues of alachlor and other herbicides (atrazine, cyanazine, metolachlor, and simazine). The investigators used a three-stage, stratified, unequal probability selection procedure to obtain samples of 1,430 wells from the estimated 6 million private, rural, domestic wells located in the 89 counties in 45 states where alachlor was sold in 1986. These wells serve 6.5 million households consisting of 20 million people. The wells located in areas of highest alachlor use and in areas of groundwater vulnerability had a higher probability of selection. Most counties were located in the midwest, northeast, and southeast, where pesticides containing alachlor are used primarily to control annual grasses and certain broadleaf weeds in corn, soybeans, and peanuts.

The results indicated that 100,000 people in the sampled area are consuming water from wells with detectable concentrations of the compound. They also suggest that an estimated 36,000 people are exposed to minimum concentrations of 0.2 μg/liter and that 3,000 people are exposed to concentrations at or exceeding 2 μg/liter—the maximum contaminant level (MCL).

The investigators found that 12.95% of the wells contained detectable residue levels of herbicides, the five highest being atrazine in 11.68%, alachlor in 0.78%, metolachlor in 1.02%, cyanazine in 0.28%, and simazine in 1.6%. Not only was atrazine detected in the highest percentage of the wells, it also exceeded the proposed MCL level by the highest percentage (0.09%), compared with alachlor (0.02%) and simazine (0.01%). Metolachlor and cyanazine were not found in levels higher than the MCL (Holden et al., 1990).

In 1990 the EPA completed a 5-year National Survey of Pesticides in Drinking Water Wells, the first survey undertaken to estimate the frequency and occurrence with which pesticides and their degradation products as well as nitrate were detected in drinking water wells (EPA, 1990). The investigators sampled 1,349 drinking water wells for 126 pesticides and products as a statistical representation of the more than 10.5 million rural domestic wells and 94,600 wells operated by the 38,300 community water systems that use groundwater. Their findings indicate that 10.4% [6.8-14.1%, 95% CI (confidence interval)] of the community water system wells and 4.2% (2.3-6.2%, 95% CI) of the rural domestic wells contain more than one pesticide.

The pesticides most frequently detected were the degradation products of the herbicide 2,3,5,6-tetrachloro-1,4-benzenedicarboxylic acid dimethyl ester (DCPA), which were found in 6.4% of the community wells sampled and in 2.5% of the rural domestic wells. Next highest was atrazine (1.7% of the community wells and 0.7% of the domestic wells). Simazine, prometon, and DBCP were also found in both sources of groundwater, but with considerably lower frequency. Other pesticides found at lower frequencies were hexachlorobenzene and dinoseb in community wells and EDB, γ-hexachlorocyclohexane (γ-HCH), ethylene thiourea (ETU), bentazon, and alachlor in domestic wells in the United States.

Reported levels of DCPA products ranged from 0.35 to 0.89 μg/liter in community systems and 0.22 to 0.63 μg/liter in domestic wells. Atrazine concentrations were 0.20 to 0.81 μg/liter and 0.18 to 1.04 μg/liter in community and domestic wells, respectively. The level of at least one pesticide was found to exceed the MCL or health advisory level (HAL) in 0.6% of the domestic wells and 0.8% of the community wells.

The EPA survey was designed to examine the relationships among contamination, groundwater vulnerability, and intensity of agriculture. Although the survey was stratified by patterns of pesticide use and groundwater vulnerability, the number of positive samples was generally too low to be considered representative of groundwater contamination or that could be used effectively by the committee in the estimation of exposure through groundwater.

Surface Water

Surface water contributes 59.9% of the water in public water systems (USGS, 1990) and supplies drinking water to approximately 47% of the U.S. population (USGS, 1988). The data on this source of the nation's water supply are even sparser than those for groundwater.

In addition to his study of groundwater, Hallberg (1989) compiled data on pesticide detections in raw and finished drinking water drawn from surface water supplies in Illinois, Iowa, Kansas, and Ohio. In most state samples, the detection rate for the herbicides alachlor, atrazine, cyanazine, metolachlor, and 2,4-D exceeded 67% in both raw and finished water. Less frequently detected were the herbicides butylate, dicamba, linuron, metribuzine, simazine, and trifluralin and the insecticides carbofuran and chlorpyrifos. Alachlor was detected in 54% of 334 raw water samples in Illinois, 67% of 15 raw water samples and 52% of 33 treated samples in Iowa, and 100% of 3 raw water and 4 treated samples in Kansas. Atrazine was found in raw water in 77% of the Illinois samples, in 93% of the Iowa samples, in 100% of the Kansas samples, and in lower percentages of the samples obtained in 12 other states (Hallberg, 1989).

Seasonal variations in pesticide concentrations in surface water are striking (Figure 6-3). Baker and Richards (1989) reported that time-weighted concentrations of atrazine, alachlor, and metolachlor peaked during the late spring and early summer months in the Maumee River, the Sandusky River, and Honey Creek—all in Ohio. Atrazine, alachlor, and metolachlor concentrations exceeded maximum contaminant levels and health advisory levels during that period but were well below them during the rest of the year (Figure 6-3). Similar seasonal patterns were noted in statewide observations of atrazine concentrations in Illinois (Good, 1988).

During the reporting period (1983 through 1988) in the Ohio study, the

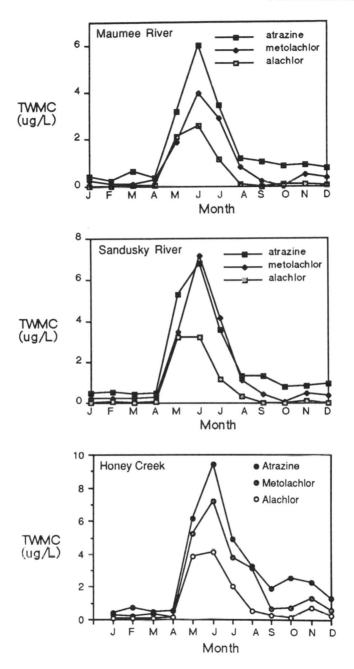

FIGURE 6-3 Monthly time-weighted mean concentrations of atrazine, alachlor, and metolachlor in 1983-1988 reported at the Maumee River, Sandusky River, and Honey Creek stations in Ohio. SOURCE: Baker and Richards, 1989.

concentrations of the three pesticides also varied extensively from year to year. In the Sandusky River, for example, a time-weighted mean atrazine concentration of 7.81 μg/liter was reported in 1986 compared with 0.90 μg/liter in 1988. In general, the highest mean concentrations for all pesticides in all rivers were highest in 1986 and 1987, dropping in 1988 to levels lower than those recorded at the beginning of the observation period in 1983. The authors attributed these differences to variations in the timing, intensity, and amount of rainfall.

Thurman et al. (1991) reported similar results from a study of a much larger watershed encompassing most of the corn belt region (parts of South Dakota, Nebraska, Kansas, Minnesota, Iowa, Missouri, Wisconsin, Illinois, Michigan, Indiana, Kentucky, and Ohio). During May and June (the planting period), median concentrations of atrazine, alachlor, cyanazine, and metolachlor were 10 times higher than levels reported in March and April (before planting) and in October and November (after harvest).

The occurrence of herbicides (primarily molinate and thiobencarb) in California's Sacramento River has been monitored extensively. Peak concentrations, which occur during May and June, have declined over the past decade because of the initiation of water management programs designed to minimize herbicide runoff from rice fields (Department of Fish and Game, State of California, 1990).

The Importance of Water Data to Infants and Children

As demonstrated in Chapter 5, water is an important component of the diets of infants and children. Water consumed by itself, water added to infant formula, and water used in the preparation of foods may represent a significant source of pesticide exposure by ingestion. However, because of the limited information on pesticide residues in water and the lack of monitoring data on water intake by infants and children, quantitative risk estimates cannot be made at this time. The committee noted, however, that water residues tend to run in the low or sub-ppb levels when present, so that the contribution of waterborne residues to ingested food prepared by using water will generally be expected to be low, except in specific locations where water contamination is far above the U.S. average.

PESTICIDES IN INFANT FORMULA

Infant formula is one of the most important processed foods fed to babies not breastfed because it is usually their sole source of nutrition during the first few months of life. Although pesticide application to some components of processed foods is likely to have occurred at some point (e.g., field applications to crops used as infant formula ingredients), measurements have consistently demonstrated that no pesticides are detected in finished infant formulas (Gelardi and Mountford, 1993). These invariably negative analytical findings are attributable to ingredient selection and processing procedures that reduce the potential for pesticide residues to appear in the finished product.

In preparing ingredients for use in infant formula, manufacturers use numerous separation and purification procedures and heat treatments that reduce pesticide residues in raw agricultural commodities (Swern, 1979; Pancoast and Junk, 1980; Considine and Considine, 1982a,b; Snyder and Kwon, 1987). Chemical and physical processes of refinement and purification include washing, solvent extraction, filtration (including carbon filtration), acidification, basic extractions, clarification (centrifugation), crystallization, deodorization, evaporation, spray drying, and heat treatments such as ultra-high temperature (UHT). Because of processing and the relatively low levels of the individual ingredients in finished products (e.g., soybean oil constitutes only 1.7% of some formulas), any likely pesticide residues in or on raw agricultural commodities are reduced below detectable limits.

Water is the principal ingredient (by weight and volume) of all liquid infant formulas. For example, ingredient water comprises approximately 87% of commercial ready-to-feed infant formulas. Ingredient water used in the processing of most infant formula is passed through activated carbon filtration columns. Analysis of influent and effluent for trihalomethanes (THMs) has shown the columns to be highly efficient at removing THMs from the water. THMs are among the most difficult compounds to remove from water by activated carbon filtration (McGuire and Suffer, 1983). Water treated in this manner is therefore considered by the manufacturers to be free of pesticide residues.

Infant formulas are broadly classified into two categories: those based on cow's milk and those based on soy protein. The manufacturing systems for producing the protein and carbohydrate ingredients used in the two formula types are quite different.

Infant Formula Based on Cow's Milk

The principal ingredients of infant formula based on cow's milk include cow's milk solids (after milk fat is removed), lactose (derived from cow's milk), and a combination of fats to provide an optimal lipid source for infants. In some cases, whey proteins derived from cow's milk are also a part of the formulation. The composition of a typical infant formula based on cow's milk is shown in Figure 6-4.

The effects of processing (e.g., fat removal, isolation of lactose, isolation of whey proteins) on any potential residues of pesticides of interest can

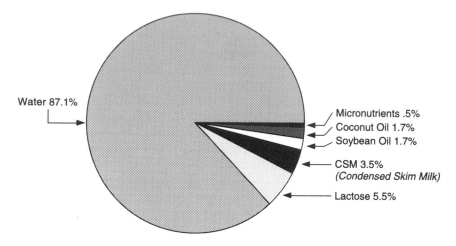

Water 87.1%

Micronutrients .5%
Coconut Oil 1.7%
Soybean Oil 1.7%

CSM 3.5%
(Condensed Skim Milk)

Lactose 5.5%

FIGURE 6-4   The composition of a typical infant formula based on cow's milk.
SOURCE: Infant Formula Council, Atlanta, Georgia, unpublished.

be logically postulated from the chemistry of the pesticides and the effects of various stages of processing. The production of lactose and whey protein concentrate is illustrated in Figure 6-5. Figure 6-6 illustrates the production of condensed skim milk and nonfat dry milk from raw cow's milk. Temperature extremes and highly effective purification processes, such as crystallization, can be expected to reduce any pesticide residues (Buchel, 1983; Hartley and Kidd, 1983; Hayes, 1975).

The likely effect of processing potential pesticide residues in raw cow's milk and the contribution of ingredients derived from cow's milk is illustrated in the following example. Assume that chlorpyrifos is present at its tolerance level of 0.020 ppm in raw cow's milk (FDA, 1989). Processing of milk to lactose could be expected to reduce that residue to 0.0002 ppm in the lactose (R.C. Gelardi, Infant Formula Council, personal commun., 1990). This is based on the assumption that 99% of any residue present would remain in the curd along with the lipids during the processing of milk into lactose (see Figure 6-5) because of the hydrophobic nature of chlorpyrifos (79% solubility in isooctane compared with 0.0002% in water). Similarly, processing is likely to reduce the residue in raw milk to 0.0005 ppm in condensed skim milk, assuming that 99% of any residue present would remain in the cream along with the lipids during the processing of milk into condensed skim milk (see Figure 6-6) because of the hydrophobic nature of chlorpyrifos (R.C. Gelardi, Infant Formula Council, personal commun., 1990). A concentration factor of 2.5 is assumed for finished condensed skim milk, resulting in a concentrating effect on the 1% of the residue assumed to reside in the skim milk fraction. Because infant formula contains 5.5% lactose and 3.5% condensed skim milk, the total possible theoretical contribution of these two milk-derived ingredients to the formula is 0.00003 ppm. This calculation is based on the preceding worst case assumptions and the percentages of lactose and condensed skim milk in a typical cow's milk-based infant formula, as noted in Figure 6-4. Thus, although condensed skim milk and lactose are major ingredients in infant formulas based on cow's milk, they are not expected to contribute greatly to potential pesticide residues.

The primary sources of lipids used in these formulas are soy oil and coconut oil. As with lactose and condensed skim milk, the contribution of these oil ingredients to pesticide residue levels in the finished product can be theoretically predicted. For example, the EPA tolerance concentration for malathion on soybeans is 8 ppm. Processing is expected to reduce the residue level in soybean oil to 4 ppm. And since infant formula based on cow's milk contains about 1.7% soybean oil, the theoretical maximum concentration of this pesticide in the soybean oil contained in the finished product would be 0.068 ppm.

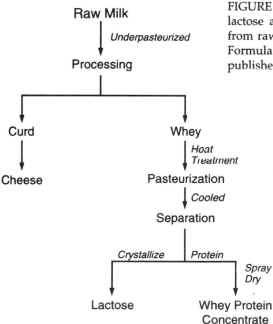

FIGURE 6-5 Steps in the production of lactose and whey protein concentrate from raw cow's milk. SOURCE: Infant Formula Council, Atlanta, Georgia, unpublished.

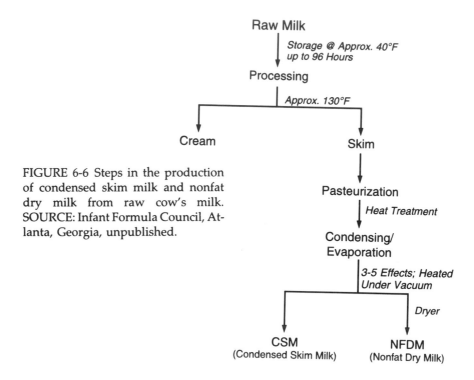

FIGURE 6-6 Steps in the production of condensed skim milk and nonfat dry milk from raw cow's milk. SOURCE: Infant Formula Council, Atlanta, Georgia, unpublished.

Soy-Based Infant Formula

The composition of a typical soy-based infant formula is shown in Figure 6-7. The principal ingredients include soy protein isolate and soy oil (the predominant ingredients derived from soybeans) as well as corn syrup solids, sucrose, and coconut oil. Soy protein isolate is a specific protein fraction derived from soybeans. Following isolation, purification and modification are required to provide a protein source that will be nutritionally beneficial to infants. The soy protein isolation process involves several physical and chemical operations that effectively decrease any pesticide concentrations (Figure 6-8). Flaking and drying (heat treatment) are used to physically alter beans to an extractable form. Hexane extraction is used to separate the oil (lipid) fraction of the soybean flakes from the nonfat portions. Any lipophilic pesticide residues that have survived previous treatment can logically be expected to be carried into the solvent phase in this operation. Residual hexane is removed from the protein-containing solids by evaporation (heat treatment). The protein is then extracted from the solids with alkali (destructive to alkaline-sensitive residues), and the proteins are then precipitated by adjusting the pH to their isoelectric point. This step reduces acid-sensitive residues, and water-soluble survivors are partitioned into the aqueous supernatant.

The acid protein precipitate constitutes the basis for the soy protein isolate. This protein fraction is further processed by UHT treatment, neutralization, and spray drying, which result in the finished soy protein isolate. It is therefore not likely that significant residues on the raw agricultural product (soybean) would still be present in the finished product.

The following example illustrates the extent to which processing can reasonably be expected to diminish residues that have survived harvest and are present at the soybean processing plant. The tolerance concentration for malathion on soybeans is 8 ppm. Processing can be expected to reduce the level of malathion in soy protein isolate to 0.016 ppm, assuming that 30% of any initial malathion residue would be extracted into hexane and an additional fraction (approximately 50% of the original residue) would be decomposed in the alkaline extraction of protein from the defatted meal. Only 1% of the remaining residue (or 0.2% of the original residue of 8 ppm at the tolerance concentration) would be expected to precipitate with the soy protein isolate (R.C. Gelardi, Infant Formula Council, personal commun., 1990). Since soy protein isolate typically constitutes 2% of soy-based infant formulas, the potential maximum possible level of

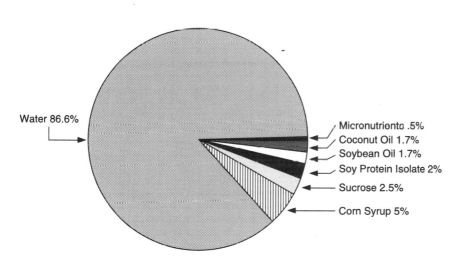

FIGURE 6-7 The composition of a typical soy-based infant formula. SOURCE: Infant Formula Council, Atlanta, Georgia, unpublished.

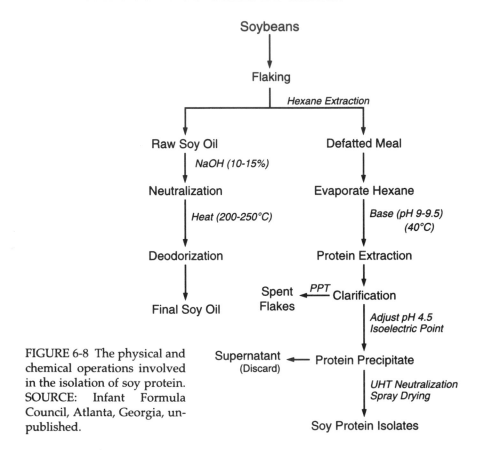

FIGURE 6-8 The physical and chemical operations involved in the isolation of soy protein. SOURCE: Infant Formula Council, Atlanta, Georgia, unpublished.

malathion (contributed by soy protein isolate) in the final product would be 0.0003 ppm.

Thus, the reductions of pesticide residues achieved by processing raw agricultural commodities combined with the relatively low concentrations of these ingredients in infant formulas account for the extremely low residue levels that can theoretically be predicted for infant formula, assuming maximum pesticide residue at EPA tolerances are initially present on the raw agricultural commodity. As stated earlier, extensive analytical testing of infant formula has failed to result in the detection of any pesticide residues.

Theoretical Maximum Residue Contributions

In numerous instances, a pesticide residue will be substantially diminished—or traces of the compound eliminated—as a result of processing

TABLE 6-4   Theoretical Maximum Residue
Contributions and Limits of Quantification

Pesticide (ppm)	Predicted Potential Maximum (ppm)	Limit of Quantification[a]
Captan	0.017	0.006-0.050
Malathion	0.068	0.002-0.050
Methylparathion	0.001[b]	0.002-0.050
Diazinon	0.002[b]	0.002-0.050
Carbaryl	0.056	0.002
Disulfoton	0.005[b]	0.020
Dimethoate	0.0002[b]	0.004
Atrazine	0.0002[b]	0.002-0.060
Dichlorvos	0.015[b]	0.020
Daminozide	0.002[b]	0.5-1.0
Alachlor	0.014	0.005
Carbofuran	0.024	0.002
Aldicarb	0.002[b]	0.050
Benomyl	0[b]	0.1
Oxamyl	0.0007[b]	0.002
Methidathion	0.002	0.002

[a] Detection limits vary with the method used and the criteria for definition.
[b] Predicted potential maximum is below limit of quantification.
SOURCE: Data from Infant Formula Council, 1993.

steps such as evaporation, distillation, partitioning, isoelectric separation, or oxidation-reduction. Such changes have been observed during the isolation of soy protein and deodorization of oil (Swern, 1979; Snyder and Kwon, 1987). Enriched samples are used to evaluate residue losses that may occur during processing. Residue levels present in unenriched samples are typically below the limit of detection, since compounds are either not used on the food or are removed through processing. Theoretical maximum residue contributions in infant formula and limits of quantification arc shown in Table 6-4.

PESTICIDES IN HUMAN MILK

Tissue concentrations of chlorinated organic pesticides are found in most people in the United States, although generally in declining levels in recent years as chlorinated pesticide use in the United States and worldwide has declined. The highest levels appear in regions where these pesticides are most commonly applied. Many of these lipid-soluble compounds are not cleared rapidly. Human milk is one route of elimination, but unfortunately that route increases the exposure of infants (Jensen,

1983; Wolff, 1983; Jensen and Slorach, 1990). In 1951, Laug et al. were the first to report an analysis of pesticides in human milk, in this case the presence of DDT and its metabolites. Since then, there have been many surveys of pesticides in human milk—some in response to episodes of food or dairy product contamination. The data from these surveys have been used to compare pesticide concentrations in human milk to allowable daily intakes.

In general, the more recent surveys of pesticides in human milk have demonstrated that the concentrations are lower than was observed in previous surveys. Despite this decrease in concentrations, more effort is needed to characterize the potential adverse effects of the low concentrations of chlorinated pesticides found in human milk. Initial attempts at estimating such effects associated with low concentrations have recently been described by Mattison (1990a,b, 1991) and Rogan et al. (1991).

DDT and Metabolites

DDT is an effective pesticide that has fairly low acute toxicity and has a long history of use worldwide (WHO, 1984; Murphy, 1986; Kurtz et al., 1989; Baker and Wilkinson, 1990). Concern over reproductive effects of DDT and its breakdown product DDE in birds and its long biological persistence led to the cessation of DDT use in the United States approximately 20 years ago (Hayes, 1975). Despite the two-decade ban, this pesticide and its metabolites continue to be found in human milk in the United States at decreasing concentrations over time, demonstrating the remarkable biological persistence of these halogenated organic compounds.

Surveys conducted before 1986 (Jensen, 1983; Wolff, 1983; Jensen and Slorach, 1990) demonstrated that p,p'-DDT or its metabolite p,p'-DDE were present in quantifiable concentrations in essentially all human milk samples assayed. Among samples analyzed in Arkansas in 1986, quantifiable concentrations of p,p'-DDT were found in 19%, and the metabolite p,p'-DDE was found in all samples surveyed (Mattison et al., 1992). The range of means of p,p'-DDT and p,p'-DDE reported in surveys of human milk in the United States before 1986 varied from 0.2 to 4.3 ppm and 1.2 to 14.7 ppm in milk fat, respectively. The mean concentrations among quantifiable samples from the survey in Arkansas were at the low end of these means. Among all samples, the mean concentrations were considerably lower than those previously reported.

The mean concentration of p,p'-DDT in all samples assayed in Arkansas was 0.039 ppm; the highest quantified level was 0.203 ppm (Mattison et al., 1992). In earlier surveys conducted in the United States, investigators found the mean concentration of p,p'-DDT to be between 0.2 and 4.3 ppm and p,p'-DDE to be between 1.2 and 14.7 ppm. Among all samples surveyed in Arkansas, the mean concentration of p,p'-DDE was 0.952

ppm. Among those with quantifiable concentrations, the mean was 0.954 ppm. This appears to show a continued decrease in DDT concentrations in human milk over the years.

Dieldrin

Dieldrin is an oxygenated metabolite of aldrin that persists in adipose tissue. Because of its persistence and toxicity, the parent compound, aldrin, has generally been removed from use in developed countries (Murphy, 1986; Kurtz et al., 1989; Baker and Wilkinson, 1990). Previous surveys conducted in the United States have identified detectable levels in 0.04% to 100% of the human milk samples analyzed (Jensen, 1983; Jensen and Slorach, 1990). Mean dieldrin concentrations ranged from 0.05 ppm to 0.24 ppm in milk fat (Jensen, 1983; Jensen and Slorach, 1990). The mean concentration in the 2% of Arkansas samples with quantifiable levels was 0.071 ppm. The mean concentration among all samples was 0.001 ppm (Mattison et al., 1992).

Lindane

Lindane is one component of a mixture of various isomers of hexachloro-cyclohexane (HCH). The composition of the commercial pesticide is α-HCH or lindane (53-70%), γ-HCH (3-14%), μ-HCH or lindane (11-18%), δ-HCH (6-10%), and other isomers (3-10%) (Rcuber, 1980). Lindane has been used as a substitute for DDT. In previous surveys conducted in the United States, HCH isomers were found in quantifiable concentrations in 4 to 68% of the human milk samples analyzed (Jensen, 1983; Wolff, 1983; Jensen and Slorach, 1990). α-HCH and γ-HCH, because of more rapid clearance, are generally found in fewer samples and in lower concentrations. This was also observed among the samples analyzed in Arkansas. However, because of its persistence, there are typically more samples with quantifiable levels of β-HCH and higher mean concentrations. For example, the β-HCH isomer was found in 27% of the human milk samples tested from Arkansas women (Mattison et al., 1992). Among those with quantifiable levels, the mean concentration was 0.12 ppm and among all samples, the concentration was 0.03 ppm. Lindane's agricultural uses in the United States have been virtually eliminated by changes in regulation over the past 20 years.

Hexachlorobenzene

Hexachlorobenzene is a persistent chemical with a variety of sources, including its previous use as a pesticide and its presence as an impurity in several other pesticide formulations. This compound can disrupt porphyrin metabolism (Jensen, 1983; Jensen and Slorach, 1990). Fatal cases of infant poisoning have been reported from ingestion of highly contaminated human milk. Because of persistence and fat solubility, hexachlorobenzene has been detected in many surveys of adipose tissue and human milk (Jensen, 1983; Jensen and Slorach, 1990). Despite the fact that hexachlorobenzene has been detected in human adipose tissues in the United States, few studies have explored the presence of this chemical in human milk. Among previous surveys conducted in the United States, the mean concentration was 0.04 ppm in milk fat (range, 0.018 to 0.063). Among the 6% with quantifiable levels in Arkansas, the mean was 0.03 ppm, and among all samples, the mean concentration was 0.002 ppm; both are lower than earlier reports (Mattison et al., 1992). Hexachlorobenzene is no longer registered for agricultural use, and its occurrence as a formulation impurity in other registered products has been greatly curtailed.

Other Cyclodiene Pesticides

Heptachlor, chlordane, and their metabolites (heptachlor epoxide, oxychlordane, *trans*-nonachlor) are closely related cyclodiene pesticides (WHO, 1984; Murphy, 1986; Kurtz et al., 1989; Baker and Wilkinson, 1990). Surveys conducted in the United States have demonstrated that between 25% and 100% of human milk samples analyzed had quantifiable concentrations of heptachlor or heptachlor epoxide ranging from 0.035 to 0.13 ppm (Jensen, 1983; Jensen and Slorach, 1990). A somewhat greater proportion of samples (46% to 100%) had quantifiable concentrations of chlordane and oxychlordane (range, 0.05 to 0.12 ppm), perhaps reflecting frequent use as a termiticide in houses (WHO, 1984; Murphy, 1986; Kurtz et al., 1989; Baker and Wilkinson, 1990). Among samples surveyed in Arkansas, 5% had quantifiable concentrations of heptachlor (mean, 0.03 ppm) and 74% had quantifiable concentrations of heptachlor epoxide (mean, 0.06 ppm) (Mattison et al., 1992). Two percent of the samples in that study contained quantifiable concentrations of chlordane; 77% and 84% had quantifiable concentrations of *trans*-nonachlor and oxychlordane, respectively. The mean concentrations among quantifiable samples measured in Arkansas for *trans*-chlordane, *cis*-chlordane, and oxychlordane were 0.18, 0.15, and 0.06 ppm in milk fat, respectively.

In most studies, heptachlor and/or heptachlor epoxide were detected in human adipose tissue samples. Curley et al. (1973) measured the concentration of chlorinated insecticides in fat samples from 241 Japanese people and found heptachlor epoxide concentrations that ranged from ≤0.01 ppm to 0.2 ppm (mean, 0.02 ppm). In Finland, Mussalo-Rauhamaa et al. (1984) demonstrated mean human adipose tissue concentrations of heptachlor epoxide as 2.7 μg/kg in males and 1.9 μg/kg in females. The difference in male and female mean concentrations may represent different volumes of adipose tissue or elimination via breastfeeding among the women studied.

In a study of approximately 1,500 women, Savage et al. (1981) explored regional differences in the pesticide content of human milk. In the southeast region of the United States, including Arkansas, 76% of the samples tested had detectable levels of heptachlor epoxide. The distribution of heptachlor epoxide concentrations in samples was also higher in that southeast region. Only 23% of the samples tested contained trace or undetectable concentrations. Half of the samples (52%) had heptachlor epoxide concentrations ranging from 0.001 ppm to 0.1 ppm. The remainder (approximately 25%) contained concentrations above 0.1 ppm. This was the highest of all regions in the United States. The mean concentration of heptachlor epoxide in these samples with detectable levels was 0.128 ± 0.209 ppm, also the highest mean level of all the regions surveyed in the United States.

Similar studies of human milk in Pennsylvania (Kroger, 1972) and Missouri (Jonsson et al., 1977) have demonstrated mean heptachlor epoxide concentrations of 0.16 ppm and 0.0027 ppm, respectively. Studies conducted in Hawaii (Takahashi et al., 1981; Takahashi and Parks, 1982) whose inhabitants have also been exposed to heptachlor and heptachlor epoxide in dairy products have demonstrated levels ranging from 0.001 ppm to 0.067 ppm (range, 0.036 ppm) among women on Oahu, and from 0.015 ppm to 0.052 ppm (mean, 0.031 ppm) among women on neighboring islands.

With the exception of endosulfan, which does not exhibit the persistence and bioconcentration characteristics of other chemicals in the group, virtually all agricultural uses of the cyclodiene pesticides have been eliminated or greatly restricted by regulatory actions over the past 20 years.

Risks to Infants from Pesticides in Human Milk

Breastfeeding has substantial benefit (Lawrence, 1989), including psychological, immunological, and general health promotion. In many countries, breastfeeding confers measurable benefit such as decreased rates of infectious disease and increased rates of growth and development. Despite the concentrations of pesticides found in human milk, no major studies have demonstrated that these pesticide concentrations have led to adverse health outcomes in the children exposed through breastfeeding. Therefore, although there is concern that exposure to pesticides in human milk may carry some potential for adverse health effects for the mother and for the nursing infant, it is important to recognize that the benefits of breastfeeding are clear and have been demonstrated. Furthermore, surveys conducted in the United States over the past four decades have shown that the number of samples with detectable pesticide concentrations have fallen—even when improved analytical methods with increased sensitivity have been used.

PESTICIDES IN FOODS

Recognizing the wide variation in sampling and analytical methods, size, and the overall design and objectives of existing residue testing programs, the committee requested residue data for its review from a variety of sources, including FDA, state regulatory agencies, the infant formula and baby food industries, the food processing industry, retail distributors, the agricultural chemical industry, and commodity associations. The most comprehensive and best recorded data were those collected through FDA's market basket sampling and analysis.

No single data bank provides ideal residue values. Residue analyses are complex, difficult to perform, and expensive. All data should be judged within this perspective. It is therefore important that samples be carefully identified and described, their processing and application history defined, and the capability and periods of analyses provided. Evaluation of the data bases provided to the committee emphasized the need for uniform sampling, collection, and reporting to reflect adequately the data's quality.

The FDA Surveillance Data

The FDA monitors pesticide residues in all food other than meat, milk, and eggs. The agency's monitoring program is *not* designed to determine dietary exposure to pesticides. Rather, its objective is to enforce compliance with EPA's tolerance levels. The usual point of sampling is the commercial warehouse, border crossings, points of importation, or some similar point after, but as close as possible to, the harvest. Table 6-5 shows the total number of samples and positive detections in FDA's surveillance program in 1988 and 1989. The ranges in both parameters are so wide that comparisons are very difficult. For example, dicrotophos samples were 100% positive, but the number sampled (15) is too small to be representative. For those pesticides with larger (>1,000) sample numbers, the highest percentages positive were found for benomyl (28.5%), EBDC (11.6%), and acephate (9.6%).

The committee reviewed FDA residue data on more than 100 different pesticides and selected a smaller subset for further analysis. The purpose was to isolate specific compounds for more extensive studies. Criteria for evaluation were

TABLE 6-5   Total of Samples and Positive Detections in FDA Residue
Data

| Chemical | Samples | | |
	Total No. Sampled	No. Positive	Percent Positive
Bromophos-ethyl	113	1	0.9
Dichlorvos	763	1	0.1
Prothiofos	1	1	100.0
Trichlorfon	1	1	100.0
Cyanophos	912	2	0.2
Ethoprop	1,927	2	0.1
Atrazine	669	4	0.6
Fonofos	290	4	1.4
Fenthion	267	6	2.2
Phenthoate	2,328	10	0.4
Carbophenothion	7,332	11	0.1
O-ethyl-O-*p*-nitrophenyl phenyl-phosphorothioate	6,912	12	0.2
Mecarbam	16	14	87.5
Dicrotophos	15	15	100.0
Ethylene Thiourea (ETU)	22	15	68.2
Fenitrothion	5,171	30	0.6
Quinalphos	40	30	75.0
Methoxychlor	5,643	36	0.6
Phorate	40	36	90.0
Phosphamidon	3,499	63	1.8
Chlorfenvinphos	9,299	66	0.7
Methomyl	2,706	69	2.5
Aldicarb	1,141	76	6.7
Phosalone	11,857	82	0.7
Profenofos	9,689	105	1.1
Disulfoton	15,121	117	0.8
Daminozide	514	125	24.3
Primiphos-methyl	4,449	176	3.9
Monocrotophos	18,617	191	1.0
Dicofol	12,430	216	1.7
Parathion-methyl	30,361	240	0.8
Benomyl	1,023	292	28.5
Ethylenebisdithiocarbamate (EBDC)	2,539	296	11.6
Phosmet	15,604	335	2.1
Methidathion	15,948	437	2.7
Azinphos-methyl	15,320	474	3.1
Parathion	40,029	591	1.5
Carbaryl	11,212	632	5.6
Diazinon	35,896	648	1.8
Ethion	30,588	699	2.3
Malathion	39,226	1,161	2.9
Mevinphos	25,639	1,320	5.1
Dimethoate	40,496	1,418	3.5
Captan	30,108	1,499	5.0
Chlorpyrifos	45,418	2,180	4.8
Acephate	39,940	3,845	9.6

SOURCE: Based on FDA Surveillance Data, 1988-1989, unpublished.

- volume of use and potential exposure,
- route of exposure (food, water),
- type of residue (surface, systemic),
- toxic potency,
- toxicity of particular concern with regard to infants and children,
- general toxic end point (cancer, cholinesterase inhibition),
- ability to characterize the potential risks that could arise from metabolic and physiologic differences between infants and adults, and
- utility of the data in evaluating different risk assessment models.

The subset of pesticides considered further by the committee includes

- total organophosphates and carbamates (high volume of use; some are easily detected by commonly used MRMs; and a common acute toxicity end point is cholinesterase inhibition);
- EBDCs and their metabolites, e.g., ETU (EBDCs are used on foods such as apples, bananas, and other fruits and vegetables consumed in large amounts by infants and young children; ETU is a possible carcinogen);
- benomyl (systemic residue; used on fresh fruits, berries, vegetables, and in fungicide problem areas; high percentage of residue occurrence in FDA surveillance program);
- captan (used on foods such as apples, peaches, and pears, which are consumed in large amounts by infants and young children; potential reproductive effects; high percentage of residue occurrence in FDA surveillance program);
- aldicarb (most acutely toxic pesticide registered; systemic residue presents high probability for exposure, potential for water-based exposure because it leaches into groundwater; used on citrus fruits, sugar beets, peanuts, pecans, and a variety of field crops and ornamentals);
- atrazine (widely applied pesticide; used on grasses, grains, corn, macadamia nuts, and guava; detected in ground and surface water; putative hormonal effects); and
- alachlor (widely applied pesticide; used on potatoes, corn, beans, peas, peanuts, and cotton; detected in groundwater; carcinogenic effects in animals).

Aldicarb, benomyl, and organophosphates were ultimately selected for the examples presented in Chapter 7.

The committee compared FDA's sampling frequency and sample sizes with the 18 foods shown to be most eaten by nursing and nonnursing infants in the USDA's 1977-1978 Nationwide Food Consumption Survey (NFCS) (see Chapter 5). As shown in Table 6-6, of the 18 foods most consumed by nursing and nonnursing infants under 1 year old, only four—apples, peaches, peas, and green beans—are among the top 18 foods sampled by the FDA. This emphasizes the importance of certain foods in the diet and, hence, their potential for contributing pesticides to the diet. A more focused and specific monitoring system is indicated for targeting foods in infant diets.

FDA monitoring is focused on crops with a history of residues exceeding EPA tolerance levels, residues with no EPA-established tolerances, and crops harboring the greatest number of different pesticides. The committee could not determine if the high number of pesticides found on some frequently sampled foods was influenced by the larger sample size and, therefore, if larger sample sizes from other crops would lead to a finding of more pesticides on them also.

The committee recognized that the 1988-1989 FDA residue data were

TABLE 6-6   Comparison of 18 Foods Most Sampled by FDA and Those Most Consumed by Nursing and Nonnursing Infants

Foods Most Consumed by Infants Starting with the Highest[a]	Foods Most Sampled by FDA[b]
Milk, nonfat solids	Peppers
Apple juice	Lettuce
Apples, fresh[c]	Tomatoes
Orange juice	*Apples, fresh*[c]
Pears, fresh	*Beans, succulent, green*[c]
Milk, fat, solids	*Peas, succulent, garden*[c]
Peaches, fresh[c]	Grapes
Carrots	Strawberries
Beef, lean	Cantaloupe
Milk sugar (lactose)	Squash
Bananas, fresh	Pineapples
Rice, milled	Potatoes
Peas, succulent, garden[c]	Oranges
Beans, succulent, green[c]	Eggplant
Oats	Broccoli
Soybean oil	Cucumbers
Coconut oil	Eggs
Wheat flour	*Peaches, fresh*[c]

NOTE: Foods are expressed as raw agricultural commodities (RACs).

[a] SOURCE: Based on data from USDA, 1983. (See Tables 5-6 and 5-7 in Chapter 5.)

[b] SOURCE: Based on FDA Surveillance Data, 1988-1989, unpublished.

[c] Foods both sampled and consumed the most by infants.

collected for compliance purposes and were therefore not appropriate for comparisons with other residue data sets and for risk calculation. Nevertheless, the FDA data were valuable in comparisons with the 1977-1978 NFCS data on food consumption and provided the committee with useful information on residues in the food supply.

The committee then reviewed FDA's sample sizes and positive detections in the 18 foods most consumed by infants for aldicarb, benomyl, captan, chlorpyrifos, dimethoate, the EBDCs and their breakdown product ETU, and parathion-methyl (see Table 6-7). In general, the pesticides most frequently detected were benomyl, captan, chlorpyrifos, dimethoate, and EBDCs. Highest residues were found for EBDC (succulent peas), captan (fresh peaches), ETU (succulent peas), and captan (fresh apples). The pesticide with no positives was aldicarb. Parathion-methyl occurred at a very low frequency at very low levels. In just a few cases did maximum residues exceed tolerance, and in only one case (EBDC in succulent peas) was the maximum residue substantially above tolerance (23 compared with 7 ppm). In a few cases, residues were found for a chemical/commodity for which no tolerance had been established; an example is dimethoate, which was found at very low frequency in fresh peaches, fresh bananas, and orange juice. The sources of occasional residues of dimethoate in food crops for which it has no tolerance was not known to the committee, although misuse or drift from neighboring fields is a possibility.

Number of FDA Samples

Adequate numbers of randomly selected samples are fundamental to developing good residue estimates for exposure and are essential to a statistically reliable basis for risk assessment. Because the FDA residue monitoring system is not designed to produce statistically reliable survey data on pesticide residues in food, sample sizes for individual foods are often small. This is true for several foods that are prominent in the diets of infants and children and for several pesticides with known toxicity.

FDA has recently begun to increase its focus on gathering information on pesticide residues in infant foods and other foods eaten by infants and children. The agency has summarized regulatory monitoring conducted in FYs 1985-1991 on eight fruit, fruit juice, and milk foods that are prominent in the diet of infants and children; incidence/level monitoring of approximately 900 prepared infant foods collected and analyzed in FY 1990-1991; and Total Diet Study (TDS) results for 33 different types of infant foods (strained/junior) collected and analyzed in FYs 1985-1991 (Yess et al., 1993). Infant foods analyzed under TDS and through the incidence/level monitoring showed that residue levels were well below EPA tolerances and that residue intakes are well below ADIs set by the

TABLE 6-7 Seven Pesticides Sampled by FDA in the 18 Foods Most Consumed by Nursing and Nonnursing Infants

Foods	Pesticide Sampled	No. Positive/No. in Sample	Residue Levels, ppm		EPA Tolerance Level
			Mean	Maximum	
Milk, nonfat solids	Captan	0/76	0	0	NT
	Chlorpyrifos	0/399	0	0	NT
	Dimethoate	0/7	0	0	0.0002[a]
	Parathion-methyl	0/394	0	0	NT
Apple juice	Dimethoate	9/110	0.0066	0.2800	NT
Milk sugar (lactose)	(No samples)				
Apples, fresh	Aldicarb	0/133	0	0	NT
	Benomyl	35/134	0.1635	2.64	7.0
	Captan	125/978	0.0352	3.4000	25.0
	Chlorpyrifos	116/968	0.0064	0.9000	1.5
	Dimethoate	10/959	0.0006	0.1900	2.0
	EBDC	2/16	0.0937	0.7700	7.0[b]
	ETU	0/1	0	0	NT
	Parathion-methyl	37/957	0.00176	0.2600	1.0
Orange juice	Chlorpyrifos	105/551	0.0248	0.5000	NT
	Dimethoate	5/537	0.0003	0.0500	NT
	Parathion-methyl	6/544	0.0008	0.1780	NT
Peaches, fresh	Aldicarb	0/34	0	0	NT
	Benomyl	13/26	0.2095	1.1100	15.0
	Captan	48/574	0.1329	9.6900	50
	Chlorpyrifos	14/581	0.0013	0.1300	0.05
	Dimethoate	2/561	0.0001	0.0875	NT
	EBDC	1/13	0.0154	0.2000	10.0[b]
	Parathion-methyl	4/570	0.0024	0.0900	1.0

TABLE 6-7 (*Continued*)

Foods	Pesticide Sampled	No. Positive / No. in Sample	Residue Levels, ppm		EPA Tolerance Level
			Mean	Maximum	
Pears, fresh	Aldicarb	0/13	0	0	NT
	Benomyl	1/23	0.0691	1.5900	7.0
	Captan	18/431	0.0155	1.0000	25
	Chlorpyrifos	6/454	0.0003	0.0400	0.05
	Dimethoate	10/455	0.0024	0.3850	2.0
	EBDC	0/3	0	0	7.0[b]
Carrots	Aldicarb	0/41	0	0	NT
	Benomyl	0/1	0	0	0.2
	Captan	1/390	0.0021	0.8300	2.0
	Chlorpyrifos	2/402	0.0001	0.0200	NT
	Dimethoate	0/375	0	0	NT
	EBDC	0/34	0	0	7.0[b]
	Parathion-methyl	1/394	0.00003	0.0100	1.0
Milk, fat solids	Captan	0/76	0	0	NT
	Chlorpyrifos	0/424	0	0	0.05[a]
	Dimethoate	0/7	0	0	0.002[a]
	Parathion-methyl	0/419	0	0	NT
Beef, lean	Aldicarb	0/1	0	0	0.01
	Captan	0/1	0	0	0.05
	Chlorpyrifos	0/1	0	0	2.0
	Dimethoate	0/1	0	0	0.02
	ETU	0/1	0	0	NT
	Parathion-methyl	0/1	0	0	NT
Soybean oil	(No samples)				

Commodity	Pesticide				
Rice, milled	Aldicarb	0/6	0	0	NT
	Captan	0/24	0	0	NT
	Chlorpyrifos	0/29	0	0	NT
	Dimethoate	0/14	0	0	NT
	Parathion-methyl	0/27	0	0	NT
Coconut oil	(No Samples)				
Bananas, fresh	Aldicarb	0/48	0	0	0.3
	Benomyl	8/72	0.0672	0.8200	1.0
	Captan	0/344	0	0	NT
	Chlorpyrifos	16/347	0.0011	0.0400	0.05
	Dimethoate	2/347	0.0058	2.0000	NT
	EBDC	4/100	0.0331	1.6000	0.50[b]
	Parathion-methyl	0/347	0	0	NT
Wheat flour	Captan	0/3	0	0	NT
	Chlorpyrifos	0/5	0	0	NT
	Dimethoate	0/1	0	0	NT
	Parathion-methyl	0/5	0	0	NT
Peas, succulent, garden	Aldicarb	0/31	0	0	NT
	Benomyl	2/6	0.1000	0.4100	NT
	Captan	2/705	0.0014	0.5900	2.0
	Chlorpyrifos	9/751	0.0017	0.4800	NT
	Dimethoate	70/720	0.0341	2.8775	2.0
	EBDC	49/124	2.5964	23.000	7.0[b]
	ETU	3/4	2.0075	4.2050	NT
	Parathion-methyl	5/732	0.0009	0.5500	1.0

TABLE 6-7 (Continued)

Foods	Pesticide Sampled	No. Positive/ No. in Sample	Residue Levels, ppm		EPA Tolerance Level
			Mean	Maximum	
Beans, succulent, green	Aldicarb	0/45	0	0	NT
	Benomyl	0/6	0	0	2.0
	Captan	3/776	0.0028	1.0000	NT
	Chlorpyrifos	23/834	0.0027	0.5650	0.05
	Dimethoate	43/805	0.0120	2.3450	2.0
	EBDC	2/24	0.1042	2.0000	10.0[b]
	ETU	1/1	0.2200	0.2200	NT
	Parathion-methyl	5/809	0.0001	0.0700	1.0
Oats	Captan	0/12	0	0	NT
	Chlorpyrifos	0/16	0	0	NT
	Dimethoate	0/4	0	0	NT
	Parathion-methyl	0/16	0	0	1.0

NOTE: NT, no tolerance level has been established by the EPA.

[a] Tolerance for milk is given in the absence of tolerances for specific components.
[b] EPA tolerance for Maneb, Zineb, and Ziram.

SOURCE: Based on FDA Surveillance Data, 1988-1989, unpublished.

Food and Agriculture Organization (FAO) and the World Health Organization (WHO). FDA's increased focus on residues in infant/children foods, exemplified by information in the cited paper, is a positive development toward improving the quality and quantity of data available to ascertain dietary exposures of infants and children to pesticides.

Milk, apple juice, and orange juice rank high in the diets of infants but do not appear in the list of foods most often sampled (Table 6-6). In fact, during the 2-year period (1988-1989) covered by the FDA data, milk constituents had been sampled only for captan, dimethoate, parathion-methyl, and chlorpyrifos in sample numbers ranging from a high of 424 for chlorpyrifos in milk fat solids to 7 for dimethoate in milk nonfat and fat solids.

FDA sampled foods (commodities) as consumed. The agency did not sample milk sugar (lactose), soybean oil, or coconut oil, which are frequently components of infant food but are not consumed as individual commodities by infants and children. The FDA program would require restructuring in order to provide the data required to estimate exposure of infants and children for use in risk assessment.

Aldicarb, ETU, EBDCs, and benomyl cannot be detected by FDA's routine multiresidue methods; they require a single residue method. As shown in Table 6-7, the sample numbers for these pesticides are small (<25) for many of the foods most consumed by infants. These low numbers of samples can be of concern when pesticides are found on a high percentage of the samples and a large percentage of the crop acreage has been treated with the pesticides. Sample numbers higher than 25 are shown in Table 6-8 along with the percentage of positive detections in those samples. Overall, the FDA data appear to be adequate for the committee's purposes, except for the low number of milk and juice samples. They are useful for comparison and discussion; however, the low number of samples does emphasize the need for careful selection and reporting. More complete and reliable information would be required for a comprehensive risk assessment.

Positive Detections

The ability of a survey to detect pesticide residues depends on at least three factors: the percentage of crop acreage treated with the pesticide, the sampling design of the survey, and the analytical limit of detection for the pesticide in that food. Other factors include the stability of the chemical; the time between pesticide application, harvest, and sampling; and the degree of postharvest processing. FDA laboratory staff that tests for pesticide residues in food have no prior knowledge of the pesticides applied to the crop being treated. The testing and sampling protocol is not designed to provide statistically valid sampling data.

TABLE 6-8   The Percentage of Positive Detections for Six Pesticides in Various Foods for Samples Larger than 25

Chemical	Foods	No. in Sample	Positive Samples No.	%
Ethylenebisdithiocarbamate	Peas	124	49	40
	Bananas	100	4	4
Benomyl	Apples	134	35	26
	Peaches	26	13	50
	Bananas	72	8	11
Captan	Apples	978	125	13
	Peaches	574	48	8
	Pears	431	18	4
	Carrots	390	1	0.3
	Peas	705	2	0.3
	Beans	776	3	0.4
Chlorpyrifos	Apples	968	116	12
	Peaches	581	14	2
	Pears	454	6	1
	Carrots	402	2	0.5
	Bananas	347	16	5
	Peas	751	9	1
	Beans	834	23	3
Dimethoate	Apples	959	10	1
	Peaches	561	2	0.4
	Pears	455	10	2
	Bananas	347	2	0.6
	Peas	720	70	10
	Beans	805	43	5
Parathion-methyl	Apples	957	37	4
	Peaches	570	4	0.7
	Beans	809	5	0.6
	Peas	732	56	0.7

SOURCE: From FDA Surveillance Data, 1988-1989, unpublished.

Nonetheless, thousands of data points are generated by FDA. For the pesticides in the infant diet examined by the committee, the data show that positive pesticide residue detections are clearly more common in fresh fruits and vegetables, specifically in apples, peaches, pears, bananas, peas, green beans, and carrots (see Table 6-7), than in other commodities. Closer examination reveals that positive detections were less than 10% in most crop-pesticide combinations with 2-year samples larger than 25 (Table 6-8). However, the percentage of positive detections for these fruits and vegetables varies by pesticide and crop. The range extends from 0.3% positive for captan on carrots and peas to 50% positive for benomyl on peaches.

Residue Levels

Ultimately, it is the levels of residues on foods that are of concern in estimating exposure and risk. FDA data indicate that most of the residues detected in food are well below the established EPA tolerance levels (Table 6-9). In general, values most dramatically below those tolerances are residues of pesticides long used, such as captan and the EBDC fungicides. Rather high tolerances were established for these chemicals in the late 1950s without the benefit of current data on the health effects of pesticide residues.

The variability and small numbers (<25) of many of the 2-year samples of the foods and pesticides of concern make it difficult to calculate residues with any certainty. When the samples were of sufficient size for calculation, the mean residues were always below the EPA tolerance level.

Following the pattern of positive detections, the infant-food crops with the highest mean residues are fresh fruit and vegetables—specifically peas, apples, peaches, pears, carrots, and green beans (Table 6-9). Sometimes, residues were found on foods for which no tolerance levels had been set—benomyl on peas, captan on beans, chlorpyrifos on carrots and peas, and dimethoate on peaches. Six of the maximum residues shown in Table 6-9 exceeded the EPA tolerance levels—EBDC on peas, chlorpyrifos on peaches and beans, and dimethoate on peas and beans. In all these cases, however, the mean concentrations were considerably below the established EPA tolerance level and therefore represented a very small number of the samples. The only chemical for which mean residues approached tolerance was EBDC on succulent peas, for which the mean (2.6 ppm) was only slightly less than half the tolerance (7.0 ppm). This demonstrates the inappropriate application of tolerances, which were intended for another purpose—not to provide a margin of safety for infants and children.

Residue Distribution

Understanding the distribution of residues in the food supply is a key factor for estimating exposure accurately. As indicated in the preceding discussions, residues are skewed toward a relatively small portion of the food supply. To better understand this distribution, the committee looked at frequency distributions for pesticide residues throughout the entire food supply and on individual crops.

Effects of Processing

Processing may exert no effect on pesticide residues in foods, or it may increase or decrease concentrations. Therefore, the effects of processing

TABLE 6-9   Six Foods Among the 18 Most Consumed by Infants and the Tolerances and Residue Levels of Six Pesticides Detected in Them

Food	Pesticide	Residue Level, ppm		
		Mean Detected	Maximum Detected	EPA Tolerance
Apples, fresh	Ethylenebisdithiocarbamate (EBDC)	0.0937	0.7700	7.0
	Benomyl	0.1635	2.6400	7.0
	Captan	0.0352	3.4000	25.0
	Chlorpyrifos	0.0064	0.9000	1.5
	Dimethoate	0.0006	0.1900	2.0
	Parathion-methyl	0.00176	0.2600	1.0
Peaches, fresh	EBDC	0.0154	0.2000	10.0
	Benomyl	0.2095	1.1100	15.0
	Captan	0.1329	9.6900	50.0
	Chlorpyrifos	0.0013	0.1300	0.05
	Dimethoate	0.0001	0.0875	NT
	Parathion-methyl	0.0024	0.0900	1.0
Pears, fresh	EBDC	0	0	7.0
	Benomyl	0.0691	1.5900	7.0
	Captan	0.0155	1.0000	25.0
	Chlorpyrifos	0.0003	0.0400	0.05
	Dimethoate	0.0024	0.3850	2.0
	Parathion-methyl	0	0	1.0
Carrots	EBDC	0	0	7.0
	Benomyl	0	0	0.2
	Captan	0.0021	0.8300	2.0
	Chlorpyrifos	0.0001	0.0200	NT
	Dimethoate	0	0	NT
	Parathion-methyl	0	0.0100	1.0
Peas, succulent, garden	EBDC	2.5964	23.0000	7.0
	Benomyl	0.1000	0.4100	NT
	Captan	0	0	2.0
	Chlorpyrifos	0.0017	0.4800	NT
	Dimethoate	0.0341	2.8775	2.0
	Parathion-methyl	0.0009	0.5500	1.0
Beans, succulent, green	EBDC	0.1042	2.0000	10.0
	Benomyl	0	0	2.0
	Captan	0.0028	1.0000	NT
	Chlorpyrifos	0.0027	0.5650	0.05
	Dimethoate	0.0120	2.3450	2.0
	Parathion-methyl	0.0001	0.7000	1.0

NOTE: NT, no tolerance level has been established by EPA.

SOURCE: Based on FDA Surveillance Data, 1988-1989, unpublished.

on pesticide residues in food must be considered in a critical evaluation of the dietary exposure of infants and children to pesticides. Earlier in this chapter, the committee discussed the effects of processing on pesticide residues on the ingredients used in infant formula.

Elkins (1989) reported NFPA data on the effects of food processing operations on the residues of pesticides permitted on raw agricultural commodities. In most instances, washing by itself was shown to reduce residues, blanching reduced them even further, and the canning process led to even further decreases. These data indicated that in tomatoes and green beans subjected to each of these three steps, levels of malathion were reduced 99% and 94%, respectively, and carbaryl concentrations decreased 99% and 73%. Levels of parathion were reduced 66% in spinach and 10% in frozen broccoli. Elkins also pointed out that some processing activities can actually increase levels in certain instances. For example, levels of ETU were increased 94.5% in frozen turnip greens as a result of maneb degradation during cooking in a saucepan.

Elkins noted that the distribution of different pesticides in a product is also an important consideration. In tomatoes, for example, malathion tends to concentrate in the peel or waste, while carbaryl, a fairly polar compound, is easily removed by washing and does not tend to concentrate in waste material. Table 6-10 lists the residues of both pesticides in the washed and unwashed product, in the peeled tomato, and in waste material.

The committee reviewed unpublished data provided by the NFPA on pesticide residues found in foods used in processed baby foods. These foods, along with infant formula, comprise a large proportion of the infant's diet. Of the 6,580 samples tested in 1987, NFPA members found residues in 165 samples (2.5%) distributed among a total of 7 foods (Table

TABLE 6-10   Pesticide Concentrations in Washed and Unwashed Tomatoes, in Peeled Tomatoes, and in Waste Material

Tomato Product	Malathion, ppm	Carbaryl, ppm
Unwashed	15.9	5.2
Washed	0.8	0.14
Peeled	0.1	Trace
Waste	5.3	0.1

SOURCE: Elkins, 1989. Reprinted from *Journal of the AOAC*, Volume 72, Number 3, pages 533-535, 1989. Copyright 1989 by AOAC International.

TABLE 6-11 Number of Samples and Detections for Foods Used in Baby Food

Product	Sample Size	No. of Samples with Detected Residues	Ratio of Maximum Detection to LOQ
Apples, fresh	1,560	27	11
Apples, juice	1,602	24	5.6
Apricots, fresh	260	0	—
Bananas, fresh	472	39	4.1
Beans, succulent, green	28	0	—
Carrots	183	0	—
Corn, sweet	2	0	—
Grapes, juice	3	0	—
Oats	36	0	—
Orange, juice	220	0	—
Peaches, fresh	356	21	13.6
Pears, fresh	328	47	140
Peas (garden), green immature	1	0	—
Pineapples, fresh, juice	1	0	—
Plums-Prunes, dried	78	0	—
Plums (Damsons), fresh	193	0	—
Rice, milled	42	0	—
Squash, summer	226	0	—
Squash, winter	174	2	3.2
Sweet potatoes (including yams)	815	5	4.2

SOURCE: National Food Processors Association, 1987, data unpublished.

6-11). As shown in the table, the largest number of positive samples and the highest ratio of detected residue to the LOQ were noted for fresh pears. The ratio represents a comparison of the highest concentration of amitraz found in pears to the relatively low (1 ppb) LOQ. The highest concentration found—140 ppb—is considerably lower than the EPA tolerance of 3,000 ppb.

A comprehensive study of the effects of processing on food residues is badly needed. This study should be undertaken by EPA, FDA, or USDA working through grants and contracts to university and private laboratories. The committee believes that the information presently available needs to be brought together in a review document and then supplemented with additional studies. However, this requirement is beyond the time and resources available to the committee. The committee therefore opted to provide some brief examples of the type of information that can be obtained on many of the foods consumed by infants and young children.

In these examples, the committee applies basic and deliberately simplified explanations of several processes involved in food preparation and processing.

In addition to infant formulas, which have already been discussed, the most critical and highly consumed foods are apple-based processed products and cereal-based foods. Following are brief overviews of the processing factors that influence pesticide residues in these foods.

Apple-Based Foods

Apple-based foods constitute a substantial portion of foods consumed by infants and young children, as shown in Chapter 5. Knowledge of the form in which these products are consumed is important to the understanding of residue data. Virtually all the foods consumed by infants are processed, and most are manufactured by a limited number of processors, who exercise stringent controls. The processing of apples for applesauce (which forms the basis for many foods) and apple juice is specific, and the controls for the finished products are extensive.

Steps in processing of apples for use in foods for infants and children involve washing, blanching, peeling, pressing (for juice), finishing (removal of fibrous or indigestible material), and heating (sterilization). The washing process removes the exterior (nonsystemic) compounds, and is effective in the removal of many pesticides. Blanching is done with steam or hot water, primarily to inactivate enzyme systems and to prevent discoloration. It involves treatment at high temperature for a relatively short time. The skin is removed by abrasion or by peeling with a knife. A substantial portion of nonsystemic pesticides is concentrated at the surface of the apple or in the peel. The calix (or core) is removed by actually cutting or by removing seeds and fibrous material with a finisher. Physical pressure, usually accompanied by heat or enzyme treatment, is used to separate clear juice from cellulose, fibrous (pectin), and protein material to yield a clear, light-colored juice. Pressing is conducted through a filter press with paper filtration or by ultrafiltration. Substantial pesticide concentrations are removed with the fibrous or protein fractions of apple solids. Thus, estimates at the farm gate are not reflective of the residue content of foods that have undergone such processing steps.

Infant Cereal

Infant cereal is consumed by a large percentage of infants. The lack of large pesticide residues detected in these products probably reflects the extensive and unique processes to which they are subjected.

The basic ingredient of infant cereal is flour, which is the dehulled and fractionated grain (rice, oat, or barley) that has been milled. The flour is formed into a slurry, treated with a hydrolyzing enzyme (α-glucosidase), then heated to cook the cereal, destroy enzyme activity, and sterilize the food. The slurry is then dried on a steam-heated drum, which effectively distills off into steam any volatile material subject to partition.

The resulting product is a sterile, precooked, partially hydrolyzed cereal-based food. Clinical trials of usage have established the acceptance and digestibility of cereal for infant consumption. There has been no relationship established between retention of components in original grain to finished, processed infant cereal.

CONCLUSIONS AND RECOMMENDATIONS

Data on residues on foods are collected by FDA, state agencies, the food industry, private organizations, and in university programs such as IR-4. Other government, industry, and academic sources were identified by the committee for specific categories such as water, infant formula, and human milk, which are particularly important in the diets of infants and children. The committee also reviewed the complex and varied methods for pesticide residue analysis and sampling performed by these groups.

Conclusions

• There is no comprehensive data source, derived from actual sampling, on pesticide residue levels in the major foods consumed by infants and children.

For example, of the foods (expressed as commodities) most consumed by nursing and nonnursing infants under 1 year of age, data showing the 18 foods most frequently tested for pesticide residues in the FDA Surveillance Program (Table 6-6) include only 4 of the 18 major foods consumed by this age group.

• Data on pesticide residues in foods are extensive, but are difficult to interpret because of variation in sample selection, analytical methods, and quality control procedures. The extensive data available from numerous testing programs for pesticide residues in food would be far more useful in profiling residues in the diet if they were presented in a more complete and coherent form.

Food samples analyzed for pesticide residues may have been selected for surveillance, compliance, or other purposes. Analytical methods, their limits of quantification, and their degrees of precision and accuracy may therefore differ among laboratories. Record-keeping practices in pesticide residue monitoring programs are generally not uniform and not well articulated.

• Many of the existing data on pesticide residues were generated for targeted compliance purposes. Although these data may be appropriate for enforcement, they have limited usefulness in generation- or population-based evaluations of actual exposure. For example, the sampling technique overrepresents suspected violators and does not adequately represent foods eaten in large quantities by infants and children.

• The limited data available suggest that pesticide residues are generally reduced by processing; however, more research is needed to define the direction and magnitude of the changes for specific pesticide-food combinations.

• Infant formulas and processed baby foods are routinely monitored to ascertain pesticide residue levels. Although sampling and analytical techniques lack the desirable degree of uniformity, residues were not generally detected in these products.

• Human milk is a food whose constituents are subject to wide variation, depending on the diet, medical history, and exposure of the mother. For some infants this may represent the primary route of exposure. Ongoing surveillance indicates that pesticide concentrations in human milk continue to decline over time, especially since organochlorine pesticide use in the United States has been reduced.

• Pesticide residues in water—both drinking water and water used in food preparation—have previously been largely overlooked in assessing dietary exposure of infants and children.

Recommendations

• A computerized data base for pesticide residue data collected by laboratories in the United States should be established.

If standardized reporting procedures were developed and adopted, pesticide residue data could be accumulated in a national data bank in a form accessible for future use.

• In future applications of residue data, consideration should be given to the development of a standardized reporting format for use by all laboratories involved in residue analyses. Since pesticide residue data are collected by a variety of laboratories using different methods for sampling and analyses, it would be desirable to maintain records of sample collection, analytical methods used, the basis of detection, and the precision and accuracy of the results obtained.

Reports of pesticide residue testing should indicate

–food commodity analyzed (and whether it is processed or unprocessed),
   –analytical method used,
   –compounds tested (including metabolites),
   –quality assurance-quality control (QA-QC) notation, and
   –limit of quantification (LOQ).

These reports should follow a standard format, should be timely and consistent, and should include not only the LOQ but also all negative and positive findings. The methods of reporting must also be consistent (e.g., using similar computer software).

• Food residue monitoring programs should target a special market basket survey designed around the diet of infants and children. The methods to be used in this survey should be validated using fortified samples circulated among the participating laboratories.

• Residue analysis methods need to be standardized in a timely manner through an independent review and validation process conducted by a government or professional organization.

• FDA, working with USDA, EPA, and state and other federal agencies, needs to create:

–a clearly explained sampling strategy that could be used to ascertain the representativeness of the results of food residue analyses;
   –guidelines for those generating, processing, and using residue data to ensure that an explanation of LOQs and nondetectables are provided with all reports and are uniformly used in data analyses (e.g., in averaging);
   –a residue data management system that will improve the quality, accessibility, and comparability of food residue data, including those generated by the commercial sector; and
   –a repository of information on the fate of compounds during food processing and preparation.

• Laboratories performing pesticide residue analysis for regulatory purposes should participate in QA-QC programs, including regular quality control checks by an independent, external organization.

REFERENCES

Baker, D.B., and R.P. Richards. 1989. Herbicide concentration patterns in rivers draining intensively cultivated farmlands of northwestern Ohio. Pp. 103-119 in Pesticides in Terrestrial and Aquatic Environments, D. Weigmann, ed. Blacksburg, Va.: Virginia Water Resources Research Center, Virginia Polytechnic Institute and State University.

Baker, S.R., and C.F. Wilkinson, eds. 1990. The Effect of Pesticides on Human Health. Advances in Modern Environmental Toxicology Ser.: Vol. XVIII. Princeton, N.J.: Princeton Scientific Publishing Co. 438 pp.

Buchel, K.H., ed. 1983. Chemistry of Pesticides. New York: John Wiley & Sons. 618 pp.

Considine, D.M., and G.D. Considine, eds. 1982a. Maize (corn) processing. Pp. 1132-1139 in Foods and Food Production Encyclopedia. New York: Van Nostrand Reinhold.

Considine, D.M., and G.D. Considine, eds. 1982b. Soybean processing. Pp. 1870-1883 in Foods and Food Production Encyclopedia. New York: Van Nostrand Reinhold.

Curley, A.V., W. Burse, R.W. Jennings, E.C. Villanueva, L. Tomatia, and K. Akazaki. 1973. Chlorinated hydrocarbon pesticides and related compounds in adipose tissue from people of Japan. Nature 242:338-340.

Department of Fish and Game, State of California. 1990. Hazard Assessment of the Rice Herbicides Molinate and Thiobencarb to Aquatic Organisms in the Sacramento River System. Environmental Services Division, Administrative Report 90-1. Sacramento: California Department of Fish and Game.

Dynamac. 1989. Anticipated Pesticide Residues in Food. Draft Final Report, Volume 1. Rockville, Md.: Dynamac Corp.

Elkins, E.R. 1989. Effect of commercial processing on pesticide residues in selected fruits and vegetables. J. Assoc. Off. Anal. Chem. 72(3):533-535.

EPA (U.S. Environmental Protection Agency). 1990. National Survey of Pesticides in Drinking Water Wells. Phase I. NTIS Doc. No. PB-91-125765. Springfield, Va.: National Technical Information Service.

FDA (Food and Drug Administration). 1989. Food and Drug Administration Pesticide Program. Residues in foods—1987. J. Assoc. Off. Anal. Chem. 72:133A-152A.

FDA (Food and Drug Administration). 1991. Food and Drug Administration Pesticide Program; Residue Monitoring 1991. Washington, D.C.: Food and Drug Administration.

GAO (General Accounting Office). 1986a. Pesticides: Better Sampling and Enforcement Needed on Imported Food. (GAO/RCED-86-219) Washington, D.C.: U.S. Government Printing Office.

GAO (General Accounting Office). 1986b. Pesticides: EPA's Formidable Task to Assess and Regulate their Risks. (GAO/RCED-86-125, Apr. 18) Washington, D.C.: U.S. Government Printing Office.

GAO (General Accounting Office). 1986c. Pesticides: Need to Enhance FDA's Ability to Protect the Public from Illegal Residues. GAO/RCED-87-7. Washington, D.C.: U.S. Government Printing Office.

GAO (General Accounting Office). 1992. Pesticides: Adulterated Imported Foods Are Reaching U.S. Grocery Shelves. Washington, D.C.: U.S. Government Printing Office.

Garner, W.Y., and M.S. Barge. 1988. Good Laboratory Practices: An Agrochemical

Perspective. ACS Symposium Series 369. Washington, D.C.: American Chemical Society.

Gelardi, R.C., and M.K. Mountford. 1993. Infant formulas: Evidence of the absence of pesticide residues. Regul. Toxicol. Pharmacol 17:181-192.

Gianessi, L.P. 1986. A National Pesticide Usage Data Base. Washington, D.C.: Resources for the Future.

Good, G. 1988. IEPA surface water monitoring support programs for pesticides. Pp. 97-107 in Pesticides and Pest Management, D. Cavanaugh, ed. Springfield, Ill.: Illinois Department of Energy and Natural Resources.

Hallberg, G.R. 1989. Pesticide pollution of groundwater in the humid United States. Agric. Ecosyst. Environ. 26:299-368.

Hance, R.J. 1989. Accuracy and precision in pesticide analysis with reference to the EC "water quality" directive. Pest. Outlook 1:23-25.

Hartley, D., and H. Kidd, eds. 1983. The Agrochemicals Handbook. Nottingham, England: Royal Society of Chemistry.

Hayes, W.J., ed. 1975. Toxicology of Pesticides. Baltimore, Md.: Williams & Wilkins.

Holden, L.R., and J.A. Graham. 1990. Project Summary for the National Alachlor Well Water Survey. Monsanto Final Report NSL-9633. St. Louis, Mo.: Monsanto.

Holden, L.R., J.A. Graham, and D.I. Gustafson. 1990. The National Alachlor Well Water Survey. Part C, Statistical Analysis. Monsanto Final Report MSL-9629. St. Louis, Mo.: Monsanto.

Immerman, F.W., and J.L. Schaum. 1990. Nonoccupational Pesticide Exposure Study (NOPES), U.S. Environmental Protection Agency Order No. PB-90-152 224/AS. Springfield, Va.: National Technical Information Service.

Infant Formula Council. 1993. Infant formula: Evidence of the absence of pesticide residues. Unpublished paper prepared for the Infant Formula Council, Atlanta, Ga.

Jensen, A.A. 1983. Chemical contaminants in human milk. Pp. 1-128 in Residue Reviews: Residue of Pesticides and Other Contaminants in the Total Environment, F.A. Gunther and J.D. Gunther, eds. New York: Springer-Verlag.

Jensen, A.A., and S.A. Slorach. 1990. Contaminants in Human Milk. Boca Raton, Fla.: CRC Press. 288 pp.

Jonsson, V., G.J.K. Liu, J. Armbruster, L.L. Kettelhut, and B. Drucker. 1977. Chlorohydrocarbon pesticide residues in human milk in greater St. Louis, Missouri, 1977. Am. J. Clin. Nutr. 30:1106-1109.

Judge, D.P., ed. 1992. The Almanac of the Canning, Freezing, Preserving Industries. Vol. 2. Westminster, Md.: Judge & Sons.

Keith, L.H. 1983. Principles of Environmental Analysis. Washington, D.C.: American Chemical Society.

Keith, L.H., ed. 1988. Principles of Environmental Sampling. Washington, D.C.: American Chemical Society.

Kroger, M. 1972. Insecticide residues in human milk. J. Pediat. 80:401-405.

Kurtz, P.J., R. Deskin, and R.M. Harrington. 1989. Pesticides. In Principles and Methods of Toxicology, 2d ed., A.W. Hayes, ed. New York: Raven.

Laug, E.P., F.M. Kunze, and C.S. Prickett. 1951. Occurrence of DDT in human fat and milk. A.M.A. Arch. Indust. Hyg. 3:245-246.

Lawrence, R.A. 1989. Breastfeeding: A Guide for the Medical Profession, 3rd ed. St. Louis: C.V. Mosby. 652 pp.

Lykken, L. 1963. Important considerations in collecting and preparing crop samples for residue analysis. Residue Rev. 3:19-34.

Mattison, D.R. 1990a. Estimate of Cancer Risk from Halogenated Pesticides in Human Breast Milk: Arkansas 1986. Society for Risk Analysis Annual Meeting, October 7-10, New Orleans, La.

Mattison, D.R. 1990b. Pesticides in Human Breast Milk: Population Based Estimates of Excess Lifetime Cancer Risk. Society for Risk Analysis Annual Meeting, October 7-10, New Orleans, La.

Mattison, D.R. 1991. Pesticides in Human Breast Milk: Changing Patterns and Estimates of Cancer Risk. Washington, D.C.: Society for Gynecologic Investigation.

Mattison, D.R., J. Wohlleb, T. To, Y. Lamb, S. Faitak, M.A. Brewster, R.C. Walls, and S.G. Selevan. 1992. Pesticide concentrations in Arkansas breast milk. J. Ark. Med. Soc. 88(11):553-557.

McGuire, M.J., and I.H. Suffer, eds. 1983. Treatment of water by granular activated carbon. In Advances in Chemistry—Series 202. Washington, D.C.: American Chemical Society.

McMahon, B.M., and J.A. Burke. 1987. Expanding and tracking the capabilities of pesticide multiresidue methodology used in the Food and Drug Administration's pesticide monitoring programs. J. Assoc. Off. Anal. Chem. 70:1072-1081.

Minyard, J.P., Jr., W.E. Roberts, and W.Y. Cobb. 1989. State programs for pesticide residues in foods. J. Assoc. Off. Anal. Chem. 72:525-533.

Murphy, S.D. 1986. Toxic effects of pesticides. Pp. 519-581 in Casarett and Doull's Toxicology: The Basic Science of Poisons, 3rd ed., C.D. Klaassen, M.O. Amdur, and J. Doull, eds. New York: Macmillan.

Mussalo-Rauhamaa, H., H. Pyysalo, and R. Moilanen. 1984. Influence of diet and other factors on the levels of organochlorine compounds in human adipose tissue in Finland. J. Toxicol. Environ. Health 13:689-704.

NACA (National Agricultural Chemicals Association). 1988. Guidelines for Conducting Agricultural Chemical Residue Field Trials in the USA. Washington, D.C.: National Agricultural Chemicals Association.

Osteen, C.D., and P.I. Szmedra. 1989. Agricultural use trends and policy issues. In Agricultural Economic Report No. 622. Resources and Technology Division, Economic Research Service. Washington, D.C.: U.S. Department of Agriculture.

OTA (Office of Technology Assessment). 1988. Pesticides in Food: Technologies for Detection. Washington, D.C.: U.S. Government Printing Office.

Pancoast, H.M., and W.R. Junk. 1980. Handbook of Sugars, 2nd Ed. Westport, Conn.: AVI Publishing.

Reuber, M.D. 1980. Carcinogenicity of benzene hexachloride and its isomers. J. Environ. Pathol. Toxicol. 4:355-372.

Rogan, W.J., P.J. Blanton, C.J. Portier, and E. Stallard. 1991. Should the presence of carcinogens in breast milk discourage breast feeding? Regul. Toxicol. Pharmacol. 13:228-240.

Savage, E.P., T.J. Keefe, J.D. Tessari, H.W. Wheeler, F.M. Applehans, E.A. Goes, and S.A. Ford. 1981. National study of chlorinated hydrocarbon insecticide residues in human milk. Am. J. Epidemiol. 113:413-422.

Snyder, H.E., and T.W. Kwon. 1987. Soybean Utilization. New York: AVI Publishing.

State of California. 1981. Pesticide Use Report, Annual. Sacramento: California Department of Food and Agriculture.

State of California. 1987. Pesticide Residue Annual Reports. Sacramento: State of California.

Swern, D., ed. 1979. Bailey's Industrial Oil and Fat Products. New York: John Wiley & Sons.

Takahashi, W., and L.H. Parks. 1982. Organochlorine pesticide residues in human tissues, Hawaii, 1968-1980. Hawaii Med. J. 41:P250-251.

Takahashi, W., D. Saidin, G. Takei, and L. Wong. 1981. Organochlorine pesticide residues in human milk in Hawaii, 1979-80. Bull. Environ. Contam. Toxicol. 27:506-511.

Thurman, E.M., D.A. Goolsby, M.T. Meyer, and D.W. Kolpin. 1991. Herbicides in surface waters of the midwestern United States: The effect of spring flush. Environ. Sci. Technol. 25:1794-1796.

USDA (U.S. Department of Agriculture). 1983. Nationwide Food Consumption Survey. Nutrient Intakes: Individuals in 48 States, Year 1977-78. Report No. I-1. Hyattsville, Md.: Consumer Nutrition Division, Human Nutrition Information Service.

USDA (U.S. Department of Agriculture). 1991. Agricultural Chemical Usage 1990. Vegetables Summary. Economic Research Service Ag Ch 1(91). Washington, D.C.: U.S. Department of Agriculture.

USDA (U.S. Department of Agriculture). 1992a. Agricultural Chemical Usage 1991. Fruits and Nuts Summary. Economic Research Service, Ag Ch 1(92). Washington, D.C.: U.S. Department of Agriculture.

USDA (U.S. Department of Agriculture). 1992b. Agricultural Chemical Usage 1991. Field Crops Summary. Economic Research Service, Ag Ch 1(92). Washington, D.C.: U.S. Department of Agriculture.

USDA (U.S. Department of Agriculture). 1992c. Pesticide Use on 1991 Fall Potatoes. Pp. 26-28 in Agricultural Resources: Inputs Situation and Outlook Report. Economic Research Service, AR-25. Washington, D.C.: U.S. Department of Agriculture.

USGS (U.S. Geological Survey). 1988. National Water Summary 1986: Hydrologic Events and Ground-water Quality. U.S. Geological Survey Water Supply Paper 2325. Denver, Colo.: U.S. Government Printing Office.

USGS (U.S. Geological Survey). 1990. National Water Summary 1987: Hydrologic Events and Water Supply and Use. U.S. Geological Survey Water Supply Paper 2350. Denver, Colo.: U.S. Government Printing Office.

WHO (World Health Organization). 1984. Environmental Health Criteria 38, Heptachlor. Geneva: World Health Organization.

Williams, W.M., P.W. Holden, and D.W. Parsons. 1988. Pesticides in Ground Water Data Base, 1988. Interim Report. NTIS No. PB-89-164230. Springfield, Va.: National Technical Information Service.

Wolff, M.S. 1983. Occupationally derived chemicals in breast milk. Am. J. Ind. Med. 4:259-282.

Yess, N.J., E.L. Gunderson, and R.R. Roy. 1993. Food and Drug Administration monitoring of pesticide residues in infant foods and adult foods eaten by infants/children. J. Assoc. Off. Anal. Chem. Int. 76:492-507.

Zweig, G. 1970. The vanishing zero: The evolution of pesticide analyses. In the Essays of Toxicology, Vol. 2. New York: Academic.

7

Estimating Exposures

T HE TWO PRECEDING CHAPTERS have reviewed data on the diets of infants and children (Chapter 5) and on pesticide residues in food (Chapter 6). This chapter addresses methods for estimating ingestion of pesticides by infants and children using the data from the preceding two chapters. Although nondietary sources of pesticide exposures such as air, soil, and consumer products are also considered, emphasis is placed on the ingestion of pesticide residues present on foods consumed by infants and children.

Dietary exposure to pesticides depends both on food consumption patterns (Chapter 5) and on residue levels on food (Chapter 6). Multiplying the average consumption of a particular food by the average residue of a particular pesticide on that food yields the average level of ingestion of that pesticide from that one food commodity:

$$\text{Consumption} \times \text{Residue} = \text{Dietary Exposure}.$$

In reality, however, estimation of dietary exposure to pesticides is more complex than this simplified equation. Since many pesticides are used on a number of food crops, determination of the total exposure to a pesticide must be based on consumption data for all such foods. Also, it may be of interest to consider the total ingestion of different pesticides such as organophosphates and carbamates that fall within related classes and may pose similar risks to health.

The data presented in Chapter 5 indicate that food consumption levels vary both among and within individuals. This variation can be represented in terms of a distribution of food consumption, reflecting both high and low consumption levels, as well as the average level of consumption. Pesticide residue levels present in food will also vary, depending on several variables including application practices in different regions, time that has elapsed since application, degradation during transportation and storage of food, and the manner in which food is prepared by the consumer. Thus, both food consumption and pesticide residue data are characterized not by a single value but, rather, by a broad distribution reflecting high, low, and average values.

The variation in food consumption and residue data produces considerable variation in dietary exposure of pesticides by infants and children. This can be represented by a distribution of exposures across individuals within a particular age group. The distribution of dietary exposures is determined by the distribution of food consumption levels and the distribution of pesticide residues in food. If both the distribution of food consumption and the distribution of residue levels are known, statistical methods can be used to infer the distribution of dietary exposures. The process for combining different distributions into one distribution is termed *convolution*. The statistical convolution methods that can be used for this purpose are discussed later in this chapter.

Since ingestion of pesticides is dependent upon both food consumption and pesticide residue levels in food, it follows that the quality of dietary exposure data is determined by the quality of consumption and residue data. Although food consumption surveys such as the Nationwide Food Consumption Survey (NFCS) provide data on consumption patterns in the population at large, these surveys have generally not targeted infants and children. Hence, they included relatively small sample sizes within the age groups of primary interest for this report. One exception is the 1985-1986 Continuing Surveys of Food Intakes of Individuals (CSFII), which did focus on food consumption patterns of children.

Determination of the distribution of pesticide residues in foods consumed by infants and children is also difficult: only a fraction of all food consumed can be tested for the presence of pesticide residues. Many of the available residue data are based on surveillance studies that because of their focus on potential problem areas may overstate residue levels in the general food supply. The detection limit of residue monitoring methods can also impart uncertainty as to the residue levels actually present on food, especially when many residues are below the limit of detection and the detection limit is relatively high.

Recognizing these data limitations, the committee has included in this chapter several examples to illustrate possible approaches to estimating the distribution of dietary exposure to pesticides for infants and children. Each of these examples is designed to illustrate different aspects of exposure estimation, including the estimation of average daily exposures for use in chronic toxicity risk assessment and the estimation of peak exposures for evaluating acute toxic effects. Examples are included to illustrate how total exposure to pesticides used on more than one food crop can be estimated, and how exposures from different pesticides falling within the same toxicological class can be combined based on their relative toxicity.

Because of the limitations in the available consumption and residue data, it must be stressed that the purpose of the examples is to identify methods for estimating exposure and not to produce representative estimates of actual exposure. The particular compounds chosen for study were selected because sufficient data were available to illustrate the approaches to exposure estimation considered by the committee. All results should be taken in the context of the limitations of the data as described in this and the previous two chapters. Application of these methods in a regulatory context will be possible only if adequate data on the distribution of both food consumption and pesticide residues in food can be obtained.

The first example deals with benomyl, a systemic fungicide that has not been permitted for postharvest use in the United States since 1989. Because of the chronic toxic effects of this compound (benomyl has been shown to cause malignant liver tumors in mice), the average daily ingestion of benomyl was considered to be most relevant for estimating long-term exposure. Note that although the focus is on the average daily ingestion by individuals over an extended period, the daily ingestion will vary from person to person, depending on their food consumption habits and the residues of benomyl in the foods consumed by each person. Since residue data were available for apples, grapes, oranges, peaches, and tomatoes, this example was used to illustrate the estimation of total exposure to a single pesticide from multiple food commodities.

Data on benomyl from different residue monitoring programs were available to the committee, permitting a comparison of exposure estimates based on different residue data. For example, field trial data derived from pesticide analysis in the manufacturer's laboratory (using a special method not adapted to multiresidue screening) usually show higher detection rates than those found by government agencies in random sampling of food shipments. Field trial data are useful only as estimates of maximum residue concentrations from field test plot trials at treatment levels proposed for registration purposes. Because field tests are generally conducted at the maximum pesticide use allowed in its registration, the residue concentrations are often higher than those found in random sampling. The results of field trials are generally used to establish farm tolerances and analytical methodology for purposes of registration. Further evaluation of field trial data is required in order to evaluate pesticide degradation following application.

The impact of residue data below the limit of quantification (LOQ), a concentration below which residues cannot be accurately measured, was also investigated in this example. For nondetectable residues, it is possible that the actual (unknown) residue could be as low as zero or as high as the LOQ itself. The limitation of data on actual residue concentrations below the LOQ imparts additional uncertainty about the level of exposure to infants and children.

Aldicarb is the subject of the second example. This acutely toxic pesticide exerts its effects by inhibiting cholinesterase enzymes in the nervous system. The example focuses on dietary exposure to aldicarb first from potatoes and bananas separately and then from potatoes and bananas combined. It serves to illustrate how estimates of exposure to a single pesticide found on more than one food can be derived. In contrast to benomyl, where average daily exposures are of interest, individual daily intakes are examined in this example because of the acute toxicity of aldicarb.

Part of the aldicarb residue data is derived from composite sampling, which may underestimate peak residues found in individual potatoes or bananas as a consequence of compositing prior to residue analysis. Composite samples are not very satisfactory in acute risk assessment for raw food commodities like potatoes and bananas. However, residue levels in processed foods can be estimated by using composite samples.

The third example addresses methods for estimating exposure to a class of pesticides inducing a common toxic effect. Specifically, the committee considered five organophosphate compounds used on different fruits and vegetables. All these compounds can inhibit plasma cholinesterase. A measure of total exposure to all five organophosphates is proposed based on their relative potencies.

Before these short examples are presented, there is a discussion of statistical methods for combining the distribution of food consumption with the distribution of residue levels in food to arrive at a distribution of dietary exposures based on the method of convolution. The chapter concludes with a brief summary of nondietary sources of exposure to pesticides.

THE USE OF FOOD CONSUMPTION AND RESIDUE DATA FOR EXPOSURE ASSESSMENT

Food Consumption Data

The most appropriate dietary exposure data for risk assessment depends on the nature of the adverse health effects of concern. In the absence of specific dose-response effects, the average level of exposure of an individual over a certain period provides a reasonable measure on which to base estimates of chronic toxic effects such as cancer. For acute toxic effects, peak exposures over shorter periods are more appropriate for risk assessment.

Average Levels of Consumption

The development of food consumption data for evaluating chronic toxicity requires careful consideration. In general, food consumption surveys yield data on the consumption of that food over all days for which data are available. The average daily consumption for children within a given age class is then obtained by averaging across all the individuals in the age class.

Estimating the average daily consumption of a particular food within a given class warrants some discussion. Since some foods will not be consumed at all by some individuals, estimates of average daily consumption based on all individuals in the sample will underestimate average consumption for the subpopulation of individuals who consume the food in question. For this reason, separate estimates of average daily consumption for "all children" and for "eaters only" are considered when estimating exposure. Average consumption levels for "eaters only" are typically 2 to 3 times higher than those for "all children."

Because food consumption data are available for only a few days each year, the proportion of children falling into the eaters-only group is underestimated. This problem is accentuated if only a 24-hour recall or 24-hour food record is used. If food consumption data were available for every day of the year, more children who consume the food of interest on an infrequent basis would be included in the eaters-only group. Thus, since the eaters-only group omits some individuals whose consumption levels are low, the average food consumption for "eaters only" calculated in this way actually overestimates the average consumption for this group. This bias does not occur when information on food consumption is obtained through food frequency questionnaires rather than 24-hour recalls or 24-hour food diaries, since food frequency tables in principle accurately identify those individuals who consume the food at any time during a given year.

Scientists working with food consumption data have long recognized that consumption by a "typical" individual will not be representative of consumption by people who eat large amounts of a particular food. This has stimulated interest in examining the distribution of average daily consumption levels across individuals in order to estimate consumption by individuals who consistently consume greater quantities of the food of interest than the average. This distribution of average daily consumption across individuals can be used to estimate upper quantiles of consumption, such as the 90th, 95th, or 99th percentile. Reliable estimates of extreme percentiles can, however, be obtained only with relatively large sample sizes. Because the distribution of average daily intakes based on a sample of food consumption records for several days includes variability both between children and among days within children, this distribution will be subject to greater dispersion than would be the case if day-to-day variability were eliminated. (In the ideal case, this could be achieved by monitoring food consumption data over a full year or by using food frequency questionnaires.) The implication of such overdispersion is that upper percentiles of consumption will be overestimated.

Peak Levels of Consumption

Although the average level of individual exposure to pesticide residues in food is an important determinant of chronic toxicity, peak levels of exposure are more relevant for evaluating acute toxicity. Episodes of relatively high exposure occurring in a single day or even during a single meal may be more pertinent for acute risk assessment, depending on the toxic effect of interest.

The 1977-1978 NFCS provides information about food consumption during individual eating occasions for 3 different days. These data permit estimation of the total ingestion of a particular pesticide for each individual in the survey on each day. Using data for different individuals in the survey, one can estimate the distribution of person-days of consumption of specific foods. By combining this information with data on the distribution of pesticide residues in the food product or products of interest, it is then possible to estimate the number of person-days each year during which exposure to pesticides in the diet will exceed a critical level such as the reference dose (RfD), as defined in Chapter 8.

Although average levels of consumption and exposure will be reasonably well estimated with this approach, upper percentiles will be underestimated since food consumption data are available for only 3 of the 365 days in a year that are of interest. This is in contrast to the case for chronic risk assessment, where upper percentiles of exposure are likely to be overestimated.

Residue Monitoring

The point at which food samples are taken will influence the residue levels found. The highest residue levels generally occur immediately following application, and are reflected in field trial data. In samples taken for surveillance or compliance purposes, the residues will generally be higher than those in samples randomly drawn from the entire stock of a particular food commodity available for sale in a particular region of the country. Market basket surveys are based on a composite sample of a limited number of commonly consumed foods after they have been cooked or prepared for consumption in the usual manner. Although market basket surveys provide residue data under conditions designed to emulate foods as consumed, they are limited because they provide only composite sampling results on a few foods included in a typical meal.

Most analytical methods for measuring pesticide residues in food are subject to an LOQ below which residue levels cannot be accurately determined. Although improved analytical methods for testing for pesticide residues in food have made it possible to detect lower and lower residue levels, even the most sensitive techniques are subject to an LOQ. When residue levels below the LOQ are reported, it is not possible to determine whether the food contains no residue of the pesticide of interest or whether there is a residue present but at a lower level than can be detected with the analytical methods used.

This uncertainty about the actual residue level with residues below the LOQ confers uncertainty on the distribution of pesticide residues in food products and, subsequently, on the distribution of dietary exposure to pesticide residues by individuals consuming those foods. For example, consider a hypothetical distribution of residue levels based on the analysis of a number of food samples that may have been treated with a particular pesticide, as shown in Figure 7-1. The residues above the LOQ will generally follow a log-normal distribution. However, an appreciable proportion of the samples will produce results below the LOQ.

What can be inferred about residue levels in samples below the LOQ? The only certain inference is that the actual residue level lies between a lower limit of zero and an upper limit equal to the LOQ. (Even this upper bound may not be entirely correct, since analytical results near the LOQ will be subject to some degree of measurement error.) Because not all crops grown in the United States are treated with pesticides approved for use on those crops, it is possible that results below the LOQ may be entirely pesticide free.

Consider, for example, the data on the use of different pesticides approved for use on apples shown in Figure 7-2. The percentages of the U.S. apple crop treated with specific pesticides varies widely, ranging from a low of 1% for malathion to a high of 90% for azinphos-methyl. Thus, most apples will not contain residues of malathion and would produce residue levels below the LOQ when tested. It is also possible tests for azinophos could yield results generally below the LOQ if residues of this widely used pesticide were present at low but nondetectable levels.

The data on pesticide use in Figure 7-2 also reveal marked regional differences in pesticide usage patterns in different regions of the country.

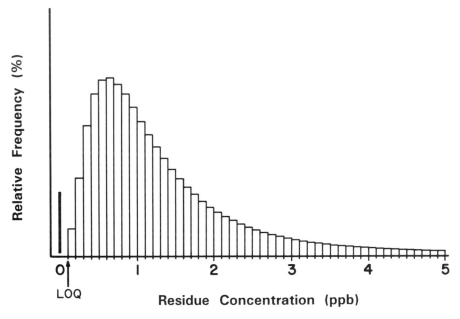

FIGURE 7-1 Hypothetical distribution of residue levels with a log-normal distribution for residues greater than zero.

Captan, for example, is widely used on apples grown in the central, northeast, and northwest regions of the United States but is virtually unused in the western regions of the country. Variation between pesticide usage patterns in different countries also warrants consideration with regard to imported food products.

In the past, results below the LOQ have been handled in different ways. A simple resolution of this uncertainty is to assume that all the results below the LOQ contain no residue and to assign them a residue level of zero. This is an optimistic approach, since the possibility of small but undetectable residues in some or all such samples cannot be excluded. A conservative approach is to assume that all residue levels are present at the LOQ. Although this may provide an upper bound on undetectable residues, it is unlikely that all the samples for which no residue was detected actually contain residues equal to the LOQ. An intermediate approach is to assume all nondetectable residues are present at one-half the LOQ. Clearly, the lower the LOQ, the less difference there will be between these different approaches, and the less uncertainty the LOQ will confer on estimates of potential human exposures.

Combining Residue and Exposure Data

Variation in food consumption patterns and in levels of pesticide residues in food leads to variation in dietary exposure to pesticides among infants and children. This variation in the ingestion of pesticide residues is characterized by a distribution of exposures, reflecting high, low, and average exposure concentrations. Statistically, the distribution of exposures can be obtained by convoluting (i.e., combining) the distribution of food consumption with the distribution of pesticide residues in food (Feldman and Fox, 1991). Thus, once the food consumption and residue distributions have been determined, the distribution of dietary exposures can be calculated (Figure 7-3).

The technical basis of convoluting two distributions can be described briefly as follows. Let C denote the consumption of a particular food by an individual, R the residue level in that food, and e the corresponding dietary intake or exposure level. The level of consumption will vary from person to person in accordance with the cumulative distribution $F_C(c)$ with corresponding density $f_C = F^1$. Note that $F_C(c)$ denotes the proportion

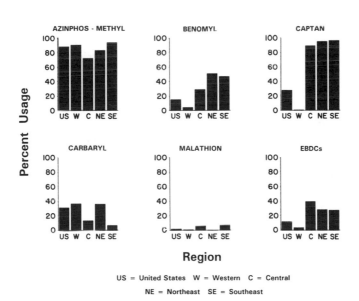

FIGURE 7-2   In the 1990 apple crop, percent of apple production treated with the following chemicals: azinphos-methyl, benomyl, captan, cabaryl, malathion, and EBDCs.

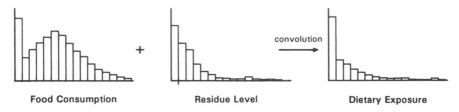

FIGURE 7-3   Convolution of food consumption distributions and residue distributions to produce dietary exposure distributions.

of the people in a given age group whose consumption C is less than a particular value c; the densities f_C and f_R reflect the relative frequency of different levels of consumption and residue, respectively, within the group. Letting F_R denote the residue distribution with density $f_R = F_R{}^1$, the distribution $F_E(e)$ of dietary intakes is defined by

$$F_E(e) = \int_0^e F_C\left(\frac{e}{r}\right)f_R(r)dr,$$

assuming that consumption C and residues R are statistically independent (Feldman and Fox, 1991, p. 349). This relationship provides the technical basis for combining the consumption distribution F_C with the residue distribution F_R to obtain the exposure distribution F_E.

In practice, estimates of consumption and residue distributions are based on survey data and are represented as histograms based on the observed sample. (If different weights are attached to the survey observations, a weighted distribution should be used.) Computationally, these two distributions can then be convoluted simply by taking the result of each point from the consumption distribution and multiplying it by each point in the residue distribution; the distribution of dietary intakes is then defined by the distribution of these products.

This empirical approach to convolution will work well, provided that the number of observations used to obtain the consumption and residue distributions is not large. With large distributions, the computation burden can be reduced by working with a random sample of both the consumption and residue data. The Monte Carlo approach (i.e., random sampling) to convolution was used by the committee in those examples where the computational effort required to convolute the two distributions was found to be excessive. The form of Monte Carlo sampling used by the committee was simply a means of reducing the amount of computational time required for convolution by using the original consumption and exposure distributions; no artificial distributional assumptions were required to implement this technique. The Monte Carlo distribution of dietary exposures will converge to that based on the entire exposure distribution as the number of Monte Carlo samples increases. As the number of samples converge, the distributions become identical.

The convolution method can be extended to more complex situations such as the estimation of total exposure to a pesticide that may be present on more than one food commodity. In this case, a single point on the exposure distribution is estimated by randomly combining points from the consumption distributions for all foods of interest with points from the corresponding residue distributions (one for each food), and then summing the total exposure across all foods. This process is repeated to generate a distribution of total exposures from all foods combined.

Total exposure to pesticides within the same class can be estimated in a similar fashion using the relative potency values for those pesticides to express the intake in toxicity equivalence factors. This is illustrated in the example of organophosphate pesticides later in this chapter.

LONG-TERM EXPOSURE TO BENOMYL

The Compound

Benomyl, or Benlate, is a systemic fungicide that was used in the United States from the time of its registration in 1972 until registration was voluntarily withdrawn for postharvest use by the manufacturer in 1989. Before then, it was the most widely used of the fungicides in the family of benzimidazole pesticides.

Benomyl is effective in preventing more than 190 fungal diseases, and it acts as a protective surface barrier while also penetrating the plant tissue to arrest infections. It was applied as a seed treatment, a transplant dip, and a foliage spray and was registered for use on more than 70 crops in 50 countries, including imported foods such as bananas and pineapples. In the United States, more than 100 EPA tolerances were established for benomyl in a variety of foods and feeds.

Benomyl has been shown to induce hepatocellular carcinomas, and combined hepatocellular neoplasms occurred in male and female mice treated with benomyl at all doses. In tests that included methyl-2-benzimidazole carbamate (MBC)—a metabolite of benomyl—investigators observed combined hepatocellular neoplasms in male mice and hepatocellular adenomas, carcinomas, and the combined hepatocellular neoplasms in female mice (NRC, 1987). Because of its carcinogenic potential, exposure assessment for benomyl is based on the distribution of average ingestion levels for different individuals.

The Consumption Data

Data from USDA's 1985-1986 Continuing Survey of Food Intake of Individuals (CSFII) were found to be more suitable for use in this example than the data from the 1977-1978 NFCS: the CSFII is more current than the NFCS, and includes consumption data collected over 6 days at 2-month intervals over an entire year, as compared to the 3-consecutive-day sample in the 1977-1978 NFCS. The CSFII included 170 1-year-olds, 195 2-year-olds, 225 3-year-olds, 191 4-year-olds, and 209 5-year-olds.

Intake data were divided into person-day food intake, and consumption was then averaged for each individual in each age class across the days of the reporting period. For example, 170 average daily intake values were recorded for each food for each of the 1-year-old children surveyed. A probability distribution of exposures could then be constructed for each yearly age class.

The Residue Data

Five different sets of residue data on benomyl residues were reviewed by the committee for this example. They were

- results of field trials conducted by the manufacturer,
- results of a market basket survey conducted by the manufacturer,
- 1988-1989 compliance and surveillance data collected by the Food and Drug Administration (FDA),
- data provided by the food industry, and
- data from tests of raw food by a certification business operating in California.

These data were collected using different sample designs, sample sizes, and analytical methods. Table 7-1 compares the number of benomyl samples and the number of detections for the FDA surveillance data, the manufacturer's field trials and market basket surveys, the food industry's data, and data supplied by the certification business. Data on apples, grapes, oranges, peaches, and tomatoes are shown in Figure 7-4 for all but the certification business.

Estimation of Exposure

Exposure was estimated using each of the five sets of residue data reviewed by the committee separately. An individual child's exposure to benomyl from a particular food was estimated by multiplying the mean

TABLE 7-1 Number of Benomyl Samples and Detections for Selected Foods Based on Data from the FDA, a Pesticide Manufacturer, the Food Industry, and a Certification Business

Food	FDA		Manufacturer's Field Trials		Manufacturer's Market Basket		Food Industry		Certification Business	
	No. of Samples	No. of Detections	No. of Samples	No. of Detections	No. of Samples	No. of Detections	No. of Samples	No. of Detections	No. of Samples	No. of Detections
Apple	134	35	138	122	26	5	68	30	127	65
Apple juice							30	16		
Apricot	72	8					6	0	19	5
Banana	5	0					4	0	38	10
Bean			35	29	30	3	19	0	14	3
Blueberry							3	0	12	1
Carrot									24	4
Celery									4	1
Cherry	21	5					7	0	4	1
Cucumber					21	1			11	5
Grape	27	12	71	65	11	4	6	0		
Watermelon	5	0					3	0		
Nectarine	14	8	18	5					39	14
Orange	6	0	18	13	12	12	1	0	6	1
Orange juice	1	0					2	0		
Peach	26	13	82	72	37	1	15	0	81	44
Pear	23	1	15	14	24	6			7	4
Pineapple	25	18							28	18
Plum	21	10								
Raisin							13	0		
Raspberry	14	0					2	0	17	6
Rice							6	0		
Squash	4	0					4	1		
Strawberry	30	2			25	1	6	0	16	11
Tomato	20	0	35	23			12	0		
Wheat										

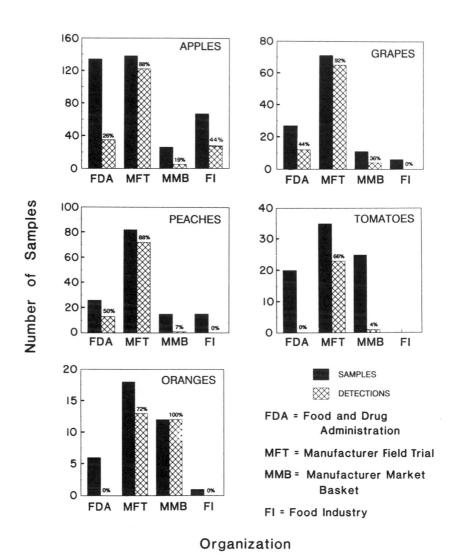

FIGURE 7-4 Number of benomyl samples and detections in apples, grapes, oranges, peaches, and tomatoes.

residue for that food by the average daily intake of that food. The exposures were then summed across up to 26 foods that 1-year-old children consume most to produce an average daily exposure estimate for each child. Note that different foods would be included with different residue data sets, depending on the availability of residue data for those foods. Finally, the distribution of average daily exposure from all foods combined across individuals was calculated.

The committee did not adjust these estimates for the percentage of the crop acreage treated. That adjustment is customarily applied by the EPA to residue data from field trials, thereby substantially reducing estimated exposures. EPA also multiplies the number of samples with no detected residues by the percentage of crop treated and assumes that residues in those samples are at the LOQ while the remainder of the undetected residues are at zero, i.e., *anticipated residues*.

In the procedure described in this chapter it is assumed that all crops are treated with benomyl. The committee notes that this is an unlikely scenario; however, the purpose of this analysis is to illustrate the probability distribution approach to estimating exposure.

The Manufacturer's Field Trials

In 1989 the manufacturer submitted to the EPA a substantial amount of residue data obtained from field trials in support of the continued registration of benomyl. These data are useful because they included information on the application rate (i.e., frequency of application [%] and amount used) and on the residue levels detected in each sample of raw agricultural commodity. Furthermore, sample sizes for single raw commodities were often large enough to permit statistical analysis. Unfortunately, many processed foods were not sampled for residues, thus forcing the EPA to rely on assumptions about the fate of residues during processing.

Data generated by the manufacturer for benomyl residues on fruit products following processing are shown in Table 7-2. Detected levels shown in those data are likely to be far higher than those in actual market basket data, due to the uneven use of the compound throughout the United States. In a nationwide market basket survey, fewer samples treated with lower amounts of benomyl would actually be found than in the manufacturer's field trials, which focused on crops known to be treated with benomyl.

As shown in Table 7-1, a total of 412 samples of eight unprocessed foods (apples, beans, grapes, nectarines, oranges, peaches, pears, and tomatoes) were tested. Of these, 343 (83%) contained residues above the detection limit. Since neither apple juice nor orange juice were sampled,

TABLE 7-2   Changes in Benomyl Concentrations
During Washing and Processing

Food	Food Form	Reduction, %
Apples	Washed	13
	Juice	69
	Applesauce	82
Peaches	Washed	73
	Canned	99
Bananas	Pulp	No detectable residue

SOURCE: Based on data from the pesticide manufacturer.

EPA must rely on the results of processing studies, such as those shown in Table 7-2, to determine the fate of residues in juices most consumed by young children.

The committee conducted two separate analyses of these data based on two different assumptions: that all reported nondetections were actually zero (Figure 7-5) and that nondetections were really residues at the LOQs, which were provided for each sample (Figure 7-6). (The actual exposure is somewhere in between those shown in the two figures.) The exposure estimates were not greatly affected by either assumption, principally due to the relatively high number of samples containing detectable residues. Estimates of young children's benomyl exposure based on the manufacturer's field trial residue data are almost identical, regardless of whether a value of zero or the LOQ is used in exposure calculations when no residue is detected. Estimates based on the manufacturer's market basket data are also comparable, regardless of the value assigned to nondetectable residues.

That portion of Figures 7-5 and 7-6 displaying the manufacturer's field trial data was constructed by combining the individual consumption reports with the mean of field trial residues for each of 10 foods: apples, apple juice, oranges, orange juice, grapes, grape juice, peaches, pears, green beans, and tomatoes. The committee could therefore produce separate exposure estimates for each food for each child from 1 to 5 years of age. Each analysis was conducted under the assumptions that only those 10 foods were consumed in a child's diet and that the juices lost no benomyl during processing.

The Manufacturer's Market Basket Survey

The sample design of a market basket survey is important, since the results can be dramatically affected by regional patterns of pesticide use and food distribution. A good design can obviate the need to make complex assumptions regarding processing, percentage of crop treated, and food distribution effects.

A limited number of foods was surveyed in the manufacturer's market basket survey. A total of 143 samples of 7 foods, and no juices, were analyzed for a benomyl. Thirty-two (22%) of the samples contained residues at levels above the LOQ. This percentage is similar to that found by the food industry but approximately 50% lower than that detected by the more focused sampling design used by the certification business.

FDA Surveillance Data

The committee used only FDA surveillance data in assessing chronic exposures, and it did not estimate exposures for any food for which there were fewer than 20 samples. Although benomyl was registered for use on many foods, sample size exceeded 20 for only 10 of the 26 foods listed in Table 7-1. Of the total of 448 samples tested, 112 (25%) had residues that exceeded the LOQ.

FDA monitoring is focused on fresh rather than processed foods. Therefore, many of the processed foods often consumed by young children are never or seldom sampled, and the utility of small samples is limited in estimating exposures. As shown in Table 7-1, a number of foods sampled by other groups were not sampled at all by FDA in 1988 and 1989. Other weaknesses of FDA surveillance data are noted in Chapter 6 on pesticide residues.

The Food Industry

A food industry association provided a large amount of data collected from its member organizations. These data identified the food, the pesticide used, the residue level, and the LOQ of the analytical method used. Since the food industry used a variety of sampling and analytical methods, the representativeness of the data for the nation's food supply is uncertain.

Despite this uncertainty, these data are useful in illustrating the method proposed here for exposure estimation. The majority of the positive findings in this data set relate to apples (with 30 residues above the detection limit observed in 68 samples) and apple juice (with 16 of 30 samples showing positive).

A Certification Business

The committee obtained residue data from a certification business operating in California, a commercial organization that guarantees to grocery store owners and consumers that any residues in produce will be below

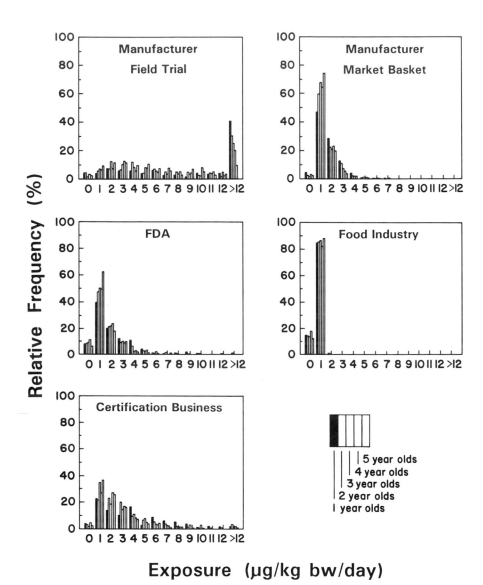

Exposure (μg/kg bw/day)

FIGURE 7-5 Daily exposure of 1- to 5-year-old children to benomyl in different combinations of foods, as shown by residue data from a certification business, FDA, the food industry, and the manufacturer (field trials and market baskets). Based on the assumption that the nondetects were equal to zero.

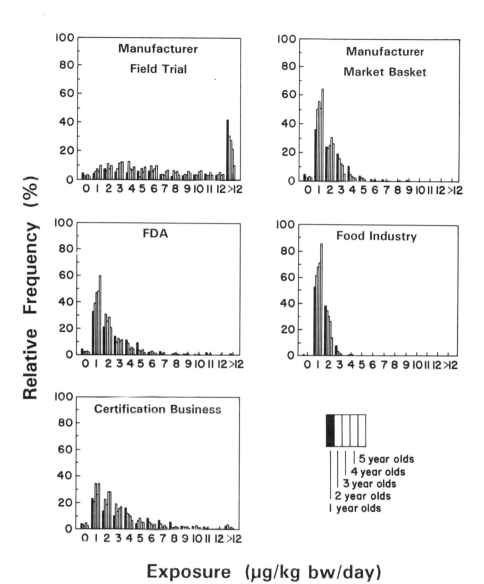

Exposure (µg/kg bw/day)

FIGURE 7-6 Daily exposure of 1- to 5-year-old children to benomyl in different combinations of foods, as shown by residue data from a certification business, FDA, the food industry, and the manufacturer (fields trials and market baskets). Based on the assumption that the nondetects were equal to the LOQ.

a detectable level. These data are subject to certain limitations such as nonrepresentative localized sampling, unidentified analytical methods, and analysis of selected produce. Nonetheless, they are used in this example to provide a range of data for comparison purposes. As shown in Table 7-1, this organization tested 447 samples of which 193 (43%) had benomyl residues above the detection limit. This compares to 448 FDA samples with 112 (25%) positive detections and to 203 food industry samples with 42 (20.6%) detections over the LOQ. The difference could be explained by the certification business's focus on produce with a history of residues of concern. Moreover, although the certification business, FDA, and the food industry tested a similar number of foods (16, 17, and 18, respectively), the selection of foods varied. For example, the food industry detected residues in 16 (53.3%) of 30 samples of apple juice, whereas the certification business tested no apple juice samples. Similarly, the certification business detected residues in 18 (64.3%) of 28 plum samples, but plums were not sampled by the food industry.

Summary

In this example, the committee examined multiple data sets reflecting concentrations of benomyl in several different foods. These data sets varied in different ways (Table 7-1). The manufacturer's field trial survey showed a much higher frequency of benomyl detections than found in the market basket survey, mainly because the study focused on crops known to be treated with benomyl. In the market place, fewer foods would actually have been treated with benomyl than in the field survey.

The effect of processing on benomyl concentrations in foods is shown in Table 7-2. These data indicate that substantial reductions in residues may occur because of processing. Figures 7-5 and 7-6 show the estimated exposure distributions for children 1 to 5 years of age to total benomyl residues for several common foods where residues below the LOQ were set at zero and at the LOQ, respectively.

The most important result of this analysis is that most children are exposed to relatively small concentrations of benomyl in their diets (less than 0.012 mg/kg bw/day), if they are exposed to any at all. Current exposures could be even less since registration of benomyl for postharvest use has been suspended. In any case, their exposures would be much below the current reference dose (RfD). In the committee's analysis, it is only the manufacturer's field trial data that suggest some children have larger exposures, and as noted above, these data would be inappropriate for this type of analysis. Both the manufacturer's market basket and the certification business's residue data show relatively small total residue exposures. Finally, Figures 7-5 and 7-6 show that assigning a value of zero or the LOQ to nondetectable residues had little effect on the overall outcome of the committee's calculations for benomyl. A principal reason for this finding is that a relatively high number of benomyl samples contained detectable residues. Certain impacts on exposure estimates would be seen for pesticides where a relatively low number of samples contained detectable residues.

SHORT-TERM EXPOSURE TO ALDICARB

The Compound

The evaluation of short-term peak exposures is illustrated using data on aldicarb residues in potatoes and bananas. Aldicarb is an acutely toxic pesticide whose use on potatoes and bananas was voluntarily suspended by the manufacturer in 1990 and 1992, respectively. It is an N-methyl carbamate that exerts its toxic effects by inhibiting the enzyme cholinesterase in the central and peripheral nervous system and at the neuromuscular junctions. Single oral doses of 25 μg/kg bw in humans produces approximately 50% inhibition of blood cholinesterase (NRC, 1986). Inhibition above 30% is usually of concern in humans.

Aldicarb is a systemic toxicant that is used primarily as an insecticide and nematocide. It is absorbed by the roots, stems, leaves, and fruits of plants. Aldicarb sulfoxide is a toxic metabolite that is distributed throughout the plant and degrades relatively slowly. Aldicarb-treated crops commonly eaten by children include potatoes, bananas, and citrus fruits. As demonstrated in Chapter 5, infants and children consume proportionately more of these foods than do adults with the exception that infants do not eat potatoes.

Acute Effects of Dietary Aldicarb Exposure

In 1970 Union Carbide gave three groups of four healthy male adult volunteers doses of aldicarb at concentrations of 25, 50, or 100 μg/kg bw. Subjects given the highest dose became acutely ill; one of those who received the lowest dose developed severe mood symptoms (i.e., anxiety reaction). Whole blood cholinesterase depression was observed in all the subjects. After reviewing the results of this study, the Safe Drinking Water Committee of the National Research Council estimated a no-observed-adverse-effect level (NOAEL) of 10 μg/kg bw/day (NRC, 1986). Applying a 10-fold uncertainty factor, the EPA established an RfD, formerly an acceptable daily intake (ADI), for aldicarb at 1.0 μg/kg bw/day.

Studies in the dog demonstrate depression of plasma cholinesterase at doses as low as 1 ppm (20 μg/kg bw/day). At doses of 50 μg/kg bw/

day, there were statistically significant increases in diarrheal stools in both sexes, along with statistically significant increases in both plasma and RBC cholinesterase inhibition in males. Applying an uncertainty factor of 100 on the lowest dose level (20 μg/kg bw/day) to account for interspecies and intraspecies variation, EPA identified an RfD of 0.2 μg/kg bw/day. EPA's Scientific Advisory Panel recommended on November 6, 1992, that the RfD for aldicarb be reestablished at 1.0 μg/kg bw/day, consistent with the 1986 NRC recommendation (EPA, unpublished data, 1993).

In July 1985, severe acute illness was observed in more than 1,000 people in the western United States a few hours after they had eaten watermelons treated with aldicarb—a nonregistered (illegal) use. The symptoms included nausea, vomiting, diarrhea, muscle fasciculations, mood changes, and other symptoms of cholinergic poisoning. The most seriously ill person was a 62-year-old woman who had eaten approximately one-fourth of a watermelon later found to contain a 2.7-ppm concentration of aldicarb sulfoxide, which presented an estimated dose of 57 μg/kg bw. She required emergency room treatment and atropine to reverse the symptoms.

The Consumption Data

Data on consumption of bananas and potatoes by children between 12 and 24 months of age were obtained from the 1977-1978 NFCS. The mean, median, and 90th and 95th percentiles of the average daily consumption of bananas and potatoes are shown in Table 7-3. Of the 529 children surveyed, 157 did not eat either bananas or potatoes on any of the days during which the survey was conducted; of those that did, fewer children ate potatoes than bananas. Of the 1,831 person-days included in the survey, there were 1,077 days on which neither bananas nor potatoes were consumed. The distribution of daily consumption of bananas and potatoes by children between 12 and 24 months of age is shown in Table 7-4.

TABLE 7-3 Average Daily Consumption by Children Between 12 and 24 Months of Age

Food	Sample Size	Number of Noneaters	Consumption, g/kg bw/day			
			Mean	Median	P90	P95
Bananas	529	208	0.90	0.16	3.01	4.17
Potatoes	529	349	0.72	0	2.25	3.41

SOURCE: Based on data from the U.S. Department of Agriculture's Nationwide Food Consumption Survey, 1977-1978.

TABLE 7-4   Daily Consumption by Children Between 12 and 24 Months of Age

Food	Sample Size	Number of Days No Food Eaten	Consumption, g/kg bw/day			
			Mean	Median	P90	P95
Bananas	1,831	1,224	0.91	0	2.52	7.47
Potatoes	1,831	1,602	0.69	0	2.38	5.20

SOURCE: Based on data from the U.S. Department of Agriculture's Nationwide Food Consumption Survey, 1977-1978.

The Residue Data

Until the early 1980s, residue sampling focused on aldicarb rather than on aldicarb sulfoxide—its persistent and more toxic metabolite. Furthermore, only composite samples were used. That is, many individual samples of a single commodity were blended and the resulting mixture was analyzed. Because aldicarb is an acute toxicant, and the foods it contaminates are often eaten individually, EPA required a survey in which individual foods were examined separately. In this survey, concentrations higher than the EPA tolerance level were found in individual bananas and potatoes, although they had not been detected previously in blended samples.

Sampling revealed that the distribution of residues in individual potatoes from treated fields followed a log-normal distribution pattern, that is, most individual sample results were clustered at low concentrations. The highest single concentration, 8.7 ppm, was found in one potato. In the event that a 20-kg child consumed that one 200-g potato, cooked by itself in a microwave oven, the child could receive an exposure of 87 µg/kg bw—an acutely toxic level. This illustrates the potential problems associated with use of composite samples for evaluation of exposure to acute toxicants.

Banana trees have been treated with aldicarb since 1977. In 1991 composite and individual samples of bananas were analyzed in five strictly controlled field trials. Half the bananas from one field were found to contain aldicarb residues higher than the tolerance level of 0.3 ppm. If a 20-kg child were to eat the 170-g edible portion of a single banana at the highest level found, 3.14 ppm, the resulting dose would be 26 µg/kg bw—again a potentially toxic dose. Even at the 0.3-ppm tolerance level for aldicarb in bananas, that child would be exposed to approximately 3 µg/kg bw— a level well above the RfD. This does not take into account exposure to other cholinesterase inhibitors in the diet, including possible aldicarb residues in citrus fruit or potatoes.

Since pesticides are usually approved for use on more than one food, it is important to consider the total exposure to a particular pesticide from

TABLE 7-5   Residues of Aldicarb

Food	Sample Size	Number Below LOQ	Mean Residue, ppm			
			Mean	Median	P90	P95
Bananas	2,697	2,442	0.008	0	0.02	0.05
Potatoes	294	6	0.239	0.085	0.510	0.840

SOURCE: Based on data from the 1987 National Aldicarb Food Survey and survey data from the manufacturer.

all dietary sources. The methods for assessing exposure to aldicarb in bananas can be extended to cover multiple foods. Although the committee used only two foods (bananas and potatoes) in its example, extension of the method to more than two foods is straightforward. In fact, the FDA data on aldicarb discussed in Chapter 6 failed to identify residue levels above the LOQ in the 350 samples tested. Consider the data on residues of aldicarb in bananas examined earlier, along with data on residues of aldicarb in potatoes given in Table 7-5. The former data are from the National Aldicarb Food Survey. The latter data are from a special survey conducted by the manufacturer of the compound in 1989. In the manufacturer's survey, representative samples were taken from 26 locations in the states of Washington, Oregon, California, Michigan, and Maine. Only three of the locations, Washington, Oregon, and Maine, had composite samples with residues above the LOQ. Residue data from these three states were also selected because they gave data for individual potatoes. Of the 294 reported residue values, 6 were below the LOQ. The mean aldicarb residue of 0.239 ppm in potatoes is much higher than the mean residue of 0.008 ppm in bananas. This difference is due largely to the use of field trial data for potatoes, which were obtained from crops known to have been recently treated with aldicarb.

Effects of Assumptions Regarding Residues Below the LOQ

The implications of results below the LOQ for exposure estimation can be illustrated using data on the levels of aldicarb in bananas obtained from the 1987 National Aldicarb Food Survey. The composite samples tested in this survey were obtained from 225 groups averaging 12 bananas each—a total of 2,700 bananas. These samples were initially tested for the presence of aldicarb with an analytical method that had an LOQ of 0.01 ppm. If any composite sample was found to have a residue greater than 0.01 ppm, each banana in that group was analyzed individually. In this survey, residues over 0.01 ppm were detected in 27 of the 225 composite samples. The investigators then conducted separate tests on the 299 bananas that were available for testing out of the 302 bananas in the 27

TABLE 7-6   Residues of Aldicarb in Bananas

Number Below LOQ/Sample Size	Value Used for Residues Below LOQ	Mean Residue, ppm			
		Mean	Median	P90	P95
2,442/2,697	LOQ	0.017	0.01	0.02	0.05
	0	0.008	0	0.02	0.05

SOURCE: Based on data from the 1987 National Aldicarb Food Survey.

samples. They found aldicarb concentrations above the LOQ in 255 of those bananas.

For risk assessment purposes, let us assume that the remaining 2,442 bananas in the sample had residue levels below the LOQ. This underestimates actual residue levels because compositing masks any unusually high residue levels on individual bananas in a given batch. Let us also assume that individual bananas testing negative do not contain residues above the LOQ. Despite these approximations, it is instructive to examine the impact of assumptions regarding residues lower than the LOQ on estimation of dietary exposures to aldicarb from bananas.

Table 7-6 presents the mean, median, and upper 90th and 95th percentiles of aldicarb on the 2,697 bananas in the survey sample. The mean residue level obtained by assigning a value of 0.01 ppm to all residues below the LOQ is 0.017 ppm—slightly more than twice the value of 0.008 ppm obtained by assigning a value of 0 to nondetectable residues. The median value of 0.01 ppm obtained by substituting the LOQ for nondetectable residues is close to the corresponding mean residue. Assigning a value of 0 to the bananas with no detectable residues leads to a median residue of 0. Since less than 10% of the detections were above the LOQ, both the 90th and 95th percentiles of the residue distribution are unaffected by the value chosen for observations below the LOQ.

Estimating Dietary Exposure

Of the 529 children between 12 and 24 months of age in the 1977-1978 NFCS, only 321 reported eating bananas on any of the 3 days during which food consumption data were recorded. The mean daily consumption of bananas among all the children surveyed was 0.90 g/kg bw/day (Table 7-7). The mean consumption by the 321 children who ate bananas on at least one occasion during the survey was 1.47 g/kg bw/day. Since 61% of the children consumed bananas at least once, the upper 90th and 95th percentiles for the subgroup of eaters are only slightly higher than the corresponding consumption percentiles for the entire sample. The 90th and 95th percentiles will be overestimated since the distribution of average daily consumption contains variability between children and among days.

TABLE 7-7 Daily Consumption of Bananas by Children Between 12 and 24 Months of Age

Subsample	Sample Size	Consumption, g/kg bw/day			
		Mean	Median	P90	P95
All	529	0.90	0.16	3.01	4.17
Eaters only	321	1.47	0.48	3.66	5.67

SOURCE: Based on data from the 1987 National Aldicarb Food Survey and USDA's Nationwide Food Consumption Survey, 1977-1978.

The aldicarb residue distribution shown in Table 7-6 may be combined with the distribution of mean daily intake of bananas shown in Table 7-7 to estimate the distribution of mean daily intakes of aldicarb residues on bananas. Statistically, this is accomplished by convoluting the two distributions by pointwise multiplication of the residue and consumption distributions to obtain an estimate of the distribution of intakes.

The results of this calculation are summarized in Table 7-8. Separate estimates of intake are presented for the entire sample and for the subsample of children who ate bananas during the survey period. Separate estimates of intake are given for residue levels below the LOQ using the assumptions that the residues were either 0 or at the LOQ.

These estimates of mean intake in Table 7-8 are identical to those obtained simply by multiplying the mean consumption of bananas by the mean residue concentration (Table 7-9). As indicated in Table 7-9, however, multiplication of the upper 90th percentile of the residue and consumption distribution in this fashion does not yield the 90th percentile of intake based on the method of convolution (Table 7-8). The discrepancy between these two values is particularly large for the subgroup of banana eaters only with nondetectable residues assigned a value of 0. Thus, esti-

TABLE 7-8 Daily Intake of Aldicarb from Bananas for Children Between 12 and 24 Months of Age

Subsample	Value Used for Residues Below LOQ	Intake, μg/kg bw/day			
		Mean	Median	P90	P95
All	LOQ	0.015	0.002	0.034	0.058
	0	0.007	0	0	0.008
Eaters only	LOQ	0.025	0.005	0.050	0.087
	0	0.012	0	0.003	0.032

SOURCE: Based on data from the 1987 National Aldicarb Food Survey and USDA's Nationwide Food Consumption Survey, 1977-1978.

TABLE 7-9   Methods for Estimating the Mean and 90th Percentiles of
Aldicarb Intake

Subsample	Chronic Values Used for Residues Below LOQ	Mean	P90	Mean	P90
		Acute			
All	LOQ	0.015	0.060	0.015	0.050
	0	0.007	0.060	0.007	0.050
Eaters only	LOQ	0.025	0.073	0.047	0.208
	0	0.012	0.073	0.022	0.208

SOURCE: Based on data from the 1987 National Aldicarb Food Survey and the U.S.
Department of Agriculture's Nationwide Food Consumption Survey, 1977-1978.

mates of upper percentiles of ingestion should be based on the more
accurate method of convolution. At the upper percentiles, estimates are
higher than the true percentiles since the average consumption distribution incorporates day-to-day variability.

This method is most appropriate for estimating the average daily ingestion of pesticide residues over an extended period. Although average
daily ingestion is an appropriate measure of exposure for chronic risk
assessment, a different approach is required for acute toxic effects caused
by short-term exposure to relatively high levels of substances.

Assume that the total intake of a particular pesticide in a single day
represents a good indicator of whether an acute toxic response will occur.
In this event, we may examine the distribution of individual daily intakes
in Table 7-10 rather than the distribution of average daily intakes shown
in Table 7-7. In the present context, this corresponds to the distribution
of individual daily intake of bananas for all days for which observations
were recorded for all children in the survey. Convolution of this distribution with the aldicarb residue distribution provides an estimate of the
distribution of the number of person-days in the sample associated with

TABLE 7-10   Daily Consumption of Bananas by Children Between 12
and 24 Months of Age

Subsample	Sample Size	Consumption, g/kg bw/day			
		Mean	Median	P90	P95
All	1,831	0.91	0	2.52	7.47
Eaters only	607	2.74	0.72	10.4	8.73

SOURCE: Based on data from the U.S. Department of Agriculture's Nationwide Food
Consumption Survey, 1977-1978.

TABLE 7-11 Individual Daily Intake of Aldicarb from Bananas for Children Between 12 and 24 Months of Age

Subsample	Value Used for Residues Below LOQ	Intake, μg/kg bw/day			
		Mean	Median	P90	P95
All	LOQ	0.016	0	0.038	0.086
	0	0.008	0	0	0
Eaters only	LOQ	0.047	0.008	0.097	0.135
	0	0.023	0	0.007	0.047

SOURCE: Based on data from the 1987 National Aldicarb Food Survey and the U.S. Department of Agriculture's Nationwide Food Consumption Survey, 1977-1978.

a given daily intake of aldicarb (Table 7-11). This distribution thus provides a basis for estimating the percentage of person-days during which exposure would exceed a health-based exposure standard, such as a reference dose based on toxicity studies. The upper percentiles will be underestimated since the food consumption data are available for only 3 to 4 days of the 365 days in the year. Although several methods for dealing with results below the detection limit of the analytical method were discussed previously, all nondetectable residues were assumed to be zero in this analysis for simplicity.

The mean, median, and 90th and 95th percentiles of average daily intake and individual daily intake of aldicarb from bananas and potatoes alone and for bananas and potatoes combined are shown in Tables 7-12 and 7-13 for children between 12 and 24 months of age. The distribution is dominated by the intake of potatoes. Figures 7-7 and 7-8 show the distribution of individual and average intakes of aldicarb from potatoes and bananas, separately and combined. Intake values greater than 0.8 g/kg bw/day represented a very small proportion and were therefore omitted from the figures.

The distribution of aldicarb intake from both bananas and potatoes

TABLE 7-12 Average Daily Intake of Aldicarb for Children Between 12 and 24 Months of Age

Food	Intake, μg/kg bw/day			
	Mean	Median	P90	P95
Bananas	0.007	0	0	0.008
Potatoes	0.172	0	0.302	0.673
Bananas and potatoes	0.179	0	0.327	0.705

SOURCE: Based on data from the U.S. Department of Agriculture's Nationwide Food Consumption Survey, 1977-1978, the 1987 National Aldicarb Food Survey, and survey data from the pesticide manufacturer.

TABLE 7-13   Individual Daily Intake of Aldicarb for Children Between 12 and 24 Months of Age

Food	Intake, µg/kg bw/day			
	Mean	Median	P90	P95
Bananas	0.008	0	0	0
Potatoes	0.164	0	0.123	0.537
Bananas and potatoes	0.172	0	0.164	0.593

SOURCE: Based on data from the U.S. Department of Agriculture's Nationwide Food Consumption Survey, 1977-1978, the 1987 National Aldicarb Food Survey, and survey data from the pesticide manufacturer.

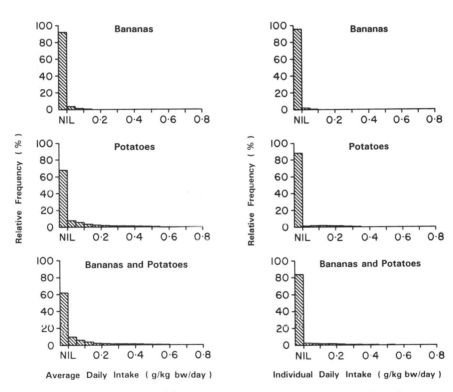

FIGURE 7-7   Distribution of the average daily intake of aldicarb from bananas and potatoes, separately and combined. SOURCE: Based on data derived from USDA, 1983, the National Aldicarb Food Survey, and survey data from the manufacturer.

FIGURE 7-8   Distribution of individual daily intake of aldicarb from bananas and potatoes, separately and combined. SOURCE: Based on data derived from USDA, 1983, the National Aldicarb Food Survey, and survey data from the manufacturer.

resembles that of aldicarb intake from bananas or potatoes alone except for a slight shift toward nonzero values; the number of zeros has decreased from 66% of the values to 63%. This is to be expected since some children ate bananas but not potatoes or potatoes but not bananas.

Summary

Aldicarb was examined by the committee because it is an acutely toxic chemical that may potentially be found in several foods consumed by children and because there were also good sampling data for aldicarb residues in several commodities. This example could also be used to illustrate approaches to estimating residue concentrations in both individual foods and in multiple foods combined in a child's diet to provide total estimated residue.

Composite samples were used until the early 1980s for measuring residues in foods. Samples of a single commodity were blended, and the resulting mixture was analyzed. Because aldicarb is an acute toxicant, new approaches do not use blended samples. Rather, single commodities are analyzed. These new residue surveys have shown log-normal distribution patterns of aldicarb residues in commodities such as potatoes. As would be expected, most individual samples show residues clustered at lower concentrations or approaching zero.

The committee estimated exposure to aldicarb from both bananas and potatoes. The estimated distribution of aldicarb residues either from the single commodity potatoes or the combination of the two commodities shows exposure above the RfD of 1.0 μg/kg bw/day. In general, when these foods are eaten in the absence of other cholinesterase inhibitors, exposures would be much lower than those that would produce toxic effects. However, in the unlikely event that a single exposure occurred at the highest residue concentrations found in either bananas or potatoes, toxic effects could occur in a child.

Since the use of aldicarb on potatoes and bananas was voluntarily withdrawn by the manufacturer in 1990 and 1991, respectively, children are not presently at risk of cholinesterase depression from residues of aldicarb on these two foods (Debra Edwards, Chief, EPA Chemistry Branch, personal commun., 1993). Nonetheless, this case study illustrates the use of food consumption and residue distributions in estimating the number of person-days on which ingestion of an acutely toxic pesticide might exceed the RfD. As in the benomyl case, total ingestion from more than one food on which residues might be present was taken into account.

MULTIPLE EXPOSURE ASSESSMENT: ORGANOPHOSPHATE INSECTICIDES

Pesticide regulation in the United States has been focused on single chemicals rather than on combinations of compounds likely to appear as mixtures in the human diet. This practice can be attributed not only to the absence of data on the residues of multiple compounds that coexist on foods but also to the lack of methods for estimating simultaneous exposures to multiple chemicals, which cannot be accomplished merely by combining mean values (or other statistical summaries) of food intake and residue data. The regulatory process has therefore progressed on a chemical-by-chemical basis without consideration of possible additive and synergistic effects that could result from exposures to mixtures.

The committee developed a method for estimating exposure to multiple pesticides with a common toxic effect: in this case, inhibition of plasma cholinesterase (ChE). This method was used to determine how many children are likely to be exposed to unsafe levels of multiple pesticides with that common effect and to express the exposures in the most desirable form—person-day exposures—using actual individual daily consumption data and actual residue data.

More than 25 compounds that inhibit cholinesterase are permitted to exist as residues in foods. Although N-methyl carbamates inhibit cholinesterase, their mechanism of action is reversible and duration of action is shorter than for organophosphates. For purposes of simplicity, therefore, the committee selected five commonly used organophosphates (acephate, chlorpyrifos, dimethoate, disulfoton, and ethion) and used actual data on their presence on eight foods (apples, oranges, grapes, beans, tomatoes, lettuce, peaches, and peas) and three juices (apple, orange, and grape) to explore the development of methods for assessing exposure to multiple chemicals.

Criteria for choosing the five chemicals included the following:

- They must each exert the same adverse effect, in this case, blood plasma ChE inhibition.
- Credible estimates of the no-observed-effect level (NOEL) for ChE inhibition must exist for each chemical.
- The chemicals must be permitted as residues on several of the eight foods analyzed.
- FDA residue data must exist for the chemical-food groups selected.

The selection of foods for analysis was driven by the availability of data on residues and on the amount of each food consumed by 2-year-old children sampled in the USDA's 1977-1978 NFCS. An attempt was made to include foods that children consume most; however, it became apparent that residues were most common on other foods, such as peaches, which were therefore included in the analysis. The frequency distributions for the foods analyzed by the committee are presented in Figure 7-9A-K.

Estimating pesticide exposure in this way was considerably constrained by the absence of residue data for certain foods and compounds, especially processed foods such as juices whose type of processing could greatly influence pesticide residue levels. There are few available residue data for processed foods; thus, little is known about the effects of processing on pesticide residues.

For this analysis, the committee assumed that exposures to ChE-inhibiting compounds should be summed across foods and compounds that induce a similar type of ChE inhibition. Although exposure to a single compound may not exceed the RfD, concurrent exposures to numerous compounds could exceed a safe level because of the increased ChE inhibition. It was also assumed that the toxic potencies of diverse compounds can be standardized by developing estimates of relative potency in the manner described below.

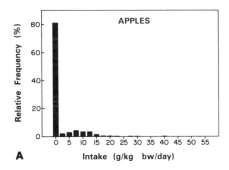

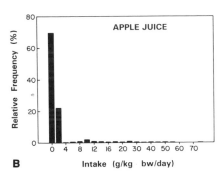

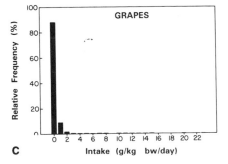

FIGURE 7-9A-K Consumption distributions for (A) apples, (B) apple juice, (C) grapes, (D) grape juice, (E) green beans, (F) lettuce, (G) oranges, (H) orange juice, (I) peaches, (J) peas, and (K) tomatoes. SOURCE: USDA, 1983.

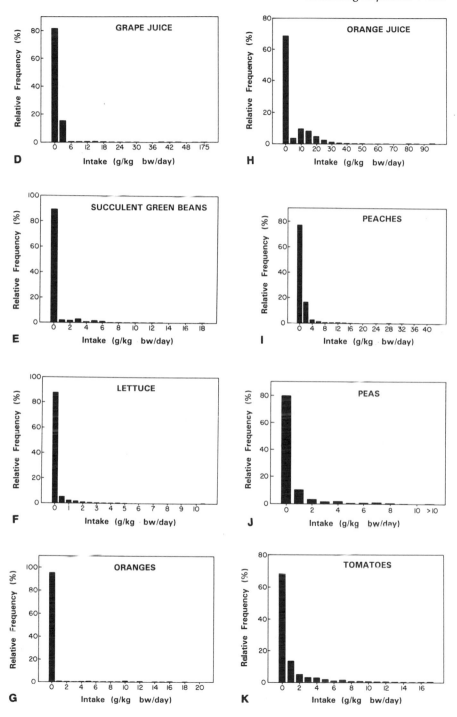

Cholinesterase Inhibition

Among pesticides, organophosphate and carbamate insecticides are the ChE-inhibiting pesticides of primary concern. These chemicals bind with cholinesterases and block their action in the hydrolysis of the acetylcholine (ACh) neurotransmitter. ACh is the principal neurotransmitter at neuromuscular junctions in the parasympathetic nervous system and in many regions of the central nervous system. High concentrations of ACh are also found in areas of the brain linked to higher cognitive functions such as learning and memory.

Organophosphate compounds, such as acephate, chlorpyrifos, dimethoate, disulfoton, and ethion, bind and phosphorylate the active site of ChE, thereby inactivating the enzyme. Carbamates, including aldicarb, lannate, methomyl, propoxur, and carbaryl, also interact with the acetylcholinesterase (AChE) receptor by reversible carbamylation of the seryl hydroxyl moiety at the active site of the enzyme (Murphy, 1986).

Some organophosphate and carbamate insecticides are acutely toxic and are frequently implicated in poisonings of humans. Exposures to high levels of AChE-inhibiting compounds may lead to severe cholinergic toxicity, with symptoms of headache, nausea and vomiting, cramps, weakness, blurred vision, pinhole pupils, chest tightness, muscle spasms, and coma (Ecobichon, 1991). Delayed neuropathy has also been associated with exposure to some organophosphorus esters (i.e., phosphate, phosphorate, and phosphoramidate esters), some of which have been used as insecticides (Amdur et al., 1991). Symptoms of acute organophosphate toxicity are difficult to recognize in the clinical setting for two major reasons: the complaints are nonspecific, and most physicians have limited familiarity with the signs and symptoms of pesticide poisoning.

Most ChE inhibitors degrade relatively rapidly in the environment and do not appear to accumulate or concentrate in the food chain (in contrast to organochlorine pesticides). In addition, these pesticides do not accumulate in the body, since they are rapidly biotransformed and excreted. Nevertheless, ChE inhibition can occur, producing signs and symptoms of poisoning after exposure to small repeated doses. Long-term effects of acute and subchronic exposures to pesticides have been reported. Some investigators have reported chronic, subtle neurologic sequelae to acute organophosphate poisoning (Savage et al., 1988). Epidemiological literature reported by the Office of Technology Assessment provides some evidence of delayed, persistent, or latent effects in humans. The literature includes case reports and studies of agricultural workers with and without histories of acute poisoning (OTA, 1990).

Organophosphates and carbamates may be converted in the environment or in vivo to form metabolites with toxicity potentially greater than that of the parent compounds. Synergism among organophosphate compounds, such as that demonstrated among malathion, O-ethyl O-p-nitrophenyl phenylphosphonothioate (EPN), and other organophosphates, may be an important variable to consider in assessing exposure to compound mixtures (NRC, 1977).

ChE inhibition is widely regarded as a good, general indicator of exposure to organophosphate pesticides except among people occupationally exposed over long periods who have developed persistently low active ChE levels. However, knowledge about this inhibitory effect is still incomplete. For example, the relationship between ChE inhibition and neurotoxicity has not been adequately demonstrated (EPA, 1988).

Neurotoxicity is defined by EPA as an adverse change in the structure or function of the nervous system following exposure to a chemical agent. The level at which ChE inhibition is associated with such changes is unclear (NRC, 1986). Furthermore, some investigators question the validity of measuring ChE inhibition in peripheral tissues, i.e., in plasma and blood, as a surrogate for measuring ChE inhibition of the central nervous system.

Further study is required to correlate ChE inhibition with identifiable changes in the central and peripheral nervous systems. Techniques for measuring neurotoxic effects include nerve conduction studies, sensory studies, evoked brain responses, electrocardiograms, and biochemical assays (NRC, 1992).

Relative Potency of Organophosphates

The EPA guidelines for the study of mixtures containing dioxins and dibenzofurans consider the relative potency of different components of the mixture (EPA, 1987). This method permits the estimation of toxicity equivalence factors (TEFs) by comparing the toxicity of the compounds of interest to a standard defined as the most thoroughly tested compound. In the present study, the committee selected as a standard chlorpyrifos— a commonly used organophosphate insecticide. A TEF may be derived by comparing the no observed-effect level (NOEL), or lowest-effect level (LEL), for any other chemical shown to produce the same type of ChE-inhibiting effect, to the NOEL (or LEL) for chlorpyrifos. The ratio of the chlorpyrifos NOEL to the NOEL of a different chemical (chemical X) provides an estimate of relative potency for chemical X and was used to adjust the laboratory-detected residue levels of the five chemicals of concern (Tables 7-14 and 7-15). The new values based on relative potency may be added to estimate cumulative exposure to chemicals believed to induce similar adverse effects, in this case ChE inhibition.

The committee assumed for this example that ChE recovers to a normal level in a 24-hour period. This may not be an appropriate assumption, however, because intake data are summed over all eating occasions for

TABLE 7-14   Estimating Toxicity Equivalence Using the LOAEL for Chlorpyrifos as Reference Standard

Pesticide	LOAEL[a]	LOAEL Ratio = TEF
Acephate	0.12 mg/kg bw Rats, ChE inhibition	0.1/0.12 = 0.83
Chlorpyrifos	0.10 mg/kg bw Humans, ChE inhibition	0.1/0.10 = 1.0
Dimethoate	0.25 mg/kg bw Rats, ChE inhibition	0.1/0.25 = 0.40
Disulfoton	0.04 mg/kg bw Rats, ChE inhibition	0.1/0.04 = 2.5
Ethion	0.075 mg/kg bw Humans, ChE inhibition	0.1/0.075 = 1.33

NOTE: LOAEL, lowest-observed-adverse-effect level; TEF, toxicity equivalence factors.

[a] From Integrated Risk Information System (IRIS): EPA, July 1992.

TABLE 7-15   Estimates of Toxicity Equivalence Factors for Five Organophosphate Insecticides Using the NOAEL for Chlorpyrifos as the Reference Standard

Pesticide	NOAEL[a]	NOAEL Ratio = TEF
Acephate	NA	NA
Chlorpyrifos	0.03 mg/kg bw Humans, ChE inhibition	1
Dimethoate	0.05 mg/kg bw Rats, Che inhibition	0.03/0.05 = 0.6
Disulfoton	NA	NA
Ethion	0.05 mg/kg bw Humans, ChE inhibition	0.03/0.05 = 0.6

NOTE: NA, not applicable. NOAEL = no–observed–adverse–effect level.

[a] From Integrated Risk Information System (IRIS): EPA, July 1992.

each day. Primary exposure on day 1 may occur during dinner, whereas primary exposure for day 2 may come at breakfast. Although both meals fall within a 24-hour window, they are presumed to be different 24-hour windows for the purposes of the present example.

Food Consumption Data

Data from USDA's 1977-1978 NFCS were used in the analysis. As mentioned above, consumption rates of eight foods (apples, oranges, grapes, beans, tomatoes, lettuce, peaches, and peas) and three juices (apple, orange, and grape) for 2-year-old children were selected for the analysis. There was a total of 1,831 person-days of data.

Residue Data

The committee used pooled FDA residue data from 1988 and 1989 compliance, surveillance, import, and domestic sampling. Residues for each chemical were converted to chlorpyrifos equivalents by multiplying each value by an equivalence ratio (chlorpyrifos LOAEL or NOEL/chemical X NOEL or LOAEL). For example, three chemicals allowed on a particular food item each have separate residue data sets consisting of individual sample results for each chemical detected on that food. Each of these data sets were summarized with regard to frequency and then for residue distribution.

To estimate cumulative exposure to the five organophosphate compounds, the committee adopted the assumption that the residue distributions for each compound are independent of one another. This approach may result in an overestimation of actual exposure, since there is likely to be some correlation among residue levels of the different compounds. In particular, substitution among chemicals would lead to scenarios in which all five compounds are never detected on the same sample.

The distribution of the cumulative exposure can be constructed by taking all possible combinations of chlorpyrifos equivalent values for the five chemicals and summing the values for each combination. This procedure will, however, yield large numbers of combinations and is thus impractical even on relatively large computers. A more practical alternative is to use strategic simulation in which the original shape of each component distribution is preserved. This is called a strategic simulation, since it is designed to reproduce the sample proportions for each subset of the original chemical distributions. The following procedure is used to create the distribution of cumulative exposure: a value is extracted randomly from each of the five residue distributions by using this strategic sampling technique; the resulting values arc summed and the result recorded; and the procedure is repeated 5,000 times, creating 5,000 possible combinations across the five chemicals. The 5,000 summed residues form the distribution for cumulative exposure, which is expressed in chlorpyrifos equivalents.

This computerized simulation was conducted for each food, creating a single residue distribution that is the random summation of residue values strategically extracted from each of the distinct residue distributions. This final distribution of summed residue values is expressed in chlorpyrifos equivalents. Eight such distributions were created, one each for apples, oranges, grapes, beans, lettuce, tomatoes, peaches, and peas.

The committee used two assumptions for nondetected residues: (1) that they were zero and (2) that they were present at the LOQ. The LOQ data used by the committee in this exercise were provided by the FDA. Since the FDA does not record the LOQ for each sample tested, it estimated an average LOQ of 0.01 ppm for all chemicals and foods analyzed in this study and proposed that this value be used in the committee's analysis.

Exposure Analysis

The objective of this exposure analysis is to produce a distribution of possible person-day exposures based on the food consumption data for 2-year-old children, including 1,831 person-day intake values for eight foods and eight separate residue distributions representing cumulative exposure—one for each of the eight foods. Person-day exposures are estimated by applying the following method.

1. Intake of food 1 by person 1 on day 1 is multiplied by some randomly extracted value from the residue data set (specific to food 1). The result is stored as an exposure value.
2. The process is repeated for n foods, still for person 1 on day 1.
3. The exposure values derived from n foods for one person-day are summed.
4. Steps 1 through 3 are repeated 5,000 times by using the strategic simulation to extract the residue data points from each summarized food-specific residue distribution of the residue data.
5. The 5,000 exposure values for person 1 on day 1 are stored and summarized as counts within exposure intervals.
6. Steps 1 through 5 are repeated for 1,831 person-days, producing 9,155,000 person-day exposure values, all expressed in chlorpyrifos equivalents.
7. The counts within exposure intervals are plotted as a frequency distribution.
8. The proportion of the sample falling above the RfD is estimated. The RfD for chlorpyrifos is 0.003 mg/kg bw/day.

A second exposure analysis was conducted to determine the sensitivity of exposure estimates to assumptions regarding the transfer of residues from raw to processed foods. Exposure was estimated for the same eight foods as above, and then for three juices (i.e., apple, orange, and grape).

The committee assumed that all residues in apples, grapes, and oranges were transferred unchanged to their juices, consistent with EPA practice (Peterson and Associates, Inc., 1992). This provides a maximum exposure estimate that is useful in the absence of statistically reliable data on the effects of processing.

Summary

The results of this analysis using five pesticides and eight selected unprocessed foods, excluding potent ChE inhibitors, and assuming that nondetectable residues are actually equal to zero are shown in Figure 7-10. (Results with nondetects set equal to the detection limit of 0.01 ppm are similar and are excluded for simplicity of presentation.) On the basis of these results, the RfD of 0.003 mg/kg bw/day (3 µg/kg bw/day) would be exceeded on approximately 1.3% of the person-days for children considered, representing approximately 120,000 of the 9.1 million person-days simulated (Figure 7-10).

Conclusions regarding the at-risk population are more difficult to reach. If one assumes that these simulated person-day exposure values are an accurate estimation of daily exposure for this population, then one must also assume that the consumption and residue data are also accurate representations. The committee does not believe that the data used do accurately represent the current status of those pesticide residues on foods because of the age of the consumption data, sample sizes, and the methods used by the FDA in interpreting residue values beneath an "action" or legal tolerance level. However, these data sets are the best now available

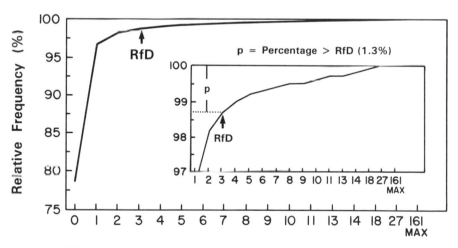

Exposure In Chlorpyrifos Equivalents (µg/kg bw/day)

FIGURE 7-10 Exposure of 2-year-old children to organophosphate pesticides— 1,831 2-year-old person-day food intake values. Foods: apples, oranges, grapes, beans, tomatoes, lettuce, peaches, and peas. Chemicals: acephate, chlorpyrifos, dimethoate, disulfoton, and ethion. Strategic simulation: 5,000 exposure values generated per person-day; cumulative distribution is summary of 9,115,000 simulated exposures.

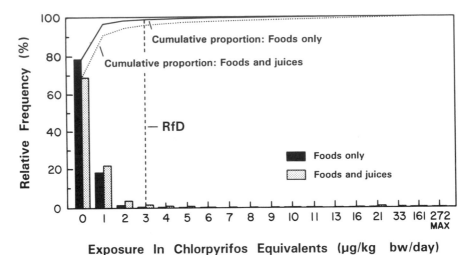

FIGURE 7-11 Exposure of 2-year-old children to organophosphate pesticides, including fruit juices. Strategic simulation: 5,000 exposure values generated per person-day; cumulative distribution is summary of 9,115,000 simulated exposures.

for a variety of chemicals and foods. (See Chapters 5 and 6 for an extensive examination of the limitations of the data.)

Although only 1.3% of the estimated person-day exposures were above the RfD, 3,584 person-days were more than ten times above the RfD, and the maximum exposure was ten times the RfD. These are clearly low probability events within a population of 3.5 million 2-year-old children; even a one-in-a-million event would occur 3.5 times per day. Even though these estimates are limited by the poor quality of the residue sampling data, they identify both a potential concern and an appropriate methodology for estimating exposure in large populations.

The results of the analysis are shown in Figure 7-11. The primary finding of this analysis is a shift in the distribution to higher residue levels. The percentage of the sample above the RfD rose from 1.3% to 4.1%, primarily because of the large intakes of apple juice and orange juice among 2-year-old children.

The committee concluded that this method is a workable and useful mechanism for assessing exposure, i.e., standardizing residue values by toxicity equivalents, combining these values based on either allowable or actual detected combinations of residues, and simulating exposures by combining residue values with actual reported intake values and summing exposures across foods within person-days. As with the other examples presented in this report, however, this discussion should be regarded as

an assessment of methodology rather than a specific attempt to characterize the proportion of children at risk. The method could also be used to
study possible combinations of residues for any class of chemicals believed
to have a common adverse effect, including cancer, where the end point
of concern is not a site-specific tumor but, rather, the probability of a
tumor occurring.

NONDIETARY EXPOSURE TO PESTICIDES

Although it was not generally within the committee's charge to examine
exposures to pesticides by routes other than dietary, the committee wishes
to point out that infants and children are subject to such exposures from
a variety of sources. These sources should not be overlooked when attempting to estimate the total exposure of infants and children to pesticides
and are therefore briefly summarized in this section.

In January 1990 EPA published the *Nonoccupational Pesticide Exposure
Study* (NOPES). One of the study's primary objectives was to assess the
relative contribution of each source to overall exposure to certain pesticides. Among their findings, the NOPES researchers concluded that (1)
"house dust may be a source of exposure to pesticides via dermal contact,
ingestion, and inhalation of suspended particulates, especially for infants
and toddlers"; (2) "acute dermal exposures that occur during application
events may contribute substantially to total exposure"; and (3) that, for
the pesticides they examined, exposure from drinking water appeared
to be minimal (EPA, 1990). Thus, exposure from all sources—not just
ingestion—must be considered when estimating total exposure and risk
to children.

Exposure via Parents

The child's first exposure to pesticides begins in utero, where chemicals
may cross over the mother's placenta. Several studies have suggested an
association between parental exposure (occupational and otherwise) to
pesticides and childhood cancers. Researchers from the Children's Cancer
Study Group (a cooperative clinical trials group with approximately 100
members and affiliate institutions in the United States and Canada) conducted a case-control study of occupational and household exposures of
parents of 204 children with acute nonlymphoblastic leukemia (ANLL)
(Buckley et al., 1989). Their most consistent finding was an association of
ANLL risk when both mother and father had been exposed to pesticides.

Human birth defects possibly associated with prenatal occupational
exposure to the organophosphate oxydemeton-methyl were published in
1989 by Romero et al. (1989). Gordon and Shy (1981) used ecologic data

to explore simultaneous maternal exposure to multiple agricultural chemicals in Iowa and Michigan and found an increased risk for facial clefts among offspring.

Lowengart and associates (1987) conducted a case-control study of leukemia patients 10 years and younger and the occupational and home exposures of their parents. With 123 matched pairs, a statistically significant association was seen between leukemia and the use of home and garden pesticides by either parent. In a cross-sectional study of 2,463 parents employed in the Central Valley area of California, limb reduction defects were observed to occur more frequently among the offspring of agricultural workers (relative risk of 2.3) at rates that were in excess of nationally established norms (Schwartz et al., 1986).

As is seen with lead and asbestos, children are at risk from toxicants that their parents unwittingly bring home on their clothes and other apparel. Children of parents employed in agricultural settings may be exposed to these "take-home" pesticides when the work clothing of their parents is washed with other family clothes (Formoli, 1990).

Exposure Through Air

Outdoor Air

In April 1990 the State of California canceled the permits for all uses of the soil fumigant 1,3-dichloropropene (commonly referred to as Telone). This action was prompted by an air monitoring program in the Central Valley that measured levels at a school in the area as high as 160 μg/m^3 (L. Baker and J. Behrmann, Air Resources Board, Sacramento, Calif., personal commun., 1990). Concentrations of 0.2 μg/m^3 of Telone in air are associated with cancer risks of 1 in 100,000 over 70 years, according to the EPA and the California Department of Health Services (CDHS).

Telone is usually applied undiluted to the soil around vegetable and tobacco crops to control nematodes and insects. Exposure to the vapor causes irritation of the mucosa and respiratory tract. The chemical is absorbed through intact skin, and systemic toxicity may follow cutaneous exposure as well as inhalation or ingestion of the compound (Flessel et al., 1978). Telone has been classified as a probable carcinogen in humans by the EPA (EPA, 1986) and is on the State of California's "Proposition 65" list of chemicals known to cause cancer.

Because pesticides are usually finely dispersed as droplets or particles at the time of application, aerial drift may cause them to be carried away from the target area where they were applied (Matsumura and Madhukar, 1984). Air sampled in areas where pesticides are used may contain residues either as vapors or bound to particles. Recent studies have found pesticide residues suspended in fog. Respiratory absorption of chemicals tends to be more rapid than absorption through other routes of exposure, because of the abundant blood supply in and the thinness of the alveolar membrane (Matsumura and Madhukar, 1984).

Exposure to pesticide residues from ambient air sources is generally higher in areas close to agricultural lands and in communities surrounding pesticide manufacturing factories. Where urban and suburban developments are interspersed within agricultural lands, movement of the more volatile chemicals present potentially significant human exposure. Episodes of illnesses in communities near agricultural areas have been reported simultaneously with applications of the fumigants methyl bromide and chloropicrin (Murray et al., 1974; CDHS, 1980; Goldman et al., 1987).

Cotton defoliants have also been associated with a higher incidence of acute health symptoms among residents living near cotton fields. Tributyl phosphorotrithioate, trade name DEF, is a defoliant of primary concern from a public health perspective (Scarborough et al., 1989). DEF is quite stable and has been detected at very low concentrations in the ambient air of residential and urban areas of cotton-growing counties, even at times of the year when cotton is not being sprayed. However, formulation impurities of butyl mercaptan or dibutyl disulfite may be the causative agents rather than the parent defoliant. In a small northern California community, an increase in health symptoms such as headaches, runny nose, and asthma attacks was reported by residents living adjacent to a potato field that had been treated with the organophosphate pesticide ethoprop. Ethoprop, like DEF, releases a strong-smelling mercaptan gas (Ames and Stratton, 1991). Organophosphates and other commonly used pesticides have been detected in ambient air in California's Central Valley, but generally in such low concentrations that they are unlikely to contribute significantly to the total exposure of humans.

Arthur et al. (1976) measured the levels of several organophosphate and organochlorine pesticides in the Mississippi delta and found that the residue levels were the highest in August and September. In rural areas, where residue levels were higher than in urban areas, the highest concentrations were found where spraying was reported (Arthur et al., 1976).

Indoor Air

The widespread use of pesticides such as flea bombs and insecticide sprays and foggers in the home exposes children to pesticides in their indoor environment. Most home-use products contain either organophosphates or carbamates as their active ingredients, both of which are cholinesterase-inhibiting compounds. These compounds affect the nervous system and at low doses may cause a variety of cholinergic symptoms such as drooling, excessive urination, or diarrhea (Berteau et al., 1989). For six pesticides (chlordane, heptachlor, aldrin, chlorpyrifos, diazinon, and gamma-BHC) analyzed in EPA's NOPES study, the mean air exposures were always or often higher than the estimated dietary exposure for the same compounds (EPA, 1990).

Fenske et al. (1990) measured chlorpyrifos (Dursban) concentrations following its application for flea treatment in a carpeted apartment. They found that the chlorpyrifos vapors measured in the infant breathing zone (25 cm above the carpet) were substantially higher than those measured in the sitting adult's breathing zone. Time-weighted averages for the 24-hour postapplication period in the infant breathing zone were 41.2 and 66.8 $\mu g/m^3$ for ventilated and nonventilated rooms, respectively; this is substantially higher than the interim guideline of 10 $\mu g/m^3$ proposed by the National Research Council's Committee on Toxicology for chlorpyrifos in indoor air following termiticide treatments (NRC, 1982). In addition, air concentrations increased from the time of application up to 5 to 7 hours later. The authors suggested that the treated carpet served as a source of volatilized chlorpyrifos and that although open windows provided mixing and dilution of air 1 m above the carpet, concentrations near the floor were affected much less (Fenske et al., 1990).

The short- and long-term health effects of exposure to commonly used home-use pesticide products are largely unknown. In assessing risk for infants in chlorpyrifos-treated homes, based on several conservative assumptions, Berteau et al. (1989) calculated an absorbed dose of 2.68 mg/kg. Fenske et al. (1990) found that the total estimated absorbed chlorpyrifos dose for infants exceeded the EPA's no-observed-effect level (NOEL) of 0.03 mg/kg/day in each case. The NOEL for chlorpyrifos is based on measurable changes in plasma acetylcholinesterase.

The indoor use of pesticides in public buildings such as schools and day-care centers leads to an additional source of exposure for children. In one episode, employees of a school for mentally handicapped children became ill within hours of entering a building that had been treated for roaches 3 days earlier and had not been ventilated. No students were admitted into the building until 14 days after the incident, when air levels of the pesticides used (dichlorvos and propoxur) had decreased to an acceptably safe level (White et al., 1987). An air analysis indicated that the levels of dichlorvos in the air were decreasing over time, but at a much slower rate than was expected from the data provided by the manufacturer.

Chlordane was the leading compound for controlling termites in homes for several years. Although it has since been canceled by the EPA for use as a termiticide, research demonstrates that chlordane air levels decline very slowly with time (Menconi et al., 1988). In a cross-sectional epidemio-logical investigation of 85 chlordane-treated households containing a total of 261 people, investigators found a dose-response relationship between chlordane levels in home indoor air and incidence of migraines, sinusitis, and bronchitis (Menconi et al., 1988). Cases of more serious health effects such as neuroblastoma, acute leukemia, and aplastic anemia have been associated with exposure to chlordane (Infante et al., 1978).

Pentachlorophenol (PCP) is commonly used as a wood preservative and has become the second most heavily used pesticide in the United States (Cline et al., 1989). It is used in wood homes and on playground equipment. Studies indicate that PCP is virtually ubiquitous in the envi-ronment, and measurable residues of PCP are found in most people (Hill et al., 1989). Hill et al. (1989) found PCP in 100% of urine samples taken from 197 Arkansas children. Researchers at the Centers for Disease Control (CDC) found that mean serum PCP levels were 10 times higher in residents of PCP-treated log homes than in the controls (40 ppb compared to 420 ppb) and that serum levels for children in the log homes were "signifi-cantly higher" than those for their parents (Cline et al., 1989).

Exposure via Contaminated Surfaces

Home-Use Products

Indoor insecticide sprays and foggers may persist on carpets, floors, and other surfaces in the home. Young children, particularly those wearing only diapers, may be exposed playing on previously sprayed surfaces. In 1980 an 11-day-old infant suffered respiratory arrest in a hospital waiting room. Pesticide poisoning was suspected because tests showed his red blood cell cholinesterase levels to be depressed to 50% of normal low baseline levels. Around the time of his birth, the child's home had been treated with chlorpyrifos; the chemical was subsequently found on dish towels, food preparation surfaces, and the infant's clothing (Dunphy et al., 1980).

Pet Products

Flea control products may persist on the pet's fur and be transferred to children during contact with the animal. Flea control products com-monly used in veterinary clinics, pet stores, and other commercial estab-lishments include carbaryl, chlorfenvinphos, chlorpyrifos, dimethyldi-chlorovinyl phosphate (DDVP), fenthion, malathion, phosmet, and propoxur (Ames et al., 1989).

Playground Equipment

Wooden playground equipment is another source of pesticide exposure because of the various kinds of wood preservatives used to prevent microbial and insect attacks. A 1987 California survey estimated that approximately 20% of all wooden structures in parks were treated with chemical preservatives. Some wood preservatives—PCP, chromium, boric acid, creosote, and arsenic—can induce adverse skin reactions such as contact dermatitis, hyperkeratosis, and, in the extreme case, skin cancer (CDHS, 1987).

Exposure via Medications and Personal Products

Another important route of exposure involves the direct application of insect repellents and pediculocides to children's skin. These include such compounds as N,N-diethyl-m-toluamide, lindane, and malathion. Lanolin, used by some breastfeeding mothers on their nipples, is also a concern because of the pesticides it can contain.

N,N-Diethyl-m-toluamide

N,N-Diethyl-m-toluamide, commonly called Deet, is the active ingredient in numerous commercially available insect repellents. Although insect repellents can provide great personal benefit, rare adverse reactions can occur. Since 1961, at least six cases of systemic toxic reactions from repeated cutaneous exposure to Deet have been reported. Six girls, ranging in age from 17 months to 8 years, developed behavioral changes, ataxia, encephalopathy, seizures, and/or coma after repeated cutaneous exposure to Deet; three died (Oransky et al., 1989). Neurobehavioral analysis showed strong correlation between Deet exposure and affective symptoms, insomnia, muscle cramps, and urinary hesitation (McConnell et al., 1987).

In August 1989 the New York State Department of Health investigated five reports of generalized seizures temporally associated with topical use of Deet. Four of the patients were boys from 3 to 7 years old (Oransky et al., 1989).

Lindane and Malathion

For almost 30 years the pesticide lindane (a chlorinated hydrocarbon) has been used in a shampoo for the treatment of head lice (Taplin and Meinking, 1988). Concern has been raised about potential central nervous system damage from exposure to lindane. In particular, cases of central nervous system toxicity have been reported from accidental ingestion as well as from single percutaneous exposures (Lee and Groth, 1977). One author reported two instances in which lindane lotion was given orally to children with scabies because of a lack of communication in one case and a language barrier in the other (Taplin and Meinking, 1988). Malathion has been recommended as a preferable treatment over lindane (Taplin et al., 1982; Fine, 1983).

Lanolin

Lanolin, a derivative of sheep's wool, is commonly used as an ointment to treat sore, cracked skin. Mothers who breastfeed frequently use it on their nipples, and it is sometimes applied directly to children's skin. The organophosphate pesticides diazinon and chlorpyrifos and several organochlorine pesticides such as dieldrin have been found at measurable levels in lanolin. The U.S. Food and Drug Administration identified 16 pesticides in lanolin it sampled in 1988. The principal source of these residues is the wool from sheep treated with a pesticide dip to control parasite infestations in the fleece (Cade, 1989). The fat-soluble organophosphate pesticide diazinon presented the greatest concern because of its frequent occurrence (21 of 25 samples) and the high levels identified (up to 29.2 ppm). (T. Levine, EPA, personal commun., 1988).

Occupational Exposures

In agricultural communities, children are often directly exposed to pesticides when they accompany their parents in the field or work there themselves (Pollack et al., 1990). In 1980, some 19 farm workers suffered organophosphate poisoning after working in a cauliflower field (Whorton and Obrinsky, 1983). Five of the workers were 18 years old or younger; three of those were between the ages of 9 and 15 years.

Exposure via Accidental Ingestion

Accidental poisonings are all too common among children. In one study of 37 children who had been hospitalized at Children's Medical Center in Dallas as a result of organophosphate or carbamate pesticide poisoning, ingestion of a liquid was the most common (73%) mechanism of exposure. Zwiener and Ginsburg (1988) reported that most poisonings took place in the home and were the result of careless storage of the original container or placement in unmarked or uncovered containers.

CONCLUSIONS AND RECOMMENDATIONS

Like other members of the general population, infants and children are exposed to pesticide residues in their diets. Estimation of dietary intakes requires information on both food consumption patterns and residue levels in food. The purpose of this chapter has been to demonstrate methods for estimating exposure to pesticides in the diet. The committee was guided by previous work on exposure estimation by the National Research Council (NRC, 1988, 1991a,b). Infants and children are also exposed to pesticides by nondietary routes, including air and contaminated surfaces such as rugs and playground equipment. Although a detailed analysis of nondietary routes of exposure to pesticides is outside the scope of this report, it is important for risk assessment purposes to consider the total exposure from all media. The following are the conclusions of the committee.

Conclusions

• Pesticide residues are present in the diets of infants and children. Estimation of dietary intakes of pesticides by infants and children requires information on both food consumption patterns and residue levels in food.

• Accurate estimation of dietary intake of pesticides by infants and children is difficult due to the limited amount of data on food consumption patterns of infants and children (Chapter 5) and limitations in the available data on pesticide residues (Chapter 6).

• Dietary exposures to pesticide residues can vary widely. Since most pesticide residues in foods are below the analytical limit of quantification, with comparatively few high residue levels, the distribution of dietary exposure to pesticides includes many low intakes. Some degree of positive skewness may be observed due to the occurrence of high consumption or high residue levels.

• To estimate dietary exposure to pesticides for infants and children, the committee combined probability distributions of food consumption with probability distributions of residue levels in order to obtain a probability distribution of individual exposures. The use of probability distributions for exposure assessment provides a more complete characterization of human exposure to pesticide residues in food than the use of summary statistics such as means or upper percentiles of exposure.

More accurate estimates of upper quantiles of the exposure distribution can be obtained by pointwise multiplication of the residue and consumption distributions than by multiplying the quantiles obtained from the residue and consumption distributions, separately. Moreover, the probability distribution approach based on 1-year age groupings of children provides useful information on differences in exposure patterns for children 1 to 5 years of age.

• Average daily ingestion of pesticide residues is an appropriate measure of exposure for chronic risk assessment, whereas actual individual daily ingestion is more appropriate for acute risk assessment.

Since chronic toxicity is often related to long-term average exposure, the average daily dietary exposure to pesticide residues may be used as the basis for risk assessment with delayed irreversible chronic toxic effects. To take into account different food consumption patterns among individuals, the distribution of average daily dietary intake of pesticides should be examined within the population of interest. Since acute toxicity is more often mediated by peak exposures occurring within a short period (e.g., over the course of a day or even during a single eating occasion), individual daily intakes are of interest for risk assessment for acute toxic effects. Examination of the distribution of individual daily intakes for persons within the population of interest reflects both day-to-day variation in pesticide ingestion for specific individuals as well as variation among individuals. This distribution can be used to estimate the number of person-days in a given period during which intake will exceed a specified level, such as the acceptable daily intake (ADI), or reference dose (RfD).

• At present, there is a relatively limited amount of information on food consumption patterns of infants and children. To obtain accurate estimates of the distribution of individual intakes, more elaborate and more intensive consumption monitoring protocols are required.

• Because residue monitoring surveys conducted for compliance purposes are expected to lead to higher residue levels than those present in the general food supply, assessment of human exposure should normally be based on surveillance surveys. In using surveillance data, however, consideration needs to be given to regional differences in pesticide use and resultant residue levels.

The committee acknowledges that pesticide food surveillance data are generated by randomly sampling food items from the distribution system. The purpose of this sampling is to ensure agricultural compliance with acceptable pesticide use practices. This sampling is broad-based and often not focused only on pesticides actually used. Pesticide field trial data are generated under strictly controlled conditions of use. These data better reflect actual levels at the time of harvest when it is known that a specific pesticide has been used. Each data source is used for purposes other than identifying actual dietary exposures, although both are useful in attempting to estimate these exposures.

• Frequently, the levels of pesticide residue in foods are below the analytical limit of quantification (LOQ). Since the actual residue level in such cases may lie anywhere between zero and the LOQ, there is some uncertainty about actual exposures in such cases.

For example, replacing the residue measurements below the LOQ with zero yields lower exposure estimates than substituting the LOQ for the unknown residue level. Notable differences occur when the analytical method is insensitive, the LOQ is high, or a large proportion of residues lies below the LOQ.

• The concentration of pesticide residues in foods may increase or decrease during food processing.

Changes in residue levels that occur during the processing of food are especially important in assessing the exposures of infants and young children, who consume large quantities of single processed foods, such as fruit juices, milk, and infant formula. In addition to the data accumulated by the food industry (Chapter 6), studies by pesticide manufacturers such as those furnished to the committee on the fate of residues during processing need to be conducted for most pesticides that produce detectable residues in food.

• Specific pesticides can be applied to more than one crop and, hence, appear on a number of food commodities. Residues of several pesticides may also appear on a single food commodity.

• Intake of multiple pesticides with a common acute toxic effect can be estimated by converting residues for each chemical to equivalent units of one of the compounds. The standardized residues can then be summed to estimate total residue levels in toxicity equivalence factors, and then combined with consumption data to construct a probability distribution of total exposure to all pesticides having a common mechanism of action.

Certain classes of pesticides such as cholinesterase inhibitors act by a common toxic mechanism. To properly evaluate the potential health effects of exposure to such pesticides, it is important to consider the total exposure to all pesticides in the class.

• Children are exposed to pesticides by nondietary routes.

Occupational exposure of the parent could result in exposure of the child in utero, in the home environment, or in the occupational setting of the parents. Pesticide residues have been detected in outdoor and indoor air, on contaminated surfaces, and in medications and personal products.

• When less than 100% of a given crop is treated with a particular pesticide, consideration might be given to adjusting exposure estimates according to the percentage of crop acreage treated. This adjustment can result in substantial reductions in estimates of exposure. This adjustment will be appropriate when the percentage of the crop treated is similar in different regions of the country, or when the crop is uniformly distributed throughout the country. Such adjustments should not be considered in the case of pesticides inducing acute toxic effects, since peak exposures are of importance in this case.

When these adjustments are used to adjust national data, they may result in averages that do not account for regional differences in pesticide use. It is therefore important that exposure estimates that have not been adjusted for acreage treated be presented and that such adjustments be critically examined.

Recommendations

The following recommendations were developed by the committee.

• Probability distributions based on actual data rather than simple summary statistics such as means or percentiles should be used to characterize human exposure to pesticide residues on food.

The advantage of using probability distributions rather than summary statistics to characterize exposure is that variation in individual food consumption patterns and residue levels in food are taken into account. This will require the collection of more detailed data on food consumption and residue levels as discussed in Chapters 5 and 6, respectively, but will provide more statistically robust estimates than the agency currently develops.

• The distribution of *average daily exposure* of individuals in the population of interest is recommended for use in chronic toxicity risk assessment; the distribution of *individual daily exposures* is recommended for evaluating acute toxic effects.

This recommendation is based on the committee's observation that chronic toxicity is typically related to long-term average exposure, whereas acute toxicity is more often mediated by peak exposures occurring within a short period, either over the course of a day or even during a single meal.

• If appropriately designed and conducted, surveillance studies of pesticide residues in food provide unbiased data on residue levels in food products. Field trials are also useful sources of information on pesticide residues in food. Such studies should be continued in order to expand the data base for evaluating dietary exposures to pesticides.

Surveillance studies based on random samples designed to provide a representative picture of residue levels in food are required to obtain unbiased information on dietary exposure to pesticides.

• The committee recommends that research to reduce the uncertainty in estimates of dietary exposure to pesticides be encouraged. Specifically, the development of improved analytical methods for residue analyses and statistical methods for imputing residue levels below the LOQ can lead to improved estimates of pesticide exposure.

All analytical methods for measuring pesticide residue levels in food are subject to an LOQ. Results below the LOQ may be as low as zero or as high as the LOQ itself, thereby imparting uncertainty regarding actual human exposure levels. This uncertainty will be reduced if more sensitive methods with lower LOQs are developed. Such technological improvements should be encouraged even in the absence of other pressures for more sensitive analytical methods.

Statistical methods for use with censored data (i.e., based on specific assumptions) can be used to impute residue levels below the LOQ, provided that the percentage of the residue data lying below the LOQ is not large. The use of such methods can reduce the uncertainty in resulting estimations of human exposure.

• When using multiresidue scans to detect different compounds in one scan of one food sample, all results should be recorded together. This will make possible more accurate evaluation of exposure distributions for multiple chemicals.

• The committee does not recommend the routine application of adjustments for the percentage of the crop treated in estimating dietary exposure to pesticides.

Adjustments for acreage treated are appropriate only under certain conditions. For example, such adjustments may be used when there is little regional variation in acreage treated, or when the crop is uniformly distributed at the national level.

• To determine total dietary exposure to a particular pesticide, intakes from all foods on which residues might be present need to be combined.

Many pesticides are approved for use on more than one crop. In addition, a single crop may be used in the production of a variety of processed foods. To estimate the total dietary exposure to a particular pesticide, it is important to consider the contribution of all foods on which residues might occur.

• To properly evaluate the potential risk from exposure to multiple pesticides with common mechanisms of action, it is necessary to develop measures of total exposure to pesticides within the same class that reflect the overall toxicity of all pesticides combined.

Since the combined effect of pesticides acting by a common mechanism can be greater than the individual effect of any single pesticide, it is important to develop risk assessment methods that address the total risk from exposure to all pesticides within the same class. One possible approach is to establish toxicity equivalence factors based on no-observed-effect levels as was done for organophosphates in this chapter.

• Because infants and children are subject to nondietary sources of exposure to pesticides, it is important to consider total exposure to pesticides from all sources combined.

REFERENCES

Amdur, M.O., J. Doull, and C.D. Klaassen, eds. 1991. Caserett and Doull's Toxicology, 4th Ed. New York: Pergamon. 1033 pp.

Ames, R.G., and J.W. Stratton. 1991. Acute health effects from community exposure to N-proply mercaptan from an ethoprop (Mocap^R)-treated potato field in Siskiyou County, California. Arch. Environ. Health 46:213-217.

Ames, R.G., S.K. Brown, J. Rosenberg, R.J. Jackson, J.W. Stratton, and S.G. Quenon. 1989. Health symptoms and occupational exposure to flea control products among California pet handlers. Am. Ind. Hyg. Assoc. J. 50:466-472.

Arthur, R.D., J.D. Cain, and B.F. Barrentine. 1976. Atmospheric levels of pesticides in the Mississippi Delta. Bull. Environ. Contam. Toxicol. 15:129-134.

Berteau, P.E., J.B. Knaak, D.C. Mengle, and J.B. Schreider. 1989. Insecticide absorption from indoor surfaces: Hazard assessment and regulatory requirements. Pp. 315-326 in Biological Monitoring for Pesticide Exposure: Measurement, Estimation and Risk Reduction, ACS Symposium Series 382, R.G.M. Wang, C.A. Franklin, R.C. Honeycutt, and J.C. Reinert, eds. Washington, D.C.: American Chemical Society.

Buckley, J.D., L.L. Robison, R. Swotinsky, D.H. Garabrant, M. LeBeau, P. Manchester, M.E. Nesbit, L. Odom, J.M. Peters, W.G. Woods, and G.D. Hammond. 1989. Occupational exposures of parents of children with acute nonlymphocytic leukemia: A report from the children's cancer study group. Cancer Res. 49:4030-4037.

Cade, P.H. 1989. Pesticide in lanolin [letter]. JAMA 262:613.

CDHS (California Department of Health Services). 1980. Ventura County Environmental Health Department, Report to the California Department of Health Services. Sacramento, Calif.: California Department of Health Services.

CDHS (California Department of Health Services). 1987. Report to the Legislature: Evaluation of Hazards Posed by the Use of Wood Preservatives on Playground Equipment. Department of Health Services, Office of Environmental Health Hazard Assessment. Berkeley, Calif.: California Department of Health Services.

Cline, R.E., R.H. Hill, Jr., D.L. Phillips, and L.L. Needham. 1989. Pentachlorophenol measurements in body fluids of people in log homes and workplaces. Arch. Environ. Contam. Toxicol. 18:475-481.

Dunphy, J., M. Kesselbrenner, A. Stevens, B. Vlec, and R.J. Jackson. 1980. Pesticide poisoning in an infant—California. MMWR 29:254-255.

Ecobichon, D.J. 1991. Toxic effects of pesticides. Pp. 565-622 in Casarett and Doull's

Toxicology: The Basic Science of Poisons, 4th Ed., M.O. Amdur, J. Doull, and C.D. Klaassen, eds. New York: Pergamon.

EPA (U.S. Environmental Protection Agency). 1986. Guidance for the Reregistration of Pesticide Products Containing 1,3-Dichloropropene (Telone II) as the Active Ingredient. NTIS No. PB-87-1117-87. Springfield, Va.: National Technical Information Service.

EPA (U.S. Environmental Protection Agency). 1987. Risk Assessment Forum: Interim Procedures for Estimating Risks Associated with Exposures to Mixtures of Chlorinated Dibenzo-p-dioxins and Dibenzofurans. Washington, D.C.: U.S. Environmental Protection Agency.

EPA (U.S. Environmental Protection Agency). 1988. Cholinesterase Inhibition as an Indicator of Adverse Toxicological Effect. Washington, D.C.: U.S. Environmental Protection Agency.

EPA (U.S. Environmental Protection Agency). 1990. Review of Cholinesterase Inhibition and Its Effects. Report of the EPA SAB/SAP Joint Study Group on Cholinesterase. Washington, D.C.: U.S. Environmental Protection Agency.

EPA (U.S. Environmental Protection Agency). 1992. Integrated Risk Information System (IRIS). Online. Office of Health and Environmental Assessment, Environmental Criteria and Assessment Office. Cincinnati, Ohio: U.S. Environmental Protection Agency.

Feldman, D., and M. Fox. 1991. Probability, The Mathematics of Uncertainty. New York: Marcel Dekker.

Fenske, R.A., K.G. Black, K.P. Elkner, C.-L. Lee, M.M. Methner, and R. Soto. 1990. Potential exposure and health risks of infants following indoor residential pesticide applications. Am. J. Public Health 80:689-693.

Fine, B.C. 1983. Pediculosis capitis [letter]. New Engl. J. Med. 309:1461.

Flessel, P., J.R. Goldsmith, E. Kahn, J.J. Wesolowski, K.T. Maddy, and S.A. Peoples. 1978. Acute and possible long-term effects of 1,3-dichloropropene—California. MMWR 27:50, 55.

Formoli, T. 1990. Pesticide Safety Information Series A-7, California Department of Food and Agriculture Worker Health and Safety Branch. HS-1228. Sacramento, Calif.

Goldman, L.R., D. Mengle, D.M. Epstein, D. Fredson, K. Kelly, R.J. Jackson. 1987. Acute symptoms in persons residing near a field treated with the soil fumigants methyl bromide and chloropicrin. West. J. Med. 147:95-98.

Gordon, J.E., and C.M. Shy. 1981. Agricultural chemical use and congenital cleft lip and/or palate. Arch. Environ. Health 36:213-221.

Hill, R.H., T. To, J.S. Holler, D.M. Fast, S.J. Smith, L.L. Needham, and S. Binder. 1989. Residues of chlorinated phenols and phenoxy acid herbicides in the urine of Arkansas children. Arch. Environ. Contam. Toxicol. 18:469-474.

Infante, P.F., S.S. Epstein, and W.A. Newton, Jr. 1978. Blood dyscrasias and childhood tumors and exposure to chlordane and heptachlor. Scand. J. Work Environ. Health 4:137-150.

Lee, B., and P. Groth. 1977. Scabies: Transcutaneous poisoning during treatment [letter]. Pediatrics 59:643.

Lowengart, R.A., J.M. Peters, C. Cicioni, J. Buckley, L. Bernstein, S. Preston-Martin, and E. Rappaport. 1987. Childhood leukemia and parents' occupational and home exposures. J. Natl. Cancer Inst. 79:39-46.

Matsumura, F., and B.V. Madhukar. 1984. Exposure to insecticides. Pp. 1-25 in Differential Toxicities of Insecticides and Halogenated Aromatics, International Encyclopedia of Pharmacology and Therapeutics Series, F. Matsumura, ed. Oxford, England: Pergamon.

McConnell, R., A.T. Fidler, and D. Chrislip. 1987. Health Hazard Evaluation Report HETA-83-085-1757. Everglades National Park, Everglades, Florida. Govt. Reports Announcements & Index (GRA&I). Issue 17. 1987.

Menconi, S., J.M. Clark, P. Langenberg, and D. Hryhorczuk. 1988. A preliminary study of potential human health effects in private residences following chlordane applications for termite control. Arch. Environ. Health 43:349-352.

Murphy, S.D. 1986. Toxic effects of pesticides. Pp. 519-581 in Casarett and Doull's Toxicology, 3rd Ed. New York: Macmillan.

Murray, R.A., L.E. Mahoney, and R.R. Sachs. 1974. Illness associated with soil fumigation, California. MMWR 23:217-218.

NRC (National Research Council). 1977. Drinking Water and Health. Washington, D.C.: National Academy Press.

NRC (National Research Council). 1982. An Assessment of the Health Risks of Seven Pesticides Used for Termite Control. Washington, D.C.: National Academy Press.

NRC (National Research Council). 1986. Drinking Water and Health, Vol. 6. Washington, D.C.: National Academy Press.

NRC (National Research Council). 1987. Regulating Pesticides in Food: The Delaney Paradox. Washington, D.C.: National Academy Press.

NRC (National Research Council). 1988. Complex Mixtures: Methods for In Vivo Toxicity Testing. Washington, D.C.: National Academy Press.

NRC (National Research Council). 1991a. Human Exposure Assessment for Airborne Pollutants: Advances and Opportunities. Washington, D.C.: National Academy Press.

NRC (National Research Council). 1991b. Frontiers in Assessing Human Exposures to Environmental Toxicants. Washington, D.C.: National Academy Press.

NRC (National Research Council). 1992. Environmental Neurotoxicology. Washington, D.C.: National Academy Press.

Oransky, S., B. Roseman, D. Fish, T. Gentile, J. Melius, M.L. Cartter, and J.L. Hadler. 1989. Seizures temporally associated with use of DEET insect repellant—New York and Connecticut. MMWR 38:678-680.

OTA (Office of Technology Assessment). 1990. Neurotoxicity: Identifying and Controlling Poisons of the Nervous System. OTA-BA-436. Washington, D.C.: U.S. Government Printing Office.

Petersen and Associates, Inc. 1992. Background: Derivation of Water Consumption Estimates Used by the Tolerance Assessment System. Paper prepared for the U.S. Environmental Protection Agency, Washington, D.C.

Pollack, S.H., P.J. Landrigan, and D.L. Mallino. 1990. Child labor in 1990: Prevalence and health hazards. Annu. Rev. Public Health 11:359-375.

Romero, P., P.G. Barnett, J.E. Midtling. 1989. Congenital anomalies associated with maternal exposure to oxydemeton-methyl. Environ. Res. 50:256-261.

Savage, E.P., T.J. Keefe, L.M. Mounce, R.K. Heaton, J.A. Lewis, and P.J. Bucar. 1988. Chronic neurological sequelae of acute organophosphate pesticide poisoning. Arch. Environ. Health 43:38-45.

Scarborough, M.E., R.G. Ames, M.J. Lipsett, and R.J. Jackson. 1989. Acute health effects of community exposure to cotton defoliants. Arch. Environ. Health 44:355-360.

Schwartz, D.A., L.A. Newsum, and R.M. Heifetz. 1986. Parental occupation and birth outcome in an agricultural community. Scand. J. Work Environ. Health 12:51-54.

Taplin, D., and T. Meinking. 1988. Infestations. In Pediatric Dermatology, 2 Vols., L.A. Schachner and R.C. Hansen, eds. New York: Churchill Livingstone. 1674 pp.

Taplin, D., P.M. Castillero, and J. Spiegal. 1982. Malathion for treatment of *Pediculus humanus* var. *capitis* infestation. JAMA 247:3103-3105.

USDA (U.S. Department of Agriculture). 1983. Food intakes: Individuals in 48 states, year 1977-78. Nationwide Food Consumption Survey 1977-78, Rep. No. I-1. Hyattsville, Md.: U.S. Department of Agriculture. 617 pp.

USDA (U.S. Department of Agriculture). 1987. CSFII, Nationwide Food Consumption Survey, Continuing Survey of Food Intakes of Individuals, 1985. Human Nutrition Information Service, 5 vols. Hyattsville, Md.: U.S. Department of Agriculture.

White, L.E., J.R. Clarkson, and S.N. Chang. 1987. Health effects from indoor air pollution: Case studies. J. Community Health 12(2-3):147-155.

Whorton, M.D., and D.L. Obrinsky. 1983. Persistence of symptoms after mild to moderate acute organophosphate poisoning among 19 farm field workers. J. Toxicol. Environ. Health 11:347-354.

Zwiener, R.J., and C.M. Ginsburg. 1988. Organophosphate and carbamate poisoning in infants and children. Pediatrics 81:121-126; erratum, 81:683.

8

Estimating the Risks

T HE PRECEDING CHAPTERS have demonstrated that infants and children may have special sensitivities to certain toxic insults. Children also consume notably more of certain foods relative to their body weight than do adults. Thus, their ingestion of pesticide residues on these foods may be proportionately higher than that of adults. For certain chronic toxic effects such as cancer, exposures occurring early in life may pose greater risks than those occurring later in life. For these reasons, risk assessment methods that have traditionally been used for adults may require modification when applied to infants and children.

To evaluate the potential risks from dietary intake of pesticide residues by infants and children, some familiarity with basic toxicological approaches to risk assessment is required. In the first part of this chapter, the committee reviews principles of toxicological risk assessment, including interspecies extrapolation, forms of toxicity, mathematical models for assessing cancer risk, low-dose linearity, and multiple exposures. Special characteristics of infants and young children that must be considered when assessing their risks from dietary exposures to pesticides are then discussed.

The fundamental purpose of regulatory toxicology is the determination of risks (including zero risk) at various possible levels of human exposure to toxicants present in the environment. For risk assessment applications, it is useful to distinguish between toxic processes that are either stochastic or nonstochastic in nature. Stochastic processes such as carcinogenesis result from the random occurrence of one or more biological events in specific individuals. Thus, whether or not a particular individual will develop a cancerous lesion under specified conditions of exposure cannot be predicted with certainty. Nonstochastic events such as enzyme inhibition following exposure to a certain toxicant are more predictable. Nonstochastic effects may not be seen if lifetime exposures are below a certain threshold concentration, whereas stochastic effects such as carcinogenesis may have no threshold. Different approaches to risk assessment are usually used for these two categories of response.

Information on health risks may be obtained directly from epidemiological and other studies of humans or indirectly through toxicological experiments conducted in animal models. Although results of laboratory studies are applicable to humans only indirectly, they can be used to predict potential health hazards in advance of actual human exposure and thus continue to be used widely to identify substances with potential toxicity.

Three major problems are inherent in the translation of results of animal tests to humans:

• Laboratory tests are conducted at relatively high doses to induce measurable rates of response in a small sample of animals. These results must often be extrapolated to lower doses that correspond to anticipated human exposure levels.

• Interspecies differences must be considered when extrapolating between the animal model and humans.

• It may be necessary to extrapolate from a route of exposure chosen for experimental practicality to a different but more likely route of human exposure.

GENERAL PRINCIPLES OF RISK ASSESSMENT

Toxicological Risk Assessment

Classical methods in toxicological risk assessment are applicable with nonstochastic toxic effects. For many kinds of agents and end points, toxicity is manifest only after the depletion of a physiological reserve. In addition, the biological repair capacity of many tissues can accommodate a certain degree of damage by reversible toxic processes (Aldridge, 1986; Klaassen, 1986). Above this threshold, however, the compensatory mechanisms that maintain normal biological function may be overwhelmed, leading to organ dysfunction. The objective of classical toxicological risk assessment, which has focused on nonstochastic end points, has been to establish a threshold dose below which adverse health effects were expected to be rare or absent.

Historically, the threshold concept was introduced by Lehman and Fitzhugh (1954), who proposed that an *acceptable daily intake*(ADI) could be calculated for chemical contaminants in human food. This concept was endorsed by the Joint FAO/WHO Expert Committee on Food Additives (JECFA) in 1961 and subsequently adopted by the Joint FAO/WHO Meeting of Experts on Pesticide Residues (JMPR) in 1962 (Lu, 1988). Formally, the ADI was defined as:

$$ADI = NOEL/SF,$$

where the NOEL is the *no-observed-effect level* in toxicity studies, usually in the most sensitive species, and SF is the *safety factor*. Weight loss, reduction in weight gain, alteration in organ weight, and inhibition of cholinesterase activity are indicative of specific adverse effects that may be considered when establishing the NOEL (Babich and Davis, 1981). The safety factor makes allowances for the type of effect, the severity or reversibility of the effect, and variability among and within species (NRC, 1970).

In 1977, the National Research Council's Safe Drinking Water Committee reviewed the methods that had evolved for establishing ADIs and made several important recommendations. First, the committee proposed that the NOEL be expressed in mg/kg of body weight rather than mg/kg of diet to adjust for differences in dietary consumption patterns. Since children consume proportionally more food relative to body weight than do adults, ADIs established on a body weight basis would be more protective of children than would ADIs established as dietary concentrations. Second, the committee explicitly supported the use of only a 10-fold safety factor (now called *uncertainty factor*) in the presence of adequate dose-response data derived from human studies. And third, the committee proposed an additional 10-fold safety factor in the absence of adequate toxicity data, for an overall safety factor of 100. The use of these factors was later supported in a report by the Safe Drinking Water Committee in a reexamination of the earlier risk assessment practices (NRC, 1986). For carcinogenesis, that committee proposed the following definitions for two levels of safety factors:

- 10: When studies in humans involving prolonged ingestion have been conducted with no indication of carcinogenicity.
- 100: When chronic toxicity studies have been conducted in one or more species with no indication of carcinogenicity. Data on humans are either unavailable or scanty.

With cholinesterase inhibitors, an uncertainty factor of 10 is applied, since toxicity in humans and animals has been shown to be similar, and because the effect is generally reversible. An uncertainty factor of 100 is more commonly used for compounds with other toxic end points, assuming that humans are 10 times more sensitive than the most sensitive test animal and that the most sensitive humans are 10 times more sensitive than the average human. The *inter*species uncertainty factor is based on studies in adult laboratory animals and, therefore, is applicable to adult human beings, but studies in adult laboratory animals are not necessarily good for predicting the response of human infants and children. The other 10-fold *intra*species uncertainty factor is meant to cover variations within human populations, including genetic predisposition, poor nutrition, disease status, and age (Babich and Davis, 1981). A factor of 10 for intraspecies variation in susceptibility may be sufficient for any one element of interpersonal difference but may not be sufficient for multiple elements. Thus, as presently determined, an uncertainty factor of 100 may not be sufficient to account for the potential increased sensitivity of infants and children.

When insufficient toxicity data are available, an uncertainty factor of 1,000 may be applied. The ADI is then converted to the *maximal permitted intake* (MPI), which is the product of the ADI and the average body weight of an adult, considered to be 60 kg, or sometimes 70 kg, by the U.S. Environmental Protection Agency (EPA) and 70 kg by the Food and Drug Administration (FDA) (Babich and Davis, 1981).

Uncertainty factors used in toxicological risk assessment have some widely recognized limitations. Since the ADI is based on an estimate of the population threshold or true *no-effect level* (NEL), it does not provide absolute assurance of safety (Crump, 1984b). Larger and better studies can demonstrate effects at lower doses, but the size of the uncertainty factor is not directly related to sample size. Therefore, smaller and poorer experiments tend to lead to larger ADIs (Schneiderman and Mantel, 1973). Although a factor of 10 is used to accommodate variation in sensitivity among species, and another factor of 10 is used for variation within a species, it cannot be guaranteed that a combined uncertainty factor of 100 will afford adequate protection in all cases. Nor are these factors of 10 based on validated biological models. There is additional uncertainty, usually small, about whether some observed biological responses are adverse effects or innocuous. Thus, the ADI is not intended to have a high degree of mathematical precision. Rather, it is a guide to human exposure levels that are not expected to present serious health risks.

The EPA has recommended using the term *uncertainty factor* (UF) rather than *safety factor* in recognition of the fact that the ADI does not guarantee absolute safety consistent with more recent recommendations of the Safe Drinking Water Committee (NRC, 1986). The agency has also adopted the *reference dose* (RfD) as a replacement for the ADI (Barnes and Dourson, 1988; EPA, 1988). In addition, EPA introduced the concept of a modifying factor (MF) to be applied to the UF in recognition of the possible special circumstances surrounding the establishment of a specific RfD. The RfD is determined by use of the equation:

$$RfD = NOAEL/(UF \times MF),$$

where the NOAEL (*no-observed-adverse-effect level*) is the highest dose at which there is no statistically significant adverse effect in the test animals beyond that exhibited by a control group, and the UF accommodates uncertainties in the extrapolation of dose-threshold data to humans. The MF is applied when scientific uncertainties in the study are not accommodated by the UF. When the data do not demonstrate a NOAEL, a LOAEL (*lowest-observed-adverse-effect level*) may be used. This is the lowest dose at which a statistically significant adverse effect is observed.

The NOAEL and the LOAEL depend on the design of the study, particularly on the selection of the experimental dose groups. If the dose groups are far apart, then the LOAEL may be significantly higher than the true concentration at which adverse effects occur and the NOAEL may be much lower than the minimum concentration producing an adverse effect. In the present context, the term *adverse effect* is defined as any effect that results in a functional impairment or pathological lesion that may affect the performance of the whole organism, or that reduces the ability of the organism to respond to additional challenges (Dourson, 1986).

By definition, a UF is a number that reflects the degree of uncertainty that must be considered when experimental data are extrapolated to humans (Dourson and Stara, 1983; Barnes and Dourson, 1988). When the critical study (the one with the best available dose-response data) is selected for calculation of the reference dose, five factors may contribute to the composite uncertainty factor:

- the need to extrapolate from animal data to humans when human exposure data are unavailable or inadequate;
- the need to accommodate human response variability to include sensitive subgroups;
- the nature, severity, and chronicity of the effect;
- the need to accommodate the necessity of using LOAEL rather than NOAEL data; and
- the need to extrapolate from a data base that is inadequate or incomplete.

The overall UF may vary from 1 to 10,000, depending on the combination of these individual factors, but usually does not exceed 100. An MF between 1 and 10 may be used to account for scientific uncertainties, either in the study or in the data base, that are not explicitly taken into consideration in any of the five factors listed above (Barnes and Dourson,

1988). For example, application of the MF may be regarded as a professional judgment that each test group contained too few animals.

Cancer Risk Estimation

The quantitative estimation of risks associated with low levels of exposure to carcinogens present in the environment is an important part of the regulatory process. At present, risk estimation methods are usually based on the assumption that the dose-response curve for carcinogenesis is linear in the low-dose region. This position is reflected in the principles proposed by the Office of Science and Technology Policy (1986), which stated:

> When data and information are limited . . . and when much uncertainty exists regarding the mechanism of carcinogenic action, models or procedures which incorporate low-dose linearity are preferred. (OSTP, 1986)

This position is also reflected in the EPA's (1986) *Carcinogen Risk Assessment Guidelines*.

Although the *Guidelines* emphasize risk estimates derived using some form of linear extrapolation, they are based on the assumption that such estimates may be more appropriately viewed as plausible upper limits on risk and that the lower limit may well be effectively zero. The *Guidelines* also state that procedures for obtaining a best estimate lying somewhere between these two extremes generally do not exist at present, but that it may be possible to move away from a linearized extrapolation to obtain an upper limit in some circumstances and instead, use a threshold approach. For example, the EPA has recently taken a step in this direction by considering the possibility of a threshold for the induction of thyroid tumors (Paynter et al., 1988).

The EPA now uses the linearized multistage model for low-dose cancer risk estimation. The most important aspect of this practice is not the choice of the multistage model itself for risk estimation purposes but, rather, the linearized form of the model. Because the model is constrained to be linear at low doses, it is expected to yield risk estimates comparable to those based on other linear extrapolation procedures, including those proposed by Gaylor et al. (1987) and Krewski et al. (1990).

Additivity to Background and Low-Dose Linearity

Additivity to background is often cited in support of the assumption of low-dose linearity in carcinogenic risk assessment. In the additive background model proposed by Crump et al. (1976) and Peto (1978), spontaneous tumors are associated with an effective background dose, to which exposure from environmental carcinogens is added. According to Crump and colleagues:

> If carcinogenesis by an external agent acts additively with an already ongoing process, then under almost any model the response will be linear at low doses. (Crump et al., 1976)

Hoel (1980) subsequently demonstrated that this result also holds for partial additivity. This is expressed as follows in the current *Guidelines*:

> If a carcinogenic agent acts by accelerating the same carcinogenic process that leads to the background occurrence of cancer, the added effect of the carcinogenic process at low doses is expected to be virtually linear. (EPA, 1986)

The basic idea behind the additive background model is illustrated in Table 8-1. Here the spontaneous response rate is considered to arise as a consequence of a background dose δ, and the effects of the test chemical administered at dose d are additive in a dosewise fashion. The linearity of the excess risk over background $P(d + \delta) - P(\delta)$ at low doses follows from the fact that the secant between doses of δ and $(\delta + d)$ converges to the tangent to the dose-response curve as the dose d of the test compound becomes small.

Within the framework of this model, the only condition required for this result to hold is that the probability of tumor occurrence be a smooth, strictly increasing function of dose. No further assumptions are required concerning either the mathematical form of the dose-response relationship or the toxicological mechanism by which tumors are induced.

The low-dose linearity implied by this model refers to the slope of

TABLE 8-1 Effect of Background Rate on Accuracy of Additivity Approximation Under the Assumption of the Multistage Model[a]

Spontaneous Background Rate	Exposure Level for Both Agents	Deviation of Additivity Approximation from True Risk After Subtracting Background, %
0.1	0.01	−0.73
	0.001	−0.10
0.01	0.01	−3.39
	0.001	−0.34
0.001	0.01	−11.23
	0.001	−1.18

[a] From NRC, 1988, p. 196.

the dose-response curve at an applied dose of zero. Without additional assumptions, no further general statements can be made about the magnitude of this slope or about the range of low doses over which this linear approximation will be sustained reasonably well. The linear approximation in the multistage model holds well, even at doses that double the background tumor rate if that rate is not trivially small (Crump et al., 1976). When attempting to fit curves in the average regulatory setting, however, the lowest dose tested under usual experimental conditions is often one-quarter or a smaller fraction of the MTD (NRC, 1993). Compared to actual dietary concentrations, one-quarter of the MTD will often be very high, and linear interpolation between a tumor response at one-quarter or less of the MTD and the tumor response of the controls may still overestimate the response at the (low) regulatory dose and the low dose to which it is anticipated that humans will be exposed.

Nonlinearity at High Doses

Linearity at low doses does not imply that the dose-response curve will also be linear at high doses. In particular, curvature at high doses can result from factors such as saturation of absorption or elimination pathways or the alteration of pharmacokinetic processes involved in metabolic activation (Hoel et al., 1983). Nonlinearity at high doses can also be attributable to saturation of DNA repair systems or the induction of cellular proliferation. Dose-response curves for chemicals that can both cause DNA damage and induce cellular proliferation can be subject to a high degree of upward curvature, as with the hockey-stick-shaped dose-response curves for tumors of the urinary bladder induced by 2-acetylaminofluorene (2-AAF) (Cohen and Ellwein, 1990). Similarly, secondary carcinogens may act only at relatively high doses.

Nonlinearity due to saturation of elimination pathways results in upward curvature (as with methylene chloride), whereas saturation of activation processes leads to downward curvature (as with vinyl chloride). For processes that saturate in accordance with Michaelis-Menten kinetics, however, the amount of the proximate carcinogen formed at low doses will be directly proportional to the administered dose, since such processes are essentially first order at low doses (Murdoch et al., 1987). But again there is no way to estimate the shape or the range of the linearity.

If the pharmacokinetic model governing metabolic activation is known, dose-response may be assessed in terms of the dose delivered to the target tissue (NRC, 1987a). This may result in a more nearly linear dose-response curve, which greatly facilitates statistical extrapolation to low doses (Hoel et al., 1983; Krewski et al., 1986). To estimate delivered dose, however, it is important to consider the uncertainty associated with the parameters used in the pharmacokinetic model (Portier and Kaplan, 1989). Several additional parameters introduced in the modeling when pharmacokinetics is considered will introduce more uncertainty.

Molecular Dosimetry

Different chemicals may induce different kinds of DNA lesions involving anywhere from two to 13 sites on molecular DNA (NRC, 1989). At the same time, there is evidence that spontaneously occurring DNA lesions (i.e., those with no known cause) can differ from those caused by exposure to alkylating agents. This suggests that fingerprinting of DNA damage in exposed and unexposed individuals may provide a practical means for distinguishing between additive and independent backgrounds.

The dose-response curve for tumor induction can be linear or nonlinear for a genotoxic agent that acts completely independently of background. If neoplastic conversion can result from a single mutagenic DNA lesion, the linearity of adduct formation at low doses implies linearity with respect to tumor induction (Lutz et al., 1990). If two or more mutagenic lesions are required to create a malignant cancer cell, however, the dose-response curve will be nonlinear with an effective threshold at low doses. This is essentially a multihit, independent background model in which the response is proportional to dose raised to a power equal to the number of DNA lesions required for neoplastic conversion.

Mathematical Modeling of Cancer Risk

Mathematical modeling has been a dominant feature of the quantitation of cancer risk. However, no model has been adequately validated in practice, and none account for all possible situations. Following are brief general descriptions of the more important risk assessment models that have been proposed and used over the years as background for the assessment of the risk to children from dietary exposures to pesticides.

Linear Extrapolation Approaches. In the absence of evidence to the contrary, existing arguments support the adoption of low-dose linearity for cancer risk assessment. Various methods for low-dose linear extrapolation have been proposed by several investigators. Gross et al. (1970) suggested discarding data starting at the highest dose until a linear model provided an adequate description of the remaining data. Van Ryzin (1980) proposed the use of any dose-response model that fits the data reasonably well to estimate the dose producing an excess risk of 1% followed by simple linear extrapolation to lower doses. Gaylor and Kodell (1980) suggested fitting a dose-response model to obtain an upper confidence limit on the probability of tumor incidence at the lowest experimental dose and then using linear extrapolation at doses between zero and that dose. Since estimates at low doses might be unduly influenced by the choice of the dose-response model, Farmer et al. (1982) recommended linear extrapolation below the lowest experimental dose, or from the dose corresponding to an excess risk of 1%, whichever is larger.

Krewski et al. (1984) proposed what is called a model-free procedure using a linear interpolation from the response at the lowest dose showing any increase, down to the response at zero dose (i.e., control response). This approach is simple, but is not strictly model free, since the use of a linear interpolation implies a model of response directly proportional to the dose or exposure. Krewski et al. (1986) modified this procedure to consider the upper confidence limits of the low-dose slopes for each dose showing no statistically significant increase in tumor incidence above background. The shallowest slope was then selected for low-dose risk estimation.

Gaylor (1987) recommended using the smallest upper confidence limit where there is evidence to believe that the lowest dose is in a portion of the dose-response curve that is likely to lie between linearity and some upward curvature. Only rarely is it known with any exactness where this region is located, and often it is not certain that it exists at all.

Because of the difficulties in specifying and implementing a fully biologically based model of carcinogenesis for purposes of risk assessment (NRC, 1993), simple model-free approaches to low-dose risk estimation are attractive, especially when little is known about the process of tumor induction. Krewski et al. (1991) compared the behavior of the upper confidence limits of the slopes obtained from the model-free extrapolation (MFX) procedure with the upper confidence limit of the low-dose slope estimate obtained from the linearized multistage (LMS) procedure of Crump (1984a), as described below in the section on "Estimates of Carcinogenic Potency." Comparisons were obtained for 572 bioassays taken from a compilation in the Carcinogenic Potency Database by Gold et al. (1984). Bioassays were restricted to rodent experiments exhibiting definite carcinogenicity. The median of the ratio of the upper 95% confidence limits on low-dose slopes for MFX:LMS was 1.3. In 433 of the 572 bioassays, the MFX estimate was within a factor of two of the LMS estimate. In eight cases, the MFX estimate exceeded the LMS estimate by a factor of 10. In these cases, there was a leveling off or decrease in the slope of the dose-response curve at higher doses, which tended to reduce the LMS values.

Biologically Based Cancer Models. The multistage model has a long history of use in theoretical descriptions of carcinogenesis (Whittemore and Keller, 1978; Brown and Koziol, 1983; Armitage, 1985). As described by Armitage and Doll (1961), a stem cell must sustain a series of mutations in order to give rise to a malignant cancer cell. This model predicts that the age-specific cancer incidence rates should increase in proportion to age raised to a power related to the number of stages in the model, and provides a good description of many forms of human cancer by allowing for two to six stages.

The biological basis for the multistage model is incomplete in that it does not incorporate tissue growth or cell kinetics. Furthermore, as many as six stages may be required to describe dose-response curves with high upward curvatures, raising questions of biological interpretation. For these reasons, dose-response models that do reflect certain biological processes (e.g., cell deaths, cell turnover) have received considerable attention in recent years. Perhaps most widely discussed is the two-stage mutation-birth-death model, which was developed by Moolgavkar, Venzen, and Knudson and is therefore known as the MVK model (Moolgavkar, 1968a,b).

The MVK model is a biologically motivated model of carcinogenesis based on the hypothesis that a tumor may be initiated following genetic damage in one or more cells in the target tissue as a result of exposure to a compound called an *initiator* (Moolgavkar, 1986a,b; Thorslund et al.,1987). The initiated cells may then undergo a further transformation to give rise to a cancerous lesion. The rate at which such lesions occur may be increased by subsequent exposure to a *promoter*, which increases the pool of initiated cells through clonal expansion. Mathematical formulations of this process have been developed by Greenfield et al. (1984) and Moolgavkar et al. (1988).

The MVK model assumes that two mutations are necessary for a normal cell to become malignant. Initiating activity may be quantified in terms of the rate of occurrence of the first mutation. The rate of occurrence of the second mutation quantifies progression to a fully differentiated cancerous lesion. Promotional activity is measured by the difference in the birth and death rates of initiated cells. In the absence of promotional effects and variability in the pool of normal cells, the two-stage mutation-birth-death model reduces to a classic two-stage model.

The MVK model provides a convenient context for the quantitative description of the initiation/promotion mechanisms of carcinogenesis (Moolgavkar, 1986a). As described above, initiator increases the rate of occurrence of the first mutation, whereas a promoter increases the pool of initiated cells. Thus, the term *progressor* is used to describe an agent that increases the rate of occurrence of the second mutation, resulting in malignant transformation to a cancerous cell (EPA, 1987). It is possible that the same agent could play two or even all three of these roles (initiator, promoter, progressor), thereby enhancing both initiation and progression.

This model was applied by Thorslund and Charnley (1988) in estimating cancer risks associated with exposure to dioxin and chlordane, two putative tumor-promoting agents. The model provided a good fit to laboratory bioassay data on these two compounds, thereby facilitating a quantitative description of the mechanism of carcinogenic action within the framework of the MVK model. Using either a simple exponential (one-hit) or logistic model to describe the dose-response relationship for promotion, these investigators developed estimates of risk that were appreciably lower than those based on the linearized multistage model. Before this model can be recommended for routine application, however, its statistical properties require further study, especially with respect to predictions of risk at low doses (Portier, 1987).

Because the MVK model can be used to describe a variety of dose-response curves, it may not always provide an adequate description of the most important events involved in tumor induction. In certain cases, more than two mutations may be required for neoplastic conversion. Similarly, nongenotoxic factors may be required to foster development of a malignant tissue mass. As additional components are incorporated, the model rapidly becomes more complex and its tractability will be reduced (NRC, 1993).

Despite this potential for further elaboration, the MVK model is viewed as biologically more meaningful than the Armitage-Doll multistage model (Armitage, 1985). To apply this model, however, it will be necessary to have data on tissue growth and cell kinetics as well as bioassay data on tumor occurrence. Attempts to estimate all the parameters involved in the model without such supplementary data may result in unstable estimates (Portier, 1987). Separate estimates of the parameters governing the two mutation rates characterizing the genotoxic process in the model will also require more elaborate bioassay protocols than those currently in use (EPA, 1987). Nonetheless, explorations of the MVK model should be encouraged in order to learn more about its utility (NRC, 1993).

Estimates of Carcinogenic Potency. Several numerical indices have been proposed to describe the potency of chemical carcinogens (Barr, 1985). This concept derives from early work by Twort and Twort (1933) and Iball (1939). Gold et al. (1984) used the results of laboratory studies of carcinogenicity to tabulate the TD_{50} for a large number of chemical carcinogens, thereby popularizing the TD_{50} index. The TD_{50} for carcinogenesis is defined as the dose that reduces the proportion of tumor-free animals by 50% (Peto et al., 1984; Sawyer et al., 1984).

Some correlation has been noted between the TD_{50} and the LD_{50} used to measure acute toxicity (Zeise et al., 1984, 1986), although the correlation is not high (Metzger et al., 1989). The TD_{50} is also highly correlated with the *maximum tolerated dose* (MTD), which is often used as the high dose in carcinogenicity bioassays (Bernstein et al., 1985). Little correlation between the TD_{50} and biological indicators of tumor pathology is apparent (Gold et al., 1986), however, indicating the impossibility of summarizing the characteristics of a chemical carcinogen in a single index (Wartenberg and Gallo, 1990).

Correlations between the TD_{50}, LD_{50}, and MTD need to be interpreted with care (Crouch et al., 1987). Since the range of possible values of the TD_{50} is limited by the choice of the MTD, the wide variation in the carcinogenic potency of different chemicals could produce a correlation between these two indices (Bernstein et al., 1985). In this regard, Rieth and Starr (1989) noted that little correlation exists between the TD_{50} and MTD after normalizing the TD_{50} values relative to the MTD. These investigators also suggest that the correlation between the LD_{50} and the TD_{50} may increase in part because an observed TD_{50} could be larger than the LD_{50}.

The TD_{50} may be of some use in representing the overall strength of carcinogenic substances, but it does not necessarily indicate the level of risk posed at low exposure levels (Wartenberg and Gallo, 1990). As discussed previously, assessment of risk at low doses is generally accomplished by extrapolating results obtained at higher doses approaching the TD_{50}.

As noted above, the EPA (1986) uses the linearized multistage model to estimate risk at low doses. A risk assessment analysis based on the LMS uses q_1^* (a quantitative carcinogenic potency factor) (Crump, 1984a), which is the estimated slope of the dose-response curve from animal tests yielding a positive carcinogenic response. The q_1^* value represents the estimated tumor incidence (the number of tumors per milligrams of pesticide per kilogram of body weight per day) at the relatively low concentrations of pesticides found in the human diet. It is based on a mathematical extrapolation of tumor incidence observed at the high doses used in tests in laboratory animals. High and low q_1^* values indicate strong and weak carcinogenic responses, respectively. For example, q_1^* values are 9.4×10^{-1} for chlordimeform, 2.3×10^{-3} for captan, and 5.9×10^{-5} for glyphosate (NRC, 1987b). Chlordimeform is thus considered a more potent carcinogen than captan (roughly 30-fold) or glyphosate (roughly 200-fold).

The q_1^* values among pesticides can vary by orders of magnitude (NRC, 1986, 1987b). The estimated q_1^* depends on whether a surface area or body weight correction is made in extrapolating risks from rodents to human beings, whether malignant and benign tumors are combined, and the extrapolation model used. The EPA uses surface area, whereas FDA uses body weight.

The EPA follows a policy in setting q_1^* values that minimizes the chance of underestimating cancer risks. As a result, EPA's estimates derived by using the q_1^* are believed by many to lead to an overestimate of the cancer risk. Use of the upper 95% confidence limit of the slope of the dose-response curve attempts to allow for variability in response among the laboratory animals and statistical stability in the estimate, and thus possibly errs on the side of caution when extrapolating risk to humans. However, the approach makes no allowance for variability among humans, such as the possible existence of highly susceptible persons, or other sources of uncertainty.

The value of q_1^* is the slope of the fitted dose-response curve in the low-dose region. It estimates the risk associated with a standard dose (such as 1 mg/kg/day). EPA uses the upper confidence limit on this slope both to introduce caution into its estimates and to provide a more stable measure of risk than would come about if the maximum likelihood (ML) estimate of the slope were used. The ML estimate is highly unstable, and the addition or removal of one or two tumor-bearing animals leads to large swings in the slope estimate and, hence, the risk estimate. The value of q_1^* derived is negatively correlated with the TD_{50} (Krewski et al., 1989, 1993). Thus, compounds with a low TD_{50} will tend to be associated with higher levels of estimated risk at low doses than will compounds with high TD_{50}s.

EPA uses a time-independent model that estimates the lifetime risk of cancer and assumes that the risk from a given dose is the same at all ages. EPA's model assumes, for example, that the excess risk from exposure at age 5 is the same as the risk from the same exposure at age 70. However, because high levels of exposure to carcinogenic pesticides may occur during childhood and because decades may pass before the cancer resulting from that exposure is manifested, the use of EPA's methodology may substantially underestimate the lifetime cancer risk at young ages of exposure.

As noted in Chapter 4 (Methods for Toxicity Testing), the toxicity data base used in either the time-independent or time-dependent models for cancer risk assessment derives from carcinogenicity studies in young adult or adult laboratory animals whose responses may not be representative of those in neonates and weanlings. Newborn animals are often found to be more sensitive to certain carcinogens than are older animals (Rice, 1979). Thus, carcinogenicity studies in young adult animals, as recommended by EPA, may underestimate risks from exposures incurred during infancy or childhood—whether the time-dependent or time-independent models are used.

Species Conversion

In the absence of direct information on the carcinogenic potential of a particular agent in humans, toxicity studies in animals are often used as the basis for carcinogenic risk assessment (Rall et al., 1987). At present, all known human carcinogens have also been shown to be carcinogenic in one or more animal species (IARC, 1987). Although this does not necessarily imply that all compounds that are clearly carcinogenic in animals will also be carcinogenic in humans, it is considered prudent to regard them as such. Empirical support for this position may be derived from the high concordance in results reported as positive and negative (74% when effects in both sexes are considered) between rats and mice exposed to 266 different chemicals tested for carcinogenicity in the U.S. National Toxicology Program (NTP) (Haseman and Huff, 1987). True concordance may be higher but observed only in small samples with imperfect laboratory methods. These same data were used by Chen and Gaylor (1987) to demonstrate good agreement between the values of q_1^* for rats and mice among chemicals that tested positive for carcinogenicity in both species. Gaylor and Chen (1986) noted a similar association for the TD_{50}.

Ideally, quantitative extrapolation of carcinogenic risks across species should take into account known species differences that might affect responses. Traditionally, dose equivalency across species has usually been estimated from average body weight. However, many physiological constants (e.g., consumption of water and food) have been shown to vary as a power function of body weight (Linstedt, 1987). Dourson and Stara (1983) noted that the metabolic rate and the toxicity of many compounds may be more accurately related to body surface area, which approximates body weight to the 2/3 power, than to body weight (Pinkel, 1958; Freireich et al., 1966). The activity of drugs in human infants and newborns has also been related to body surface area (Wagner, 1971; Homan, 1972). On the basis of an empirical study of the carcinogenic potency of chemotherapeutic agents, Travis and White (1988) recommended an intermediate scaling factor based on body weight to the 3/4 power.

In an attempt to directly address the question of species conversion for carcinogens, Allen et al. (1988) examined the relationship between the TD_{25} estimated from human and animal data and the TD_{25} expressed on both a body-weight and a surface-area basis. A correlation coefficient exceeding 70% carcinogenic potency in animals and humans was obtained by using both scales of measurement. The correlation was slightly higher when the dose was expressed in relation to body weight.

Bernstein et al. (1985) suggested that the apparent interspecies correlation in carcinogenic potency is due to the high correlation between the MTDs for rats and mice. Other investigators argued that while the correlation in carcinogenic potency between rats and mice may be due in part to the correlation between the corresponding MTDs, at least part of the correlation is not artifactual (Crouch et al., 1987; Shlyakhter et al., 1992). In a recent review of the use of the MTD in animal cancer tests, the NRC (1993) concluded that the implications of the MTD for quantitative risk assessment were unclear.

Benchmark Dose

The concept of the benchmark dose approach was proposed by Crump (1984b) as an alternative to other existing quantitative risk assessment techniques. In 1991, the EPA proposed using this approach in its recent guidelines for developmental toxicity risk assessment (EPA, 1991). A benchmark dose is defined as the lower confidence limit for the dose corresponding to a specific increase in the response rate over the background rate. The dose is estimated using a dose-response model fit to experimental data, typically with a 5 to 10 percent response. Crump (1984b) suggests using a likelihood (statistical) approach to estimating the lower confidence limit.

The advantages of using the benchmark dose approach over the NOAEL approach have been discussed by Crump (1984b) and more recently by Kimmel et al. (1991). The benchmark dose approach takes into consideration the dose-response model, which utilizes all the data available in the experiment. The NOAEL approach ignores the shape of the dose-response curve and focuses only on the experimental dose group, which was estimated to be the NOAEL. Thus, the true NOAEL could be anywhere between the estimated NOAEL and the first dose group at which adverse effects are observed. Moreover, smaller and poorly designed studies could result in larger NOAELs due to insufficient statistical power. Thus, the benchmark dose approach provides a consistent basis for calculating the RfD.

Two limitations of the benchmark dose approach are that the computations are more laborious than the NOAEL and that a dose-response model may not readily fit the data. The NOAEL approach also allows for the identification of the LOAEL.

Benchmark doses provide an integrated approach for risk assessment for both carcinogenic and noncarcinogenic health effects. Rather than using two different methods of estimating risk (e.g., the NOAEL approach for threshold toxicants and low dose extrapolation for nonthreshold toxicants), the method can be applied in a similar manner to both.

RISK ASSESSMENTS FOR INFANTS AND CHILDREN

Pharmacokinetics

Carcinogens may require some form of metabolic activation to exert their effects, whereas others are often deactivated. Pharmacokinetic models may be used to describe such processes as well as the uptake, absorption, distribution, and elimination of compounds that enter the body (O'Flaherty, 1981). This may in turn permit estimation of the amount of the reactive metabolite reaching the target tissue. Tissue dosimetry is important for risk assessment purposes when one or more steps in the process of metabolic activation and biological action are saturable, since this can lead to a relationship between the dose of the parent compound administered to the test subjects and the dose of the reactive metabolite reaching the target tissue (Hoel et al., 1983), which is not directly proportional.

Compartmental pharmacokinetic models have often been used to describe biological systems in terms of a small number of conceptual compartments (Godfrey, 1983). More recently, physiologically based pharmacokinetic (PBPK) models that imply a larger number of physiologically meaningful compartments have been used to accommodate metabolic activation (Menzel, 1987). The application of PBPK models requires extensive information on the anatomy and physiology of the test animals, the solubility of the test chemical in various organs and tissues, and biochemical parameters governing metabolism (Andersen et al., 1987).

If metabolic activation can be adequately characterized in terms of a suitable pharmacokinetic model, the dose delivered to the target tissue may be used in place of the administered dose for purposes of dose-response modeling. This may sometimes lead to more accurate predictions of cancer risk (Krewski et al., 1987), but accuracy may be reduced when the uncertainty associated with the many parameters involved in a PBPK model is taken into account (Portier and Kaplan, 1989).

Pregnancy, Lactation, and Nursing

Many changes that occur during pregnancy can have a significant impact on the toxicodynamics of a particular chemical. For example, the changes in body weight, total body water, plasma proteins, body fat, and cardiac output will alter the distribution of many xenobiotic compounds (Hytten and Chamberlain, 1980; Mattison, 1986; Mattison et al., 1991).

PBPK models have been used to describe the kinetics and disposition of the drugs tetracycline, morphine, and methadone (Olanoff and Anderson, 1980; Gabrielsson et al., 1983; Gabrielsson et al., 1985a,b). Two recent reports described the pharmacokinetics and disposition of trichloroethylene and its principal metabolite, trichloroacetic acid, in the pregnant rat as well as in the lactating rat and nursing pup. A PBPK model, consisting of eight compartments for trichloroethylene in the pregnant rat and nine compartments for trichloroacetic acid, was used (Fisher et al., 1989). Both models accommodated multiple routes of exposure and repeated dosing. The model provided for a variable litter size from 1 to 12 pups per rat as well as placental growth.

Values of the maximum rate of metabolic removal velocity (V_{max}) (rate at which a chemical is removed from the organism) in unmated and pregnant rats were 10.98 and 9.18 mg/kg/hr, respectively. This reduction is significant, and has been related to a decrease in the cytochrome P-450 monooxygenase activity due to altered steroid hormones. A high substrate affinity was demonstrated for metabolism and elimination of chemical substances. Although the value of the Michaelis constant (K_m) was low (0.25 mg/liter), it was similar in both groups of rats. Fetal exposure to trichloroethylene was estimated to range from 67 to 76% of the maternal exposure; fetal exposure to the trichloroacetic acid metabolite was 63 to 64% of that of the dam.

The model fitted data obtained after exposure by inhalation, oral gavage, or via drinking water. Other kinetic parameters predicted by the model (such as the relative volumes of distribution, the peak blood concentration following oral gavage, fetal concentrations following inhalation exposure, absorption and elimination rates) agreed well with data previously reported in the literature. These results demonstrated that fetal exposure to both the parent compound and its principal metabolite was significantly elevated in relation to the maternal exposure.

Fisher et al. (1990) examined the transfer of trichloroethylene and trichloroacetic acid to nursing pups from lactating dams exposed to trichloroethylene by inhalation of 610 ppm 4 hours per day, 5 days per week from days 3 to 14 of lactation. A further study involved exposure of the lactating dam to 333 µg/ml of trichloroethylene in drinking water from days 3 to 21 of lactation. The pups were exposed to trichloroethylene solely from ingested maternal milk; however, their exposure to trichloroacetic acid came from maternal milk and from metabolism of ingested trichloroethylene. The model provided for different published values for compartmental volumes, blood flows, and milk yield during lactation. Metabolic and other kinetic parameters were determined experimentally.

The value of V_{max} = 9.26 mg/kg/hr in the lactating rat was similar to that in the pregnant rat. However, the value of V_{max} = 12.94 mg/kg/hr obtained for male and female pups indicated that the ability of the pups to metabolize trichloroethylene was greater than that in the adult. The plasma half-life (16.5 hr) of trichloroacetic acid in the pups was also substantially greater than that in the mature rat. Unlike most other physiological distribution processes that are flow limited, the distribution of trichloroacetic acid to mammary tissue is diffusion limited. The exposure of the pups to trichloroethylene from maternal milk was small, representing only about 2% of the exposure of the dam. Pup plasma levels of trichloroacetic acid, however, were as high as 30 and 15% of the maternal exposure for drinking water and inhalation exposures, respectively.

Neurotoxicity

Two classes of insecticides, the organophosphates and the carbamates, are acetylcholinesterase inhibitors, which interfere with nerve transmission. Chemically induced functional alterations of the nervous system are often assessed with neurobehavioral tests. However, it has been very difficult to integrate observations of behavior with other aspects of neurotoxicological testing (NRC, 1992). This has been especially true for assessing neurodevelopmental effects in young animals. For example, some studies in laboratory animals have shown that exposure to organophosphates and carbamates before or immediately after birth may alter neurological development and cause subtle and long-lasting neurobehavioral impairments. (See Chapter 4 for a discussion of neurotoxicity testing.) Therefore, additional laboratory animal data are needed to evaluate the effects of low-level chronic exposures on subtle neurotoxicities, including alterations of behavior. The recent NRC report *Environmental Neurotoxicology* (NRC, 1992)describes in detail the research necessary to improve the neurotoxicological testing of chemical substances, including pesticides. There is currently no validated testing system that satisfies all the necessary requirements for a screening program to detect the neurotoxic potential of chemical substances (NRC, 1992).

Multiple Exposures

Because humans are exposed to multiple chemicals simultaneously, the potential risks associated with joint exposures require consideration (Krewski and Thomas, 1992). Factorial experiments with pairs of chemicals have demonstrated both synergistic and antagonistic effects as well as a lack of interaction, depending on the pair of agents used, the tissues in which the cancers appear, and the time of appearance of different primary site tumors (Elashoff et al., 1987; Fears et al., 1988, 1989). Epidemiological studies have also demonstrated the existence of synergism between two agents, such as tobacco and asbestos (NRC, 1988). Brown and Chu (1989) demonstrated that the multistage model can lead to a spectrum of effects, ranging from additive to multiplicative age-specific relative risks, depending on the patterns of joint exposure to the two agents of interest. Kodell et al. (1991) reported similar findings for the two-stage clinical expansion model of carcinogenesis, including the possibility of supra-multiplicative relative risks in the case of joint exposure to two promoters.

The existence of synergism at low levels of exposure cannot be assessed directly. It must be inferred by other means. For example, the multistage model predicts additivity of excess risks at low doses, even in the presence of interactive effects (NRC, 1988). Similarly, the multiplicative relative risk model leads to additivity of excess risks at low doses (Krewski et al., 1989). Although these models imply that interactions will be negligible at low levels of exposure under certain conditions, the possibility of synergy at low doses under other conditions cannot be ruled out. It is conceivable, for example, that two compounds, innocuous by themselves, might interact chemically even at low doses to form a new substance that is toxic.

Since more than one pesticide may be applied during the growing season of some crops, multiple pesticide residues are possible. Those pesticides can interact synergistically to increase the total toxic potential. For example, synergistic effects between malathion and O-ethyl O-p-nitrophenylphenylphosphonothioate (EPN) have been noted since 1957 (NRC, 1977). Nevertheless, 33 of the 35 crops with tolerances for EPN also have tolerances for malathion (AOR, 1988).

Inert Ingredients

A pesticide product is generally formulated by combining one or more active ingredients with one or more inert ingredients. The active ingredients are selected because their potency can achieve the intended pesticidal action; the inert ingredients are designed not to kill the target pest but to provide bulk to the pesticide formulation and to improve its spreading or sticking ability. Inert ingredients are not regulated, and although many of them are innocuous, others are themselves potent toxicants and may cause adverse health effects. Examples of the latter include asbestos fibers, benzene, epichlorohydrin, methylene chloride, 2-methoxyethanol, pentachlorophenol, formaldehyde, and vinyl chloride. Dietary exposure to these compounds may be significant, since they usually appear in pesticide formulations in much greater quantity than the active ingredients. Furthermore, EPA does not require registrants to submit analytic methods to detect inert ingredients in foods (AOR, 1988).

In 1987, EPA divided the 1,200 inert ingredients currently contained in pesticide products into four toxicity categories based on available hazard information: ingredients of toxicological concern (57); substances that are potentially toxic and have a high priority for testing (67); ingredients whose potential for toxicity is unknown (800); and those considered to be innocuous (300) (Peach, 1987). In an attempt to encourage the use of less toxic inert ingredients, EPA gave pesticide registrants until October 20, 1988, to replace ingredients in the first category with compounds that do not appear in the first two classifications. After that deadline, any compounds in the first category still present in pesticide products had to be identified on the product label and were subject to data call-in for chronic health effects (AOR, 1988).

Considerations Specific to Children

Certain populations of children may be more sensitive to the effects of pesticides because of physiological and biochemical factors such as genetic predisposition, chronic medical conditions, interactions (or additive effects) of pesticides with medications, and general health status. Other factors that may make certain children more susceptible include increased exposure through farm work or parental occupational exposure and low socioeconomic status.

Exposure and Low Socioeconomic Status

Children living in poverty may constitute an additional sensitive population. Illness, when it occurs, is more severe among children of poor families (Starfield, 1982). For example, common childhood conditions such as asthma tend to be more severe in these children (Mitchell and Dawson, 1973). Furthermore, children of poor families are 2 to 3 times more likely to contract illnesses such as rheumatic fever, twice as likely to contract illnesses such as bacterial meningitis, 9 times more likely to have elevated levels of lead in their blood, and 2 to 3 times as likely to have iron-deficiency anemia (Starfield and Budetti, 1985). In the poor, rural agricultural community of McFarland, California, studied because of a childhood cancer cluster, 23.8% of children from 1 to 12 years of age were anemic. In the 1- to 4-year-old age group, 30.6% were anemic (McFarland Child Health Screening Project, 1991).

One factor contributing to compromised health status is poor nutrition. Preschool children of low socioeconomic status have been found to have lower dietary intakes, lower biochemical indices, and smaller physical size for their age than children of higher socioeconomic status (Owen and Frankle, 1986).

Because of compromised health status, poor children are probably more susceptible to any toxic insult, including pesticide exposure. Children of poor families are more likely to live in highly polluted neighborhoods and thus to have greater exposure to environmental toxicants. In the Los

Angeles area, many immigrant poor may be fishing in contaminated waters to supplement their source of protein.

The combined effect of poorer health status and of likely higher exposure to environmental toxicants suggests that the further burden of pesticide exposure could lead to toxic effects at levels that do not produce effects in other children. Therefore one might expect that adverse effects of pesticides, whether acute or chronic, might be magnified in this subpopulation.

Age-Dependent Toxicity

Traditional risk assessment methods generally do not make specific allowances for any unique features of infants and children. Species conversion of dietary intakes per unit of body weight are normally based on adult body weights and food consumption data (McColl, 1989).

Infants and children are unique in a number of ways (see Chapters 2 and 3). Babich and Davis (1981) noted that children may be hypersusceptible to food toxicants, especially heavy metals and pesticides. The realization that the process of atherosclerosis begins in childhood has led to recommendations that total fat intake of children over 2 years of age not exceed 30% of calories and that cholesterol intake not exceed 300 mg daily (American Health Foundation, 1989). Gaines and Linder (1986) estimated the LD_{50}s for 57 pesticides in adult and weanling rats and found four cases (leptophos, methidathion, pyrazon, and sulfoxide) in which the weanlings were more sensitive than the adults and several in which they were less sensitive.

Infants and children may be more or less susceptible to exposure to toxic chemicals than adults (see Chapters 2 and 3). Drugs such as amikacin and the aminoglycosides are less toxic to children than adults, whereas children seem to be more sensitive to salicylates than adults (Mendelson and Grisolia, 1975; Brown et al., 1982; Faden et al., 1982). Comparative analyses of the acute lethality of drugs (Goldenthal, 1971), chemotherapeutic agents (Glaubiger et al., 1982), insecticides acting through cholinesterase inhibition (Brodeur and DuBois, 1963), and pesticides (Gaines and Linder, 1986) indicate that infants and children may exhibit a higher or lower LD_{50} than adults.

Collectively, these data provide information on the magnitude of the effects of age on susceptibility to toxic chemicals. Specifically, Brodeur and Dubois (1963) observed that the LD_{50}s for adult rats were 2 to 4 times greater than the corresponding LD_{50}s for weanling rats for 15 of 16 insecticides examined.

Potential carcinogenic risks may depend on the unique physiological sensitivities of infants and children as well as the specific age at which exposure occurs. In a study of the carcinogenic effects of N-nitroso compounds in the diets of rats conducted by the British Industrial Biological Research Association, for example, different tumor response rates were observed in animals exposed throughout their lives beginning at 3, 6, or 12 weeks of age (Peto et al., 1986). Other investigators have identified perinatal exposure as a critical factor in neoplastic development. During the 1970s, three independent two-generation cancer bioassays involving *in utero* exposure showed a clear increase in the incidence of malignant tumors in the urinary bladders of male rats, whereas this effect was not observed in single generation studies in which exposure began at the time of weaning (Arnold et al., 1983). These results confirm an important role of *in utero* exposure in bladder tumor induction. In a study conducted by the International Research and Development Corporation (IRDC), animals were exposed only from birth onward (Food and Chemical Toxicology, 1985). Tumor yields were similar to those observed in the two-generation studies. Although this association seems not to have been studied directly, these data are consistent with the view that perinatal exposure can be a critical determinant in the initiation and extent of carcinogenic response.

Recently, both the multistage and two-stage models have been extended to accommodate the age at which exposure occurs (Kodell et al., 1987; Chen et al., 1988; Gaylor, 1988; Murdoch and Krewski, 1988; Murdoch et al., 1992). These results indicate that when an early stage of carcinogenesis is dose dependent, early exposures will be of greater concern than later exposures. In this case, equivalent exposures will present greater risks to infants and children than to adults. Estimates of the size of this excess risk in infants and children depend on the particular assumptions made concerning the mechanism of carcinogenesis. In the classical interpretation of the multistage model, the increase cannot exceed a factor of k, the number of stages in the biological system under study. With the two-stage model, however, the increased risk can be substantial when the chemical greatly increases the proliferation rate of the initiated cell population.

Genetic Susceptibility

Polymorphism in Metabolism

In recent years a number of genetic polymorphisms of metabolic enzymes have been identified in humans. Several cytochrome P-450 polymorphisms have been established (Guengerich et al., 1986). Polymorphisms of N-acetyltransferase (Weber, 1987), glutathione-S-transferase (Board, 1990), and paraoxonase (Ortigoza-Ferado et al., 1984) have also been documented. High biological activity of an enzyme(s) that catalyzes the activation of a pesticide or low activity of an enzyme that mediates deactivation may increase the sensitivity of an individual to pesticide exposure.

Parathion is activated to paraoxon by mixed-function oxygenase enzymes. Hydrolysis (and deactivation of the active paraoxon) is mediated by plasma paraoxonase (arylesterase). The activity of this arylesterase is polymorphic with respect to paraoxon (La Du and Eckerson, 1984). There are two, possibly three, paraoxonase phenotypes with activity toward paraoxon varying by 11-fold (Furlong et al., 1988). Low arylesterase activity could potentiate the cholinergic effects of this pesticide; thus, children with the low activity phenotype would be at greater risk of adverse health effects because of a decreased ability to hydrolyze paraoxon.

The specificity of this arylesterase toward most other aromatic organophosphate esters is not clear. Activity toward chlorpyrifos, however, has been studied. Arylesterase activity toward chlorpyrifos is not polymorphic but follows a Gaussian distribution with a four- to fivefold variation. Even though polymorphism was not observed, individuals with the lowest paraoxonase activity also had the lowest chlorpyrifos-oxonase levels (Furlong et al., 1988). Other arylester organophosphates that may be substrates for arylesterase include guthion, EPN, and diazinon.

Fish-eye (Tangier's) disease, a rare genetic disorder characterized by severe corneal opacities and abnormal plasma lipoproteins, has been associated with decreased paraoxonase activity. Mackness et al. (1987) reported that the mean value of paraoxonase activity in two patients with this disease was only 11 percent of the mean value for 55 control subjects. One could expect children with this disorder to be more prone to adverse health effects if exposed to parathion or other organophosphates that are substrates for this arylesterase.

A toxic encephalopathy has been associated with exposure to N,N-diethyl-m-toluamide (Deet) (Gryboski et al., 1961; Zadikoff et al., 1979; Heick et al., 1980). At least four cases of Deet-induced toxicity have been documented after heavy but routine use of Deet-containing products (Edwards and Johnson, 1987). All cases involved young girls. The most recent report linked the toxicity with ornithine carbamoyltransferase (OCT) deficiency (Heick et al., 1980)—a sex-linked disorder that is fatal in males during the neonatal period. In females, severity varies. There are certain characteristic syndromes common with OCT deficiency, such as feeding problems during infancy and hepatic syndromes including hyperammonia, oroticaciduria, and increased serum glutamine. OCT deficiency is extremely rare and may go undetected. If these cases are identified, then parents and other adults concerned can be notified of the possible toxic effects that may be encountered with Deet exposure.

Increased Sensitivity of Unknown Origin. A case of acrodynia, a form of mercury poisoning that occurs in infants and young children, was recently reported after exposure to latex paint containing the fungicide phenylmercuric acetate (Agocs et al., 1990). Mercury exposure occurs when vapor is released from painted surfaces after the paint has dried. Although EPA now forbids the addition of phenylmercuric acetate to interior latex paint (as of August 20, 1990), mercury-containing paint manufactured before this date may still be sold.

There is wide interindividual variability in susceptibility to acrodynia. In 1980 in Buenos Aires, Argentina, three cases of acrodynia were reported when up to 12,000 infants were exposed to a phenylmercuric fungicide used by a commercial diaper service (Gotelli et al., 1985). The reason why particular children are more sensitive than others is unknown.

Chronic Medical Conditions

Exposure to pesticides could exacerbate common childhood conditions. For instance, the increased cholinergic response induced by organophosphates may increase contraction of bronchial muscles and stimulation of bronchial glands. Therefore, asthmatic children might be affected more easily and more severely than normal children.

Interactions with Medication

Children taking medication may be at increased risk from pesticide exposure. Many drugs compromise nutritional status, and this in itself may increase sensitivity to pesticides. In addition, acetylcholinestrase (AChE) inhibitors, antiepileptic drugs, other drugs that act on the central nervous system (centrally acting drugs), propranolol and digoxin, and drugs that alter hepatic blood flow might be of concern.

Acetylcholinesterase Inhibitors

Any child taking AChE inhibitors would be at increased risk from organophosphate exposure. For instance, increased levels of organophosphates would increase the cholinergic effects observed during the use of the cholinesterase inhibitor phospholine iodide (echothiophate) for glaucoma patients.

Antiepileptic Pharmaceuticals

Phenobarbital and phenytoin induce cytochrome P-450 isozymes. Phenobarbital is also known to induce a number of other hepatic drug metabolizing enzymes. With regard to pesticide exposures, these drugs might induce a cytochrome P-450-dependent isozyme involved in bioactivation of a pesticide, while the deactivation pathways (that is, hydrolysis or conjugation) may be induced to a lesser degree or not induced at all.

Alterations in the activation/deactivation ratio depend on the pesticide as well as the inducing agent. For example, pretreatment with phenobarbital induced both cytochrome P-450-mediated activation of malathion and carboxylesterase-linked hydrolysis in mice (Ketterman et al., 1987). In this case, phenobarbital did not affect the toxicity of malathion. In studies with parathion, on the other hand, pretreatment with phenobarbital decreased acute toxicity in mice (Sultatos, 1986). Phenobarbital pretreatment did increase the toxicity of diazinon in calves (Abdelsalam and Ford, 1986). The effect of chronic phenobarbital or phenytoin treatment on the activation and deactivation of any of these pesticides in humans is unknown.

Other Pharmaceuticals

Fenyvesi et al. (1985) reported that the fungicide thiram potentiated the effects of promethazine in behavioral studies in rats. Promethazine is an H_1 (type 1 histaminergic receptor) antagonist. Two H_1 antagonists commonly given to children are diphenhydramine (Benadryl®) and dimenhydrinate (Dramamine®).

Because of structural similarities, thiram may act similarly to its ethyl analog, disulfiram, a well-studied drug. Disulfiram inhibits hepatic microsomal, drug-metabolizing enzymes and interferes with the metabolism of many drugs, including phenytoin (Gillman and Sandyk, 1985). In addition, disulfiram has been found to greatly increase the cancer-causing effects of the pesticide ethylene dibromide. Whether low levels of thiram would have any of the above effects is not known.

Disulfiram has also been found to interact with centrally acting drugs including amphetamine (Sharkawi et al., 1978). In this regard, a potential interaction between thiram exposure and methylphenidate (Ritalin®), a drug with pharmacological properties similar to those of amphetamine, might be worth investigating. Although thiram residues in food are likely to be low, the potential susceptibility for children taking methylphenidate should be of concern.

Children taking medication for heart disease might also be at increased risk. Propranolol, a β-adrenergic blocker, decreases the heart rate, a response that could be magnified by pesticides that increase cholinergic activity. Digoxin is another medication commonly given to children with heart disease. Pesticides that increase cholinergic activity might potentiate digoxin-induced bradycardia and depressed conduction at the sinoatrial and arterioventricular nodes. One toxic effect of digoxin is complete arterioventricular blockage. There may also be a cholinergic response. Therefore, the pharmacological actions induced by pesticides may reduce the margin of safety of these drugs.

Many drugs alter hepatic blood flow and consequently may alter biotransformation or excretion of xenobiotic compounds that are readily cleared by the liver. For example, propranolol decreases hepatic blood flow. This may be an advantage in exposures to pesticides activated by metabolism, but it could cause greater toxicity in cases of exposure to pesticides that are deactivated by metabolism. Phenobarbital, on the other hand, increases liver blood flow and, therefore, speeds up delivery of pesticides to the liver. For pesticides that are extracted by the liver, extraction would be increased. Thus, for a pesticide that is bioactivated by liver enzymes, activation would increase.

The effects of altered blood flow would not be significant for orally absorbed pesticides. However, altered blood flow would have an effect on pesticides absorbed dermally. Theoretically, then, children taking medication that alters hepatic blood flow would be more affected than other children. In addition, if these children had damaged skin (such as cuts and scrapes), the increased absorption could potentiate the effects of increased (or decreased) hepatic blood flow.

Carcinogenesis

Factors That Might Increase Risk of Cancer Among Infants and Children

When subjected to the same low levels of pesticide residues in food measured as parts per million (ppm), infants and children may be at greater cancer risk than adults for several reasons. First, exposures to an early-stage carcinogen early in life can be more effective than exposures to the same carcinogen in later years (Day and Brown, 1980; Kaldor and Day, 1987). A multistage model of carcinogenesis shows that completion of the first stages of neoplastic transformation of normal cells early in life increases chances that subsequent stages will be completed within a normal lifespan. Early exposures are not always associated with increased risk, however, since exposures to late-stage carcinogens are more effective in later years.

Second, tissues undergoing rapid growth and development may be more susceptible than mature tissues to the effects of chemical carcinogens, but empirical evidence to support this is mixed. In studies of the $LD_{50}s$ of pesticides in rats, weanlings were found to be more sensitive than adults to some carcinogens but less sensitive to others (Gaines and

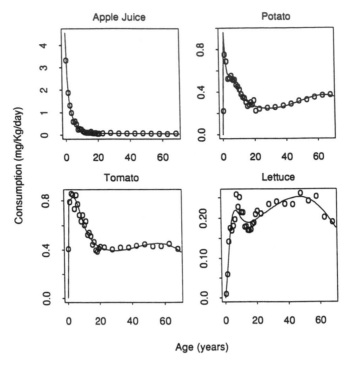

FIGURE 8-1 Consumption rates for apple juice, potato, tomato, and lettuce by age. SOURCE: Murdoch et al., 1992.

Linder, 1986). Drew et al. (1983) demonstrated that rats, mice, and hamsters are more sensitive to the carcinogenic effects of vinyl chloride when
exposure occurs early in life, compared to later exposures of the same
duration. And as noted in the preceding section, Peto et al. (1986) reported
that the incidence of liver tumors decreased appreciably with advanced
age among rats exposed to nitrosodimethylamine (NDMA) beginning at
3, 6, and 12 weeks of age. Perinatal exposure has also been shown to be
necessary for the induction of urinary bladder tumors in male rats exposed
to saccharin in the diet. *In utero* exposure to ethylenethiourea has recently
been found to increase the incidence of malignant thyroid tumors (NTP,
1991).

Third, infants and children may be at greater risk than adults because
of their greater consumption (on a kilogram of body weight basis) of
certain foods. As shown in Chapter 5, children consume fewer foods than
do adults but many of these are consumed at much higher levels when
expressed on a body weight basis. To illustrate, Figure 8-1 shows the

consumption of apple juice as reported in the 1977-1978 Nationwide Food Consumption Survey (NFCS) (Murdoch et al., 1992). The average intake of apple juice for survey respondents is indicated by open circles; the solid lines represent smooth curves fitted to the raw data. On a body weight basis, 1-year-old children consume more than 30 times as much apple juice per day as adults.

The data for the three other foods depicted in Figure 8-1 reflect different consumption patterns with age. These particular foods were chosen to illustrate a variety of food consumption patterns. The maximum consumption of potatoes and tomatoes occurs among children 1 to 6 years of age. The consumption of lettuce increases throughout childhood and is higher among adults than among children.

These consumption data have important implications for assessing the risk of carcinogenesis in infants and children. In laboratory studies of carcinogenicity in which the dose is expressed relative to body weight (in mg/kg bw/day), the dose is generally held constant throughout the study. Such studies are used to estimate carcinogenic potency in terms of a unit risk factor (q_1^*). Multiplication of the anticipated level of exposure by the unit risk factor leads to an estimate of cancer risk.

Effects of Age-Dependent Exposures

Methods for estimating lifetime cancer risk when the level of exposure varies over time have been investigated by Crump and Howe (1984) and Kodell et al. (1987). One simple approach is to calculate a lifetime average daily dose (LADD) by distributing the cumulative lifetime dose equally over a lifetime. This amortization of total lifetime exposure is suggested by the EPA (1986) in the absence of evidence of dose-rate effects. This method is especially appealing when considering carcinogens that remain in the body permanently or have very long half lives, such as some fat-soluble organic compounds and heavy metals.

Murdoch and Krewski (1988) showed that except in very special cases, the use of the LADD will not lead to the same lifetime risk as the actual time-dependent exposure pattern. These authors defined a lifetime equivalent constant dose (LECD) in terms of the relative effectiveness of dosing at different ages, which leads to the same lifetime risk as the time-dependent dosing pattern. (The LECD is a weighted average of the time-dependent dose; weights are proportional to the relative effectiveness of dosing at each point.) The ratio $C = LECD/LADD$ then provides a measure of error of risk estimates based on the LADD.

Murdoch et al. (1992) used this measure to investigate the extent risk estimates based on average adult levels of exposure may underestimate the risks for infants and children as a consequence of their higher exposures on a body weight basis. This analysis relies on the recent interpretations of the classical Armitage-Doll multistage model of carcinogenesis (Armitage, 1985) and the MVK two-stage mutation-birth-death model (Moolgavkar and Luebeck, 1990).

Application of the approximate form of the two-stage model requires specification of the number of cells at risk $N(t)$ in the target tissue at time t as well as the net birthrate $\delta(t)$ of initiated cells that have sustained the first mutation as a function of time t. Application of the exact form of the two-stage model, which is necessary when the lifetime probability of tumor formation is high (Moolgavkar et al., 1993), further requires specification of the birth rate $\alpha(t)$ and death rate $\beta(t)$ of initiated cells, rather than just the difference $\delta(t) = \alpha(t) - \beta(t)$.

In some applications of the two-stage model, it has been assumed that $N(t)$ and $\delta(t)$ are constant over time. The assumption that $N(t)$ is constant is also implicit in most applications of the multistage model. Assumptions of the constancy of $N(t)$ and $\delta(t)$ expedite the mathematical analysis of the two-stage model, but are oversimplifications of reality. For example, Enesco and Leblond (1962) found that the number of cells in most rat tissues increase with age up to approximately 34 to 48 days after birth, well past weaning in these species, at which point tissue size starts to stabilize.

Tissue growth can result from an increase in either cell size (hypertrophy) or in cell number (hyperplasia). Hyperplasia is the dominant feature in initial tissue growth, but the rate of cell division decreases with age (Brasel and Gruen, 1978) and reaches zero at maturity in tissues that are not self-renewing. Some tissues such as the brain are fully developed in early childhood, whereas others such as the skeletal system do not achieve maturity until after adolescence.

Growth in cell mass is sometimes assessed by comparing tissue weight to DNA content (Brasel and Gruen, 1978). However, this approach to distinguishing between increases in cell number and cell size is difficult to apply in mammalian liver because of the formation of multinucleate cells and cells with single nuclei, but with diploid or tetraploid amounts of DNA (polyploidy).

The interpretation of $N(t)$ as the number of normal stem cells may also be an oversimplification. Moolgavkar and Luebeck (1990) noted that this is inconsistent with the assumption that mutation occurs during cell division, because the mitotic index (the per-cell rate of cell division) varies with age. For example, Ellwein and Cohen (1988) report that mitotic indices in stem cells in 1-week-old rats are approximately 100 times higher than in rats 8 weeks of age. The estimation of mitotic rates based on DNA incorporation of tritiated thymidine during cell division is discussed by Moolgavkar and Luebeck (1992).

In general, little information is available about the net birth rate $\delta(t)$ of initiated cells and even less about $\alpha(t)$ and $\beta(t)$ separately. Initiation/promotion assays may provide some direct information on the birth and death rates of initiated cells. In the mouse liver system described by Pitot and Dragan (1991), foci believed to correspond to clones of premalignant initiated cells can be identified histologically and interpreted in terms of numbers of initiated cells. Such assays have been used to study cancers of the liver, urinary bladder, and skin, but are not yet well developed for other tissues. Moolgavkar et al. (1990) and Luebeck et al. (1991) discussed the statistical use of data on the kinetics of initiated cells in a two-stage model.

Murdoch and Krewski (1988) explored the results obtained by assuming a constant rate of proliferation of intermediate cells, independent of age, while Krewski and Murdoch (1990) assumed that the net birth rate was a constant multiple of the birth rate of normal cells. Murdoch et al. (1992) considered an intermediate approach:

$$\delta(t) = \delta_N(t) + \delta,$$

where $\delta_N(t)$ is the net birth rate of normal cells at time t calculated on the assumption that all changes in tissue weight are due to changes in cell number rather than cell size, and the constant δ is the excess net birth rate of intermediate cells. This model implies that intermediate cells respond to their microenvironment in essentially the same way as normal cells do, but may be slightly more ($\delta > 0$) or less ($\delta < 0$) successful at proliferation.

When specific information on $N(t)$ is unavailable, a rough surrogate for the number of stem cells in the target tissue is gross organ weight. This latter indicator has the advantage of being readily available in humans, but does not take into account hypertrophic effects or changes in mitotic indices over time. For example, Murdoch et al. (1992) assumed that $N(t)$ is proportional to human liver weight (Snyder et al., 1974).

Murdoch et al. (1992) calculated the values of $C = \text{LECD}/\text{LADD}$ for apple juice, potatoes, tomatoes, and lettuce, assuming that exposure is characterized by the amount of food consumed. The values of C shown in Tables 8-2 and 8-3 for the multistage and two-stage models, respectively, indicate the degree to which the use of adult exposure data may underestimate or overestimate the actual risks, taking into account the greater levels of food consumption by infants and children. Consider, for example, the value of $C = 3.74$ for a six-stage model in which only the first stage is dose-dependent, based on the actual levels of apple juice consumed. This implies that the actual lifetime cancer risk due to the presence of residues of a pesticide found in apple juice will be more than 3 times that calculated on the basis of a LADD obtained by amortization of the cumulative lifetime exposure. Assuming a constant daily lifetime expo-

TABLE 8-2 Values of the Ratio $C = LECD/LADD$ for the Multistage Model for Pesticide Residues in Apple Juice, Potatoes, Tomatoes, and Lettuce[a]

Food	Stage Affected (r)	Number of Stages (k)					
		1	2	3	4	5	6
Apple Juice	1	1.00	1.64	2.23	2.77	3.27	3.74
	2		0.36	0.48	0.61	0.76	0.93
	3			0.30	0.34	0.38	0.43
	4				0.28	0.31	0.34
	5					0.27	0.30
	6						0.27
Potato	1	1.00	1.05	1.13	1.21	1.27	1.33
	2		0.94	0.90	0.90	0.94	0.99
	3			0.97	0.89	0.85	0.84
	4				1.00	0.92	0.86
	5					1.02	0.95
	6						1.04
Tomato	1	1.00	1.08	1.16	1.22	1.27	1.31
	2		0.92	0.94	0.97	1.01	1.06
	3			0.90	0.91	0.91	0.92
	4				0.90	0.91	0.90
	5					0.90	0.91
	6						0.90
Lettuce	1	1.00	0.95	0.90	0.85	0.82	0.79
	2		1.05	1.05	1.02	0.99	0.96
	3			1.05	1.09	1.07	1.04
	4				1.04	1.10	1.10
	5					1.03	1.10
	6						1.01

[a] From Murdoch et al. (1992).

sure equal to that of adults would lead to a slight underestimation of risk.

Examination of Tables 8-2 and 8-3 reveals several important findings. First, the largest correction factors C increase with the number of stages k in the multistage model and the cumulative net birth rate of initiated cells in the two-stage model. The values of C also increase as the stage affected r decreases. Values of C less than unity correspond to cases in which the use of the LADD will overestimate rather than underestimate risk.

The largest values of C encountered under the conditions considered in Tables 8-2 and 8-3 range from 3- to 4-fold for apple juice. Since apple juice was selected to show the wide differences between childhood and adult food consumption patterns, it is not unreasonable to suggest that the use of adult exposure levels will not underestimate actual lifetime

TABLE 8-3   Values of the Ratio C = LECD/LADD for the Two-Stage Mutation-Birth-Death Model for the Pesticide Residues in Apple Juice, Potatoes, Tomatoes, and Lettuce[a]

Food	Stage Affected (r)	Cumulative Net Birth Rate of Initiated Cells ($t\delta$)				
		−10	−5	0	5	10
Apple Juice	1	0.72	0.90	1.53	3.27	5.25
	2	0.39	0.36	0.32	0.28	0.25
Potato	1	0.94	0.96	1.04	1.26	1.46
	2	0.91	0.91	0.94	1.01	1.06
Tomato	1	0.97	0.99	1.07	1.26	1.39
	2	0.92	0.91	0.91	0.90	0.89
Lettuce	1	1.03	1.01	0.96	0.83	0.71
	2	1.05	1.05	1.06	1.03	0.97

[a] From Murdoch et al. (1992).

cancer risks more than 5-fold, taking into account the increased consumption of these foods by infants and children.

Percent Contribution to Lifetime Risk.   The contribution to lifetime risk made by exposures at different ages can be characterized by applying the methods developed by Murdoch et al. (1992) and described above. The percent contribution to lifetime risk (PCLR) made by exposure at age t can be written as:

$$PCLR_t = [(d_t r_t f)/LECD] \times 100\%,$$

where d_t is exposure at age t, r_t denotes the effectiveness of exposure at age t, and LECD (effectively the sum or integral of d_t r_t over t) is a normalizing factor. Here, f is the fraction of the lifespan composed by the tth time interval. (For example, f is 1/70 for yearly age groups with an expected lifespan of 70 years.) In this example, exposure depends on both food consumption patterns and pesticide residue levels, whereas relative effectiveness is determined by the model of carcinogenesis used.

For illustration purposes, if one assumes that pesticide applications produce constant levels of residues in food over time, then the only factors that affect exposure levels are changes in food consumption. Such changes could involve changes in the sources of food (e.g., more imported foods with different levels of residues than domestically grown foods) as well as changes in quantities consumed. However, this is an unlikely scenario since the amount of pesticide residues in food is affected by change in regulations, climate, and application practices.

Figure 8-2 shows the percent contribution to lifetime risk from a hypo-

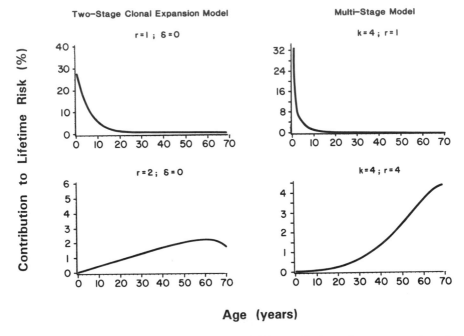

FIGURE 8-2 Annualized contributions to lifetime risk from the consumption of apple juice. SOURCE: Murdoch and Krewski, 1988.

thetical pesticide residue in apple juice by using a two-stage model with no proliferation of initiated cells ($\delta = 0$). When only the first stage is dose dependent, the PCLR is highest early in life. Specifically, more than 27% of the lifetime risk accrues during the first year of life. This increases to more than 81% by age 10. This occurs because both the highest exposures and the greatest relative effectiveness occur early in life. Calculations (not shown) for $\delta \neq 0$ indicate that the PCLR early in life increases when $\delta > 0$ and decreases when $\delta < 0$. Even for $t\delta = -10$, however, the PCLR up to age 10 exceeds 56%. (The parameter $t\delta$ represents the cumulative net birth rate of initiated cells up to $t = 70$ years of age, and provides a convenient indicator of the rate of proliferation of initiated cells.)

When the second stage is affected, the relative effectiveness of dosing is low early in life, and only about 5% of the lifetime risk accrues by age 10. After age 40, more than 60% of the lifetime risk accrues. The PCLR up to age 10 decreases when $\delta > 0$ and increases when $\delta < 0$. For $t\delta = -10$, the PCLR up to age 10 is only about 18%.

The PCLR for pesticide residues in lettuce (not shown) are notably different from those for apple juice (Figure 8-2). Because children consume less lettuce than do adults, the PCLR for up to age 10 is comparatively lower. However, with a first-stage carcinogen and a high net birth rate of initiated cells ($t\delta$ = 10), the PCLR up to age 10 is nearly 70%.

These results indicate that the PCLR depends strongly on the form of the underlying two-stage model of carcinogenesis. Which of the two mutation rates is dose dependent appears to be a more critical factor than the proliferation rate δ of initiated cells. Similar calculations (see Figure 8-2) for the multistage model with k = 4 stages indicate that assumptions about which stage or stages in the model are dose dependent have a marked effect on the PCLR, since earlier dose-dependent stages lead to a higher PCLR from exposures experienced early in life.

Age-Specific Risks. The preceding analysis of the relative contribution to lifetime risk indicates only the percentage of lifetime risk that accrues each year—not the actual lifetime risk. The lifetime risk (LR) can be expressed as follows:

$$LR = \sum_{t=1}^{70} LR_t,$$

where $LR_t = (d_t r_t P)/70$ denotes the contribution to the lifetime risk at t years of age and P is a measure of carcinogenic potency such as the unit risk factor q_1^* based on the linearized multistage model. The weights r_t defined by the relative effectiveness function represent age-specific multipliers that can be used to estimate the risk LR_t due to exposure d_t at t years of age.

The age-specific multipliers for the two-stage model with δ = 0 and the multistage model with k = 4 stages are shown in Figure 8-3. When the last stage is dose dependent, the age-specific multipliers decrease with age, reflecting the fact that earlier exposures are more effective than later exposures (Murdoch and Krewski, 1988). When only the second stage is dose dependent, the age-specific multipliers increase with age.

Consumption and Exposure Data

The data on food consumption and pesticide residues were extremely limited, as described in Chapters 5 and 6, and the limitations were especially severe for infants and children. Furthermore, when determining the acceptability of a food tolerance, EPA compared the theoretical maximum residue contribution (TMRC; see Chapter 6) and anticipated residues to

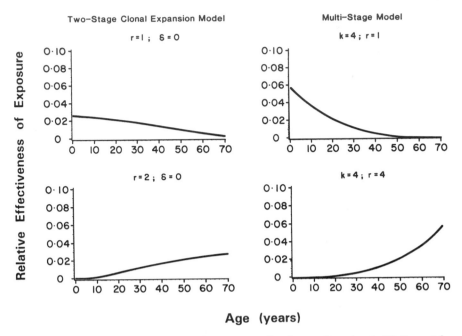

FIGURE 8-3 Multipliers used to obtain annualized contributuion to lifetime risk.
SOURCE: Murdoch and Krewski, 1988.

the RfD and if the values were less than the RfD, it was assumed that no adverse health effects would result. However, the TMRC methodology is limited to pesticide exposures from residues in food alone. Thus, although the TMRC values may be less than the RfD, an individual may also be exposed to nondietary sources of that particular pesticide, so that the *total* exposure exceeds an RfD and is detrimental to health. Although such nondietary exposures of both children and adults are of concern, children may experience greater exposures because of their tendency to ingest foreign objects (a practice called pica). Furthermore, breastfed infants may be expected to be exposed to low levels of lipophilic pesticides (Babich and Davis, 1981). Lactating women may be, or may have been, exposed to lipophilic pesticides from various sources, including air, food, water, cosmetics, and occupational and household environments, and they may have substantial stores of some that can be mobilized during lactation. Within a few months of birth, the concentration of halocarbons in the body fat of the breastfed infant may be equivalent to that in the fat of the lactating mother. Although the period of breastfeeding is short compared to the total life span, the lipophilic chemicals accumulated in the body during these early months may be retained for years (WHO, 1986).

Another source of exposure to pesticides is groundwater (Russell et al., 1987), which supplies drinking water to 97% of rural Americans and 53% of all Americans (USGS, 1988). Water intake should be considered, although an EPA survey of groundwater (EPA, 1990,1992) indicates that in most cases this will be an insignificant source of pesticide exposure. Neither nondietary exposures nor exposures in drinking water were considered in setting food tolerances using the "food factor" system. Because of this limitation, total exposures of infants and children may be underestimated.

CONCLUSIONS AND RECOMMENDATIONS

The evaluation of potential risks to infants and children due to pesticide residues present in the diet requires consideration of a number of factors. In particular, the level of risk depends on individual food consumption patterns, levels of pesticide residues in foods as consumed by infants and children, and the toxicological potency of the pesticide. A comprehensive analysis of the potential risks to infants and children exposed to pesticides in their diets requires consideration of all these factors, as well as any unique characteristics of infants and children relative to adults. The conclusion and recommendations presented here address each of these components of risk assessment.

Conclusions

• The health risks experienced by infants and children as a result of their exposure to pesticide residues in the diet may differ from those experienced by adults both in nature and severity.

• Infants and children may exhibit unique susceptibility to the toxic effects of pesticides because they are undergoing rapid tissue growth and development, but empirical evidence to support this is mixed.

• For some pesticides, infants and children are at greater risk than adults. For others they may exhibit less risk.

–Infants and children consume much greater quantities of certain foods than do adults on a body weight basis and may thus be subjected to higher levels of exposure to certain pesticides than adults.

–Exposures occurring earlier in life can lead to greater or lower risk of chronic toxic effects such as cancer than exposures occurring later in life.

• Traditional approaches to toxicological risk assessment may not always adequately protect infants and children.

Although current uncertainty factors used to extrapolate toxicological data to humans provide for 10-fold variations between species and within the human population, additional protection for developmental toxicity may be required, depending on the toxicant of interest and the amount of testing that has been conducted.

• It is known that socioeconomic, nutritional, and health status factors influence the vulnerability of human infants and children to environmental toxicants. The committee acknowledges that these are adequately protected by the existing 10-fold uncertainty factor for intraspecies variability.

• An additional uncertainty factor of up to 10-fold is currently used when prenatal developmental effects are observed.

As a result of its consideration of toxicity to organs that may be particularly susceptible to developmental effects—such as the central nervous system, immune surveillance system, and endocrine system—the committee encourages the development and validation of animal test systems to evaluate toxicity to these organ systems.

• There is a need for validated animal models for predicting the risks to infants and children.

• There exist only limited data on the food consumption patterns of infants and children that are appropriate for use in risk assessment.

The data used by the committee to evaluate food consumption patterns of infants and children were derived primarily from the 1977-1978 NFCS. This survey involved relatively small numbers of people in the younger age categories and may not reflect current food consumption patterns.

Recommendations

• Additional data on the food consumption patterns of infants and children are needed.

Current data indicate that infants and children consume much more of certain foods on a body weight basis than do adults. Because such higher exposures can lead to higher risks, it is important to have accurate data on the food consumption patterns of infants and children. The available data are based on relatively small samples and may not reflect current trends in food consumption by infants and children.

• All sources of exposure to pesticides—dietary and nondietary— need to be considered when evaluating the potential risks to infants and children.

Since many pesticides are applied to more than one crop, residues of a particular pesticide may be found on different foods. The total intake from all foods on which residue may be present should be calculated when estimating exposure of infants and children. Pesticides may also be present in drinking water due to contamination of groundwater. Nondietary environmental sources, including air, dirt, surfaces, lawns, and pets, should also be considered.

• Physiological and biochemical characteristics of infants and children that influence metabolism and disposition need to be considered in risk assessment.

Physiological parameters such as tissue growth rates and biochemical parameters such as enzyme induction may affect the response of infants and children to pesticide residues in food.

• Pharmacokinetic models that provide for the unique physiologic characteristics of infants and children should be developed.

Physiologically based pharmacokinetic models can be used to estimate the dose of toxic metabolites reaching target tissues. Such models can be extended to allow for such factors as lactation and development that are unique to infants and children.

• Three 10-fold uncertainty factors are now applied to the NOEL to develop the RfD: 10 to account for interspecies differences, 10 to account for intraspecies difference, and 10 when there is evidence of developmental effects as demonstrated by toxicological testing and metabolic/disposition studies. Thus, a 10-fold factor has been applied by the EPA whenever toxicity studies have shown fetal developmental effects. Because of specific periods of vulnerability that exist during development, the committee recommends that an uncertainty factor up to the 10-fold factor traditionally used for fetal developmental toxicity should also be considered for postnatal developmental toxicity and when data from toxicity testing are incomplete. The committee wishes to emphasize that this is not a new, additional uncertainty factor but, rather, an extended application of a uncertainty factor now routinely used by the EPA for a narrower purpose.

Data on anticancer agents and other drugs, insecticides, and pesticides have indicated that infants and children may be more susceptible or less susceptible to the effects of these agents than adults. Additional data are required to better quantify the magnitude of the differences in susceptibility between infants and children and adults, especially with agents causing neurological or immunological disorders. Systematic studies of age-dependent toxicity are needed to define the magnitude of such differences more precisely.

• The committee recommends development and validation of animal test systems to evaluate toxicity to developing organ systems.

These tests should be designed to evaluate the unique developmental waypoints of infants and children, including assessment of the central nervous system, immune system, endocrine system, and reproductive system.

• Estimates of cancer risk should take into account both the higher exposures of infants and children to certain pesticides and the earlier age at which these exposures occur in comparison to adults.

These factors can be taken into account using cancer risk estimation methods that allow for time-dependent exposure patterns and toxicological testing paradigms that include early exposures.

• The use of biologically based models of carcinogenesis that take into account the special physiological characteristics of infants and children should be developed.

Biologically based models such as the two-stage clonal expansion model of carcinogenesis provide a realistic approach to cancer risk estimation. If the rate of clonal expansion of initiated cells could be determined as a function of age, this information could be used in developing biologically based models of carcinogenesis for use in risk estimation.

• The use of the benchmark dose for risk assessment applications involving infants and children should be explored.

Although not yet widely applied in risk assessment, the benchmark dose approach offers certain advantages over the NOEL. In particular, this approach uses all the available dose-response data, better reflects the slope of the dose-response curve, and provides an explicit indication of risk at doses at or below the benchmark dose. The benchmark dose may also be used as a means of integrating toxicological and carcinogenic risk assessment methodologies.

• The use of risk distributions rather than a point estimate such as a mean, median, or outer bound should be used where possible to provide a more complete characterization of risk.

In this report, the notion of combining distributions of individual food consumption levels with distributions of pesticide residue levels in food to obtain a distribution of exposure levels was used extensively and is illustrated by the examples presented in Chapter 7. Data on the potency of a particular pesticide (expressed either in terms of a reference dose or measure of carcinogenic potency) can then combined with data on the distribution of exposures to estimate a distribution of risks across a population of infants and children. Successful application of this method will require more elaborate food consumption survey data than are currently available. To avoid serious bias in estimates of upper and lower percentiles of distributions of individual daily intakes, repeated observations of food consumption on several different days will be required.

REFERENCES

Abdelsalam, E.B., and E.J. Ford. 1986. Effect of pretreatment with hepatic microsomal enzyme inducers on the toxicity of diazinon in calves. Res. Vet. Sci. 41:336-339.

Agocs, M.M., R.A. Etzel, R.G. Parrish, D.C. Paschal, P.R. Campagna, D.S. Cohen, E.M. Kilbourne, and J.L. Hesse. 1990. Mercury exposure from interior latex paint. New Engl. J. Med. 323:1096-1101.

Aldridge, W.N. 1986. The biological basis and measurement of thresholds. Annu. Rev. Pharmacol. Toxicol. 26:39-58.

Allen, B.C., K.S. Crump, and A.M. Shipp. 1988. Correlation between carcinogenic potency of chemicals in animals and humans. Risk Anal. 8:531-544.

American Health Foundation. 1989. Coronary artery disease prevention: Cholesterol, a pediatric perspective. Prev. Med. 18:323-409.

Andersen, M.E., H.J. Clewell III, M.L. Gargas, F.A. Smith,and R.H. Reitz. 1987. Physiologically based pharmacokinetics and the risk assessment process for methylene chloride. Toxicol. Appl. Pharmacol. 87:185-205.

AOR (Assembly Office of Research). 1988. Pp. 35-36 in The Invisible Diet. Sacramento: Joint Publications Office, State of California.

Armitage, P. 1985. Multistage models of carcinogenesis. Environ. Health Perspect. 63:195-201.

Armitage, P., and R. Doll. 1961. Stochastic models for carcinogenesis. Pp. 19-38 in Fourth Berkeley Symposium on Mathematical Statistics and Probability. Berkeley Calif.: University of California Press.

Arnold, D.L., D. Krewski, and I.C. Munro. 1983. Saccharin: A toxicological and historical perspective. Toxicology 27:179-256.

Babich, H., and D.L. Davis. 1981. Food tolerances and action levels: Do they adequately protect children? BioScience 31:429-438.

Barnes, D.G., and M.L. Dourson. 1988. Reference dose (RfD): Description and use in health risk assessments. Regul. Toxicol. Pharmacol. 8:471-486.

Barr, J.T. 1985. The calculation and use of carcinogenic potency: A review. Regul. Toxicol. Pharmacol. 5:432-459.

Bernstein, L., L.S. Gold, B.N. Ames, M.C. Pike, and D.G. Hoel. 1985. Some tautologous aspects of the comparison of carcinogenic potency in rats and mice. Fundam. Appl. Toxicol. 5:79-86.

Board, P. 1990. Genetic polymorphisms of glutathione S-transferase in man. Pp. 232-241 in Glutathione S-Transferase and Drug Resistance, J.D. Hayes et al., eds. London: Taylor and Frances.

Brasel, J.A., and R.K. Gruen 1978. Cellular growth: Brain, liver, muscle, and lung. Pp. 3-19 in Human Growth: Postnatal Growth, Vol. 2, F. Falkner and J.M. Tanner, eds. New York: Plenum Press.

Brodeur, J., and K.P. DuBois. 1963. Comparison of acute toxicity of anticholinesterase insecticides to weanling and adult male rats. Proc. Soc. Exp. Biol. Med. 114:509-511.

Brown, A.E., O. Quesada, and D. Armstrong. 1982. Minimal nephrotoxicity with cephalosporin-aminoglycoside combinations in patients with neoplastic disease. Antimicrob. Agents Chemother. 21:592-594.

Brown, C.C., and K.C. Chu. 1989. Additive and multiplicative models and multistage carcinogenesis theory. Risk Anal. 9:99-105.

Brown, C.C., and J.A. Koziol. 1983. Statistical aspects of the estimation of human risk from suspected environmental carcinogens. SIAM Rev. 25:151-181.

Chen, J.J., and D.W. Gaylor. 1987. Carcinogenic risk assessment: Comparison of estimated safe doses for rats and mice. Environ. Health Perspect. 72:305-309.

Chen, J.J., R.L. Kodell, and D. Gaylor. 1988. Using the biological two-stage model to assess risk from short-term exposures. Risk Anal. 8:223-230.

Cohen, S.M., and L.B. Ellwein. 1990. Proliferative and genotoxic cellular effects in 2-acetylaminofluorene bladder and liver carcinogenesis: Biological modeling of the ED01 study. Toxicol. Appl. Pharm. 104:79-93.

Crouch, E., R. Wilson, and L. Zeise. 1987. Tautology or not tautology? J. Toxicol. Environ. Health 20:1-10.

Crump, K.S. 1984a. An improved procedure for low-dose carcinogenic risk assessment from animal data. J. Environ. Pathol. Toxicol. Oncol. 6:339-348.

Crump, K.S. 1984b. A new method for determining allowable daily intakes. Fundam. Appl. Toxicol. 4:854-871.

Crump, K.S., and R.B. Howe. 1984. The multistage model with a time-dependent dose pattern: Applications to carcinogenic risk assessment. Risk Anal. 4:163-167.

Crump, K.S., D.G. Hoel, C.H. Langley, and R. Peto. 1976. Fundamental carcinogenic processes and their implications for low dose risk assessment. Cancer Res. 36:2973-2979.

Day, N.E., and C.C. Brown. 1980. Multistage models and primary prevention of cancer. J. Natl. Cancer Inst. 64:977-989.

Dourson, M.L. 1986. New approaches in the derivation of acceptable daily intake (ADI). Comments Toxicol. 1:35-48.

Dourson, M.L., and J.F. Stara. 1983. Regulatory history and experimental support of uncertainty (safety) factors. Regul. Toxicol. Pharmacol. 3:224-238.

Drew, R.T., G.A. Boorman, J.K. Haseman, E.E. McConnell, W.M. Busey, and J.A. Moore. 1983. The effect of age and exposure duration on cancer induction by a known carcinogen in rats, mice and hamsters. Toxicol. Appl. Pharmacol. 68:120-130.

Edwards, D.L., and C.E. Johnson. 1987. Insect-repellent-induced toxic encephalopathy in a child. Clin. Pharm. 6:496-498.

Elashoff, R.M., T.R. Fears, and M.A. Schneiderman. 1987. Statistical analysis of a carcinogen mixture experiment. I. Liver carcinogens. J. Natl. Cancer Inst. 79:509-526.

Ellwein, L.B., and S.M. Cohen. 1988. A cellular dynamics model of experimental bladder cancer: Analysis of the effect of sodium saccharin in the rat. Risk Anal. 8:215-221.

Enesco, M., and C.P. Leblond. 1962. Increase in cell number as a factor in the growth of the organs and tissues of the young male rat. J. Embryol. Exp. Morphol. 10:530-562.

EPA (U.S. Environmental Protection Agency). 1986. Guidelines for carcinogen risk assessment. Fed. Regist. 51:33992-34003.

EPA (U.S. Environmental Protection Agency). 1987. Report of the EPA Workshop on the Development of Risk Assessment Methodologies for Tumor Promoters. EPA/600/9-87/03. Washington, D.C.: U.S. Environmental Protection Agency.

EPA (U.S. Environmental Protection Agency). 1988. Reference Dose (RfD): Description and Use in Health Risk Assessment. Integrated Risk Information System (IRIS). Online. Office of Health and Environmental Assessment, Environmental Criteria and Assessment Office. Cincinnati, Ohio: U.S. Environmental Protection Agency.

EPA (U.S. Environmental Protection Agency). 1990. National Survey of Pesticides in Drinking Water Wells. Phase I. NTIS Doc. No. PB-91-125765. Springfield, Va.: National Technical Information Service.

EPA (U.S. Environmental Protection Agency). 1991. Guidelines for developmental toxicity risk assessment: Notice. Fed. Regist. 56:63798-63826.

EPA (U.S. Environmental Protection Agency). 1992. Another Look: National Survey of Pesticides in Drinking Water Wells - Phase II Report. Office of Pesticides and Toxic Substances. EPA 579/09-91-020. Washington, D.C.: U.S. Environmental Protection Agency.

Faden, H., G. Deshpande, and M. Grossi. 1982. Renal and auditory toxic effects of amikacin in children with cancer. Am. J. Dis. Child. 136:223-225.

Farmer, J.H., R.L. Kodell, and D.W. Gaylor. 1982. Estimation and extrapolation of tumor probabilities from a mouse bioassay with survival/sacrifice components. Risk Anal. 2:27-34.

Fears, T.R., R.M. Elashoff, and M.A. Schneiderman. 1988. The statistical analysis of a carcinogen mixture experiment. II. Carcinogens with different target organs, N-methyl-N'-nitro-N-nitrosoguanidine, N-butyl-n(4-hydroxybutyl) nitrosamine, dipentylnitrosamine and nitrilotriacetic acid. Toxicol. Ind. Health 4:221-255.

Fears, T.R., R.M. Elashoff, and M.A. Schneiderman. 1989. The statistical analysis of a carcinogen mixture experiment. III. Carcinogens with different target systems, aflatoxin B_1, N-butyl-n(4-hydroxybutyl)nitrosamine, lead acetate, and thiouracil. Toxicol. Ind. Health 5:1-23.

Fenyvesi, G., M. Botos, and J. Iván. 1985. Pesticide-drug interaction in rats. Arch. Toxicol. (Suppl.) 8:269-271.

Fisher, J.W., T.A. Whittaker, D.H. Taylor, H.J. Clewell III, and M.E. Andersen. 1989. Physiologically based pharmacokinetic modeling of the pregnant rat: A multiroute exposure model for trichloroethylene and its metabolite, trichloroacetic acid. Toxicol. Appl. Pharmacol. 99:395-414.

Fisher, J.W., T.A. Whittaker, D.H. Taylor, H.J. Clewell III, and M.E. Anderson. 1990. Physiologically based pharmacokinetic modeling of the lactating rat and nursing pup: A multiroute exposure model for trichloroethylene and its metabolite, trichloroacetic acid. Toxicol. Appl. Pharmacol. 102:497-513.

Food and Chemical Toxicology. 1985. Saccharin-Current status. Food Chem. Toxicol. 23:543-546.

Freireich, E.J., E.A. Gehan, D.P. Rall, L.H. Schmidt, and H.E. Skipper. 1966. Quantitative comparison of anticancer agents in mouse, rat, hamster, dog, monkey and man. Cancer Chem. Rep. 50:219-248.

Furlong, C., R. Richter, S. Seidel, and A. Motulsky. 1988. Role of genetic polymorphism of human plasma paraoxonase/arylesterase in hydrolysis of the insecticide metabolites chlorpyrifos oxon and paraoxon. Am. J. Hum. Genet. 43:230-238.

Gabrielsson, J.L., and L.K. Paalzow. 1983. A physiological pharmacokinetic model for morphine disposition in the pregnant rat. J. Pharmacokinet. Biopharm. 11:147-163.

Gabrielsson, J.L., P. Johansson, U. Bondesson, and L.K. Paalzow. 1985a. Analysis of methadone disposition in the pregnant rat by means of a physiological flow model. J. Pharmacokinet. Biopharm. 13:355-372.

Gabrielsson, J.L. Paalzow, S. Larsson, and I. Blomkvist. 1985b. Constant rate of infusion-improvement of tests for teratogenicity and embryotoxicity. Life Sci. 37:2275-2282.

Gaines, T.B., and R.E. Linder. 1986. Acute toxicity of pesticides in adult and weanling rats. Fundam. Appl. Toxicol. 7:299-308.

Gaylor, D.W. 1987. Linear-nonparametric upper limits for low dose extrapolation. Pp. 63-66 in Proceedings of the Biopharmaceutical Section of the American Statistical Association. Alexandria, Va.: American Statistical Association.

Gaylor, D.W. 1988. Risk assessment: Short-term exposure at various ages. In Phenotypic Variation in Populations: Relevance to Risk Assessment, A.D. Woodhead, M.A. Bender, and R.C. Leonard, eds. New York: Plenum.

Gaylor, D.W., and J.J. Chen. 1986. Relative potency of chemical carcinogens in rodents. Risk Anal. 6:283-290.

Gaylor, D.W., and R.L. Kodell. 1980. Linear interpolation algorithm for low dose risk assessment of toxic substances. J. Environ. Pathol. Toxicol. 4:305-312.

Gillman, M.A., and R. Sandyk. 1985. Phenytoin toxicity and co-trimoxazole. Ann. Intern. Med. 102:559.

Glaubiger, D.L., D.D. von Hoff, J.S. Holcenberg, B. Kamen, C. Pratt, and R.S. Ungerleider. 1982. The relative tolerance of children and adults to anticancer drugs. Front. Radiat. Ther. Oncol. 16:42-49.

Godfrey, K. 1983. Compartmental Models and Their Application. London: Academic Press.

Gold, L.S., C.B. Sawyer, R. Magaw, G.M. Backman, M. deVeciana, R. Levinson, N.K. Hooper, W.R. Havender, L. Bernstein, R. Peto, M.C. Pike, and B.N. Ames. 1984. A carcinogenic potency database of the standardized results of animal bioassays. Environ. Health Perspect. 58:9-19.

Gold, L.S., J.M. Ward, L. Bernstein, and B. Stern. 1986. Association between carcinogenic potency and tumor pathology in rodent carcinogenesis bioassays. Fundam. Appl. Toxicol. 6:677-690.

Goldenthal, E.I. 1971. A compilation of LD_{50} values in newborn and adult animals. Toxicol. Appl. Pharmacol. 18:185-207.

Gotelli, C.A., E. Astolfi, C. Cox, E. Cernichiari, and T.W. Clarkson. 1985. Early biochemical effects of an organic mercury fungicide on infants: "Dose makes the poison." Science 227:638-640.

Greenfield, R.E., L.B. Ellwein, and S.M. Cohen. 1984. A general probabilistic model of carcinogenesis: Analysis of experimental urinary bladder cancer. Carcinogenesis 5:437-445.

Gross, M.A., O.G. Fitzhigh, and N. Mantel. 1970. Evaluation of safety for food additives: An illustration involving the influence of methyl salicylate on rat reproduction. Biometrics 26:181-194.

Gryboski, J., D. Weinstein, and N.K. Ordway. 1961. Toxic encephalopathy apparently related to the use of an insect repellant. New Engl. J. Med. 264:289-291.

Guengerich, F.P., L.M. Distlerath, P.E. Reilly, T. Wolff, T. Shimada, D.R. Umbenhauer, and M.V. Martin. 1986. Human liver cytochromes P-450 involved in polymorphisms of drug oxidation. Xenobiotica 16:367-378.

Haseman, J.K., and J.E. Huff. 1987. Species correlation in long-term carcinogenicity studies. Cancer Lett. 37:125-132.

Heick, H., M. Shipman, M. Norman, and W. James. 1980. Brief clinical and laboratory observations: Reye-like syndrome associated with use of insect repellent in a presumed heterozygote for ornithine carbamoyl transferase deficiency. J. Pediatr. 97:471-473.

Hoel, D.G. 1980. Incorporation of background in dose-response models. Fed. Proc. 39:73-75.

Hoel, D.G., N.L. Kaplan, and M.W. Anderson. 1983. Implication of nonlinear kinetics on risk estimation in carcinogenesis. Science 219:1032-1037.

Homan, E.R. 1972. Quantitative relationships between toxic doses of anti-tumor chemotherapeutic agents in animals and man. Cancer Chemother. Rep. (3):13-19.

Hytten, F.E., and G. Chamberlain. 1980. Clinical Physiology in Obstetrics. Oxford: Blackwell.

IARC (International Agency for Research on Cancer). 1987. IARC Monographs on the Evaluation of Carcinogenic Risks to Humans. Overall Evaluations of Carcinogenicity: An Updating of IARC Monographs, Vols. 1-42, Supplement 7. Lyon, France: International Agency for Research on Cancer.

Iball, J. 1939. The relative potency of carcinogenic compounds. Am. J. Cancer 35:188-190.

Kaldor, J.M., and N.E. Day. 1987. Interpretation of epidemiological studies in the context of the multistage model of carcinogenesis. Pp. 21-57 in Mechanisms of Environmental Carcinogenesis: Vol. II—Multistep Models of Carcinogenesis, J.D. Barrett, ed. Boca Raton, Fla.: CRC Press.

Ketterman A.J., S.M. Pond, and C.E. Becker. 1987. The effects of differential induction of cytochrome P-450, carboxylesterase and glutathione S-transferase activities on malathion toxicity in mice. Toxicol. Appl. Pharmacol. 87:389-392.

Kimmel, C.A., B.C. Allen, R.J. Kavlock, and E.M. Faustman. 1991. Alternatives to the NOAEL/UF approach for calculating reference dose. Unpublished.

Klaassen, C.D. 1986. Principles of toxicology. Pp. 11-32 in Casarett and Doull's Toxicology: The Basic Science of Poisons, 3rd ed, C.D. Klaassen, M.O Amdur, and J. Doull, eds. New York: Macmillan. 974 pp.

Kodell, R.L., D.W. Gaylor, and J.J. Chen. 1987. Using average lifetime dose rate for intermittent exposures to carcinogens. Risk Anal. 7:339-345.

Kodell, R.L., D. Krewski, and J.M. Zielinski. 1991. Additive and multiplicative relative risk in the two-stage clonal expansion model of carcinogenesis. Risk Anal. 11:483-490.

Krewski, D., and D.J. Murdoch. 1990. Cancer modeling with intermittent exposures. Pp. 196-214 in Scientific Issues in Quantitative Cancer Risk Assessment, S.H. Moolgavkar, eds. Cambridge, Mass.: Birkhauser.

Krewski, D., and R.D. Thomas. 1992. Carcinogenic mixtures. Risk Anal. 12:105-113.

Krewski, D., S. Brown, and D. Murdoch. 1984. Determining "safe" levels of exposure: Safety factors or mathematical models? Fundam. Appl. Toxicol. 4:S383-S394.

Krewski, D., D. Murdoch, and A. Dewanji. 1986. Statistical modeling and extrapolation of carcinogenesis data. Pp. 259-282 in Modern Statistical Methods in Chronic Disease Epidemiology, S.H. Moolgavkar and R.L. Prentice, eds. New York: John Wiley and Sons.

Krewski, D., D.J. Murdoch, and J.R. Withey. 1987. The application of pharmacokinetic data in carcinogenic risk assessment. Pp. 1-468 in Drinking Water and Health, Vol. 8: Pharmacokinetics and Risk Assessment. Washington, D.C.: National Academy Press.

Krewski, D., T. Thorslund, and J. Withey. 1989. Carcinogenic risk assessment of complex mixtures. Toxicol. Ind. Health 5:851-867.

Krewski, D., D.W. Gaylor, and M. Szyszkowicz. 1991. A model-free approach to low dose extrapolation. Environ. Health Perspect. 90:279-285.

Krewski, D., D.W. Gaylor, A.P. Soms, and M. Szyszkowicz. 1993. Correlation between carcinogenic potency and the maximum tolerated dose: Implications for risk assessment. Pp. 111-171 in Issues in Risk Assessment. Washington, D.C.: National Academy Press.

La Du, B.N., and H.W. Eckerson. 1984. The polymorphic paraoxonase arylesterase isozymes of human serum. Fed. Proc. 43:2338-2341.

Lehman, A.J., and O.G. Fitzhugh. 1954. 100-fold margin of safety. Assoc. Food Drug Off. Q. Bull. 18:33-35.

Lindstedt, S.L. 1987. Allometry: Body size constraints in animal design. Pp. 65-79 in Drinking Water and Health, Vol. 8: Pharmacokinetics in Risk Assessment. Washington, D.C.: National Academy Press.

Lu, F.C. 1988. Acceptable daily intake: Inception, evolution, and application. Regul. Toxicol. Pharmacol. 8:45-60.

Luebeck, E.G., S.H. Moolgavkar, A. Buchmann, and M. Schwartz. 1991. Effects of polychlorinated biphenyls in rat liver: Quantitative analysis of enzyme-altered foci. Toxicol. Appl. Pharmacol. 111:469-484.

Lutz, W.K., P. Buss, A. Baertsch, and M. Caviezel. 1990. Evaluation of DNA binding *in vivo* for low dose extrapolation in chemical carcinogenesis. In Genetic Toxicology of Complex Mixtures, M.D. Waters, S. Nesnow, J. Lewtas, N.M. Moore, and F.B. Daniels, eds. New York: Plenum.

Mackness, M., C. Walker, and L. Carlson. 1987. Low A-esterase activity in serum of patients with fish-eye disease. Clin. Chem. 33:587-588.

Mattison, D.R. 1986. Drug and Chemical Action in Pregnancy, S. Fabro and A.R. Scialli, eds. New York: Marcel Dekker.

Mattison, D.R., E. Blann, and A. Malek. 1991. Symposium: Pharmacokinetics in Developmental Toxicity. Physiological alterations during pregnancy: Impact on toxicokinetics. Fundam. Appl. Toxicol. 16:215-218.

McColl, S. 1989. Biological Safety Factors in Toxicological Risk Assessment. Ottawa: Health & Welfare Canada.

McFarland Child Health Screening Project. 1991. McFarland Child Health Screening Project: 1989. Environmental Epidemiology and Toxicology Branch, California Department of Health Services, Emeryville, Calif.

Mendelson, J., and S. Grisolia. 1975. Age-dependent sensitivity to salicylate. Lancet 2(7942):974.

Menzel, D.B. 1987. Physiological pharmacokinetic modeling. Environ. Sci. Technol. 21:944-950.

Metzger, B., E. Crouch, and R. Wilson. 1989. On the relationship between carcinogenicity and acute toxicity. Risk Anal. 9:169-177.

Mitchell, R., and B. Dawson. 1973. Educational and social characteristics of children with asthma. Arch. Dis. Child. 48:467-471.

Moolgavkar, S.H. 1986a. Carcinogenesis modelling from molecular biology to epidemiology. Annu. Rev. Public Health 7:151-169.

Moolgavkar, S.H. 1986b. Hormones and multistage carcinogenesis. Cancer Surv. 5:635-648.

Moolgavkar, S.H., and E.G. Luebeck. 1990. Two-event model for carcinogenesis: Biological, mathematical, and statistical considerations. Risk Anal. 10:323-341.

Moolgavkar, S.H., and E.G. Luebeck. 1992. Interpretation of labeling indices in the presence of cell death. Carcinogenesis 13:1007-1010.

Moolgavkar, S.H., A. Dewanji, and D.J. Venzon. 1988. A stochastic two-stage model for cancer risk assessment. I. The hazard function and the probability of tumor. Risk Anal. 8:383-392.

Moolgavkar, S.H., E.G. Leubeck, M. de Gunst, R.E. Port, and M. Schwarz. 1990. Quantitative analysis of enzyme-altered foci in rat hepatocarcinogenesis experiments-I. Single agent regimen. Carcinogenesis 11:1271-1278.

Moolgavkar, S.H., E.G. Luebeck, D. Krewski, and J.M. Zielinski. 1993. Radon, cigarette smoke and lung cancer: A reanalysis of the Colorado plateau uranium miners' data. Epidemiology 4:204-217.

Murdoch, D.J., and D. Krewski. 1988. Carcinogenic risk assessment with time-dependent exposure patterns. Risk Anal. 8:521-530.

Murdoch, D.J., D.R. Krewski, and K.S. Crump. 1987. Quantitative theories of carcinogenesis. Pp. 61-89 in Cancer Modeling, J. R. Thompson and B. W. Brown, eds. New York: Marcel Dekker.

Murdoch, D.J., D. Krewski, and J. Wargo. 1992. Cancer risk assessment with intermittent exposure. Risk Anal. 12:569-577.

NRC (National Research Council). 1970. Evaluating the Safety of Food Chemicals. Washington, D.C.: National Academy of Sciences.

NRC (National Research Council). 1977. Drinking Water and Health. Washington, D.C.: National Academy Press.

NRC (National Research Council). 1986. Drinking Water and Health, Vol. 6. Washington, D.C.: National Academy Press.

NRC (National Research Council). 1987a. Drinking Water and Health, Vol. 8: Pharmacokinetics in Risk Assessment. Washington, D.C.: National Academy Press.

NRC (National Research Council). 1987b. Regulating Pesticides in Food: The Delaney Paradox. Washington, D.C.: National Academy Press.

NRC (National Research Council). 1988. Complex Mixtures: Methods for In Vivo Toxicity Testing. Washington, D.C.: National Academy Press.

NRC (National Research Council). 1989. Diet and Health: Implications for Reducing Chronic Disease Risk. Washington, D.C.: National Academy of Sciences.

NRC (National Research Council). 1992. Environmental Neurotoxicology. Washington, D.C.: National Academy Press.

NRC (National Research Council). 1993. Issues in Risk Assessment. Washington, D.C.: National Academy Press.

NTP (National Toxicology Program). 1992. NTP Technical Report on the Perinatal Toxicity and Carcinogenicity Studies of Ethylene Thiourea in F/344 rats and B6C3F$_1$ mice (Feed Studies). NTP Technical Report 388. Research Triangle Park, N.C.: National Toxicology Program.

O'Flaherty, E.J. 1981. Toxicants and Drugs: Kinetics and Dynamics. New York: John Wiley and Sons.

Olanoff, L.S., and J.M. Anderson. 1980. Controlled release of tetracycline III: A physiologically pharmacokinetic model of the pregnant rat. J. Pharmacokinet. Biopharm. 8:599-620.

Ortigoza-Ferado, J., R. Richter, C. Furlong, and A.G. Motulsky. 1984. Biochemical genetics of paraoxonase. Pp. 177-188 in Banbury Report No. 16: Genetic Variability in Responses to Chemical Exposure, G.S. Omenn and H.V. Gelboin, eds. New York: Cold Spring Harbor Laboratory.

OSTP (Office of Science and Technology Policy). 1986. Chemical carcinogenesis, A review of the science and its associated principles. Environ. Health Perspect. 67:201-282.

Owen, Y., and R. Frankle. 1986. Nutrition and the Community, 2d ed. St. Louis, Mo.: Times Mirror/Mosby College Publishing.

Paynter, O.E., G.J. Burin, R.B. Jaeger, and C.A. Gregorio. 1988. Goitrogens and thyroid follicular cell neoplasia: Evidence for a threshold process. Regul. Toxicol. Pharmacol. 8:102-119.

Peach, J.D. 1987. Statement of J. Dexter Peach, General Accounting Office. Pp. 14-43 in Pesticides in Food. Hearing before the Subcommittee on Oversight and Investigations of the Committee on Energy and Commerce, House of Representatives, April 30, Serial No. 100-17. Washington, D.C.: U.S. Government Printing Office.

Peto, R. 1978. Carcinogenic effects of chronic exposure to very low levels of toxic substances. Environ. Health Perspect. 22:155-159.

Peto, R., M.C. Pike, L. Bernstein, L.S. Gold, and B.N. Ames. 1984. The TD$_{50}$: A proposed general convention for the numerical description of the carcinogenic potency of chemicals in chronic-exposure animal experiments. Environ. Health Perspect. 58:18.

Peto, R., R. Gray, P. Brantom, and P. Grasso. 1986. Nitrosamine carcinogenesis in 5120 rodents: Chronic administration of sixteen different concentrations of NDEA, NDMA, NPYR and NPIP in the water of 4440 inbred rats, with parallel studies on NDEA alone of the effect of age of starting (3, 6 or 20 weeks) and of species (rats, mice or hamsters). Pp. 627-665 in N-Nitroso compounds: Occurrence, Biological Effects and Relevance to Human Cancer, I.K. O'Neill, R.C. Von Borstel, C.T. Miller, J.I. Long, and H. Bartsch, eds. IARC Scientific Publication No. 57. Lyon, France: International Agency for Research on Cancer.

Pinkel, D. 1958. The use of body surface area as a criterion on drug dosage in cancer chemotherapy. Cancer Res. 18:853-860.

Pitot, H.C., and Y.P. Dragan. 1991. Facts and theories concerning the mechanisms of carcinogenesis. FASEB J. 5:2280-2286.

Portier, C.J. 1987. Statistical properties of a two-stage model of carcinogenesis. Environ. Health Perspect. 76:125-131.

Portier, C.J., and N.L. Kaplan. 1989. Variability of safe dose estimates when using complicated models of the carcinogenic process. A case study: Methylene chloride. Fundam. Appl. Toxicol. 13:533-544.

Rall, D.P., M.D. Hogan, J.E. Huff, B.A. Schwetz, and R.W. Tenant. 1987. Alternatives to using human experience in assessing health risks. Annu. Rev. Public Health 8:355-385.

Rice, J.M. 1979. Perinatal period and pregnancy: Intervals of high risk for chemical carcinogens. Environ. Health Perspect. 29:23-27.

Rieth, J.P., and T.B. Starr. 1989. Chronic bioassays: Relevance to quantitative risk assessment of carcinogens. Regul. Toxicol. Pharmacol. 10:160-173.

Russell, H.H., R.J. Jackson, D.P. Spath, and S.A. Bock. 1987. Chemical contamination of California drinking water. West. J. Med. 147:615-622.

Saunders, D.S. 1987. Briefing Paper on the Tolerance Assessment System (TAS) for Presentation to the FIFRA Science Advisory Panel, Hazard Evaluation Division. Washington, D.C.: Environmental Protection Agency, Office of Pesticide Programs.

Sawyer, C., R. Peto, L. Bernstein, and M.C. Pike. 1984. Calculation of carcinogenic potency from long-term animal carcinogenesis experiments. Biometrics 40:27-40.

Schneiderman, M.A., and N. Mantel. 1973. The Delaney Clause and a scheme for rewarding good experimentation. Prev. Med. 2:165-170.

Sharkawi, M., D. Cianflone, and A. El-Hawari. 1978. Toxic interactions between disulfiram and some centrally acting drugs in rats. J. Pharmaceut. Pharmacol. 30:63.

Shlyakhter, A., G. Goodman, and R. Wilson. 1992. Monte Carlo simulation of rodent carcinogenicity bioassays. Risk Anal. 12:73-82.

Snyder, W.S., M.J. Cook, E.S. Nasset, L.R. Karhausen, G.P. Howells, and I.H. Tiptar. 1974. Report of the Task Group on Reference Man. Oxford: Pergamon.

Starfield, B. 1982. Family income, ill health, and medical care of U.S. Children. J. Public Health Policy 3:244-259.

Starfield, B., and P.P. Budetti. 1985. Child health status and risk factors. Health Serv. Res. 19:817-886.

Sultatos, L.G. 1986. The effects of phenobarbital pretreatment on the metabolism and acute toxicity of the pesticide parathion in the mouse. Toxicol. Appl. Pharmacol. 86:105-111.

Thorslund, T., and G. Charnley. 1988. Quantitative dose-response models for tumor promoting agents. Pp. 245-255 in Banbury Report No. 31: Carcinogen Risk Assessment: New Directions in the Qualitative and Quantitative Aspects, R.W. Hart and F.D. Hoerger, eds. New York: Cold Spring Harbor Laboratory.

Thorslund, T., C.C. Brown, and G. Charnley. 1987. Biologically motivated cancer risk models. Risk Anal. 7:109-119.

Travis, C.C., and R.K. White. 1988. Interspecies and scaling of toxicity data. Risk Anal. 8:119-125.

Twort, C.C., and J.M. Twort. 1933. Suggested methods for the standardization of the carcinogenic activity of different agents for the skin of mice. Am. J. Cancer 17:293-320.

USGS (U.S. Geological Survey). 1988. National Water Summary 1986: Hydrologic Events and Ground-water Quality. U.S. Geological Survey Water Supply Paper 2325. Denver, Colo.: U.S. Government Printing Office.

Van Ryzin, J. 1980. Quantitative risk assessment. J.Occup. Med. 22:321-326.

Wagner, J.G. 1971. Biopharmaceutics and Relevant Pharmacokinetics, Chapter 4. Hamilton, Ill.: Drug Intelligence Publications.

Wartenburg, D., and M.A. Gallo. 1990. The fallacy of ranking possible carcinogenic hazards using the TD_{50}. Risk Anal. 10:609-613.

Weber, W.W. 1987. The Acetylator Genes and Drug Response. New York: Oxford University Press.

Whittemore, A., and J.B. Keller. 1978. Quantitative theories of carcinogenesis. SIAM Rev. 20:1-30.

WHO (World Health Organization). 1986. Environmental Health Criteria 59. Principles for Evaluating Health Risks from Chemicals during Infancy and Early Childhood. The Need for a Special Approach. Geneva: World Health Organization.

Zadikoff, C.M. 1979. Toxic encephalopathy associated with use of insect repellant. J. Pediatr. 95:140-142.

Zeise, L., E.A.C. Crouch, and R. Wilson. 1984. Use of acute toxicity to estimate carcinogenic risk. Risk Anal. 4:187-199.

Zeise, L., E.A.C. Crouch, and R. Wilson. 1986. A possible relationship between toxicity and carcinogenicity. J. Am. Coll. Toxicol. 5:137-152.

Index

A

Acceptable daily intake (ADI), 132, 315, 324-326
Accidental poisoning, 313
Acephate, 244
Acetaminophen, 57
Acetylcholinesterase, 54, 310, 341
  inhibitions by organophosphates, 347
Acrodynia, 347
Acute exposure and effects, 50-53, 103, 104, 271
  of aldicarb, 287-288
  and daily intakes, 6, 11, 315, 317
  of neurotoxins, 63-64
  pharmaceutical studies, 54-60
  testing of, 130, 133, 136
  and tolerances, 8, 18
  see also Cholinesterase inhibition
Acute nonlymphoblastic leukemia (ANLL), and parental exposure, 307
Additive background model, cancer risk, 328-330

Adolescence, see Age-related differences
Adults, see Age-related differences
Adverse effect, definition, 327
Age groups, in data analyses, 5, 6, 315
Age-related differences, 2-7, 11-12, 19, 105-107
  carcinogenesis, 75, 351
  cholinesterase inhibition, 53-54
  dermal absorption, 77-78
  dietary composition, 4, 38, 103, 181-187, 193-194, 196
  gastrointestinal absorption, 82
  metabolism, 38-39, 55-57
  pharmacokinetics, 55-59, 106
  physiology, 3, 36, 37, 39-43
  protein binding, 85-86
  pulmonary exposure, 79-80
  risk assessment, 7, 11-12, 344-345
Agricultural application, see
  Application rates and methods;
  Field trials
Agricultural Marketing Service, 216